PEARSON EDEXCEL INTERNATIONAL A LEVEL

FURTHER PURE MATHEMATICS 1

Student Book

Series Editors: Joe Skrakowski and Harry Smith

Authors: Greg Attwood, Jack Barraclough, Ian Bettison, Lee Cope, Charles Garnet Cox, Keith Gallick, Daniel Goldberg, Alistair Macpherson, Anne McAteer, Bronwen Moran, Su Nicholson, Laurence Pateman, Joe Petran, Keith Pledger, Cong San, Joe Skrakowski, Harry Smith, Geoff Staley, Dave Wilkins

Published by Pearson Education Limited, 80 Strand, London, WC2R 0RL.
www.pearsonglobalschools.com

Copies of official specifications for all Pearson qualifications may be found on the website: https://qualifications.pearson.com

Text © Pearson Education Limited 2018
Designed by © Pearson Education Limited 2018
Typeset by Tech-Set Ltd, Gateshead, UK
Edited by Eric Pradel
Original illustrations © Pearson Education Limited 2018
Illustrated by © Tech-Set Ltd, Gateshead, UK
Cover design © Pearson Education Limited 2018

The rights of Greg Attwood, Jack Barraclough, Ian Bettison, Lee Cope, Charles Garnet Cox, Keith Gallick, Daniel Goldberg, Alistair Macpherson, Anne McAteer, Bronwen Moran, Su Nicholson, Laurence Pateman, Joe Petran, Keith Pledger, Cong San, Joe Skrakowski, Harry Smith, Geoff Staley and Dave Wilkins to be identified as the authors of this work have been asserted by them in accordance with the Copyright, Designs and Patents Act 1988.

First published 2018

25 24 23
10 9 8 7

British Library Cataloguing in Publication Data
A catalogue record for this book is available from the British Library

ISBN 978 1 292244 64 8

Printed in Slovakia by Neografia

Picture Credits
The authors and publisher would like to thank the following individuals and organisations for permission to reproduce photographs:

Alamy Stock Photo: Paul Fleet 92; **Getty Images:** Anthony Bradshaw 36, David Trood 1, Dulyanut Swdp 49, gmutlu 116, jamielawton 76, Martin Barraud 127; **Paul Nylander:** 28

Cover images: *Front*: **Getty Images:** Werner Van Steen
Inside front cover: **Shutterstock.com:** Dmitry Lobanov

All other images © Pearson Education Limited 2018
All artwork © Pearson Education Limited 2018

Endorsement Statement
In order to ensure that this resource offers high-quality support for the associated Pearson qualification, it has been through a review process by the awarding body. This process confirms that this resource fully covers the teaching and learning content of the specification or part of a specification at which it is aimed. It also confirms that it demonstrates an appropriate balance between the development of subject skills, knowledge and understanding, in addition to preparation for assessment.

Endorsement does not cover any guidance on assessment activities or processes (e.g. practice questions or advice on how to answer assessment questions) included in the resource, nor does it prescribe any particular approach to the teaching or delivery of a related course.

While the publishers have made every attempt to ensure that advice on the qualification and its assessment is accurate, the official specification and associated assessment guidance materials are the only authoritative source of information and should always be referred to for definitive guidance.

Pearson examiners have not contributed to any sections in this resource relevant to examination papers for which they have responsibility.

Examiners will not use endorsed resources as a source of material for any assessment set by Pearson. Endorsement of a resource does not mean that the resource is required to achieve this Pearson qualification, nor does it mean that it is the only suitable material available to support the qualification, and any resource lists produced by the awarding body shall include this and other appropriate resources.

ABOUT THIS BOOK

The following three themes have been fully integrated throughout the Pearson Edexcel International Advanced Level in Mathematics series, so they can be applied alongside your learning.

1. Mathematical argument, language and proof

- Rigorous and consistent approach throughout
- Notation boxes explain key mathematical language and symbols

2. Mathematical problem-solving

- Hundreds of problem-solving questions, fully integrated into the main exercises
- Problem-solving boxes provide tips and strategies
- Challenge questions provide extra stretch

The Mathematical Problem-Solving Cycle

specify the problem → collect information → process and represent information → interpret results → specify the problem

3. Transferable skills

- Transferable skills are embedded throughout this book, in the exercises and in some examples
- These skills are signposted to show students which skills they are using and developing

Finding your way around the book

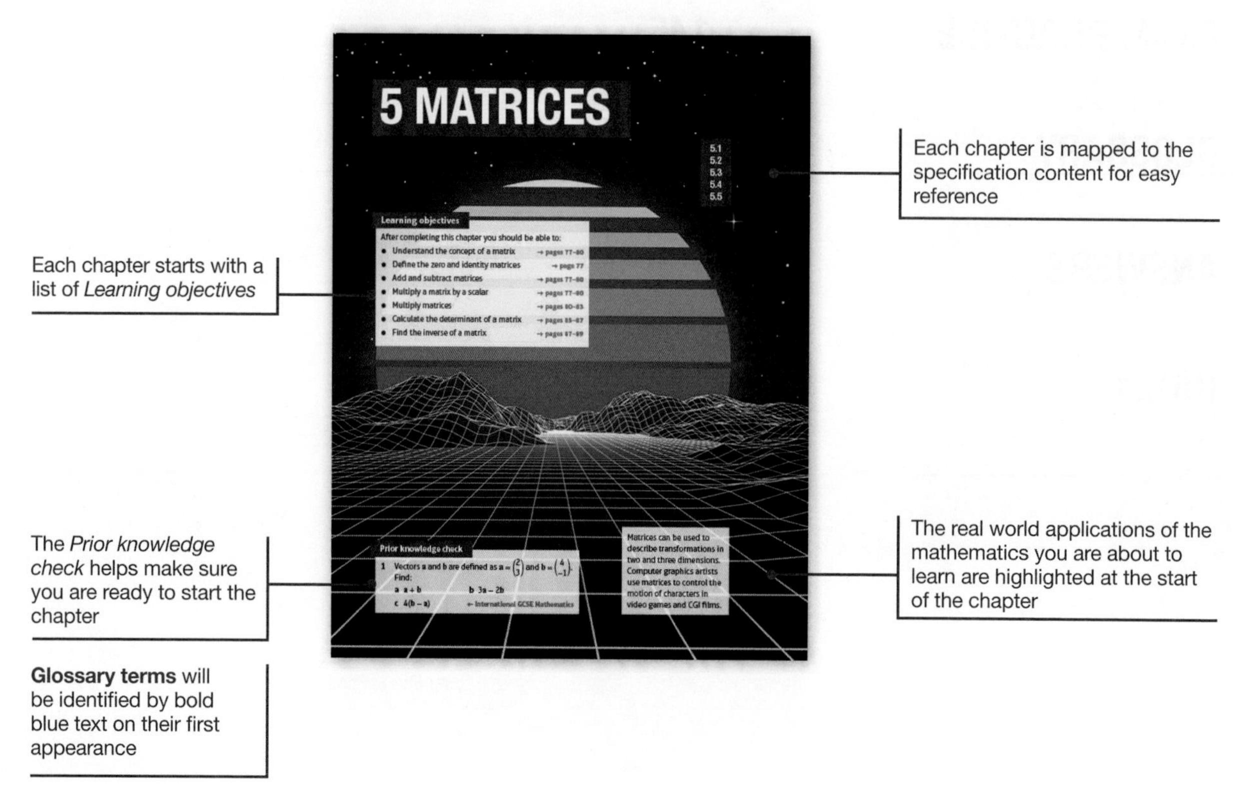

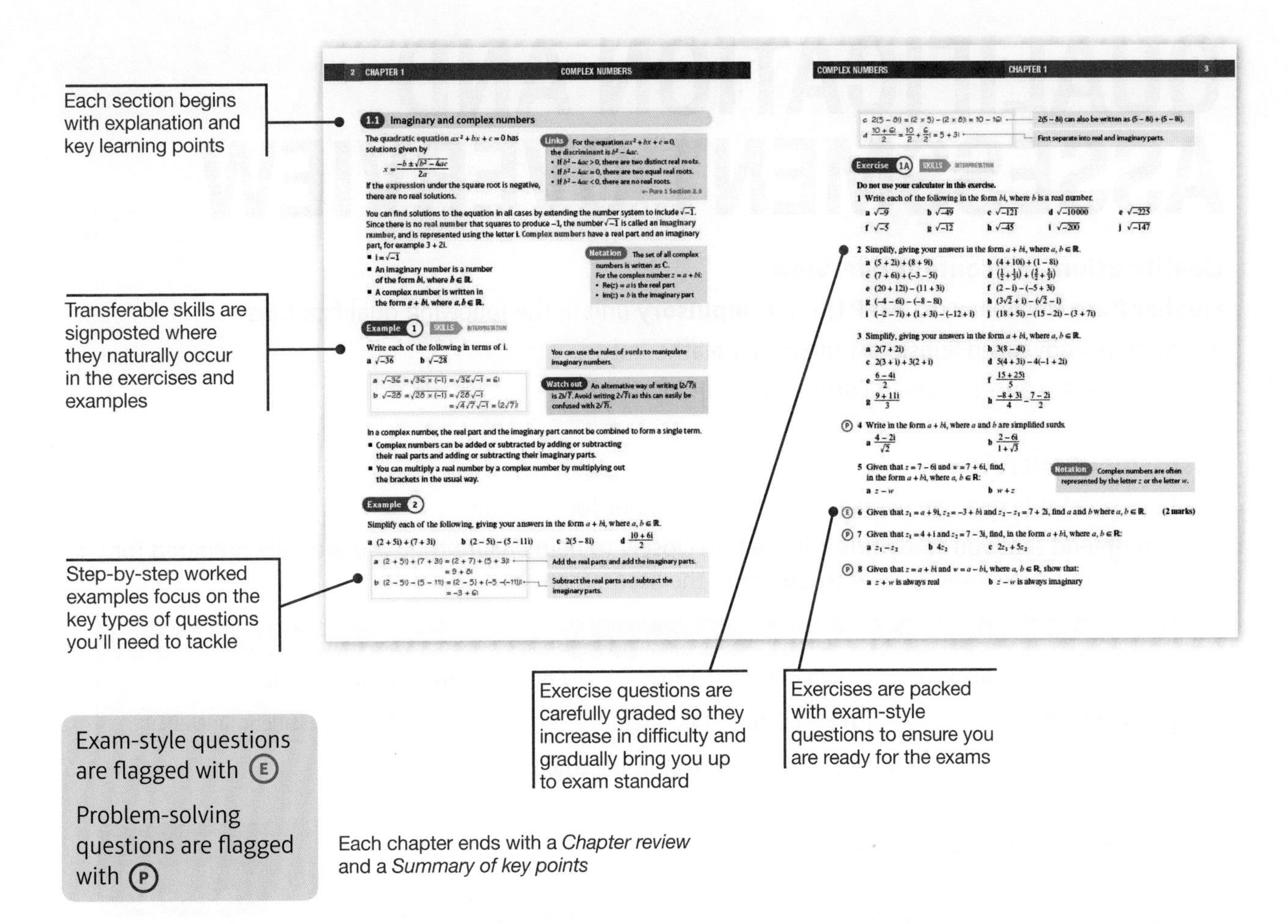

Each section begins with explanation and key learning points

Transferable skills are signposted where they naturally occur in the exercises and examples

Step-by-step worked examples focus on the key types of questions you'll need to tackle

Exercise questions are carefully graded so they increase in difficulty and gradually bring you up to exam standard

Exercises are packed with exam-style questions to ensure you are ready for the exams

Exam-style questions are flagged with Ⓔ

Problem-solving questions are flagged with Ⓟ

Each chapter ends with a *Chapter review* and a *Summary of key points*

After every few chapters, a *Review exercise* helps you consolidate your learning with lots of exam-style questions

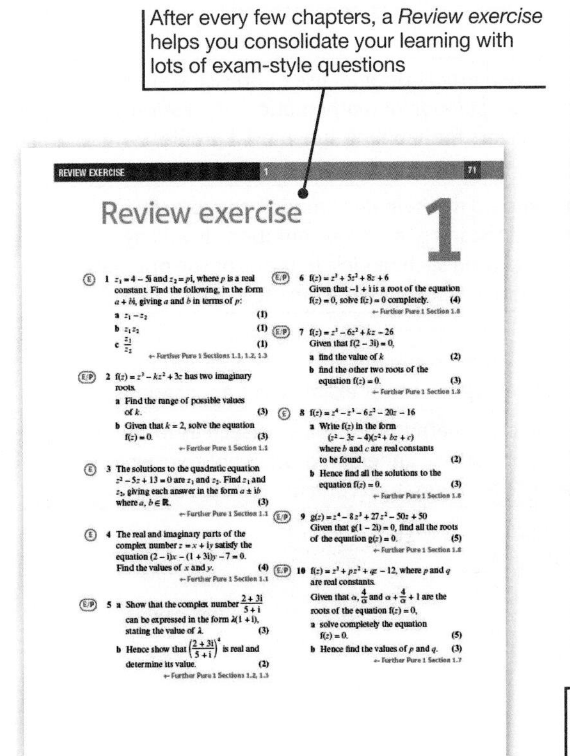

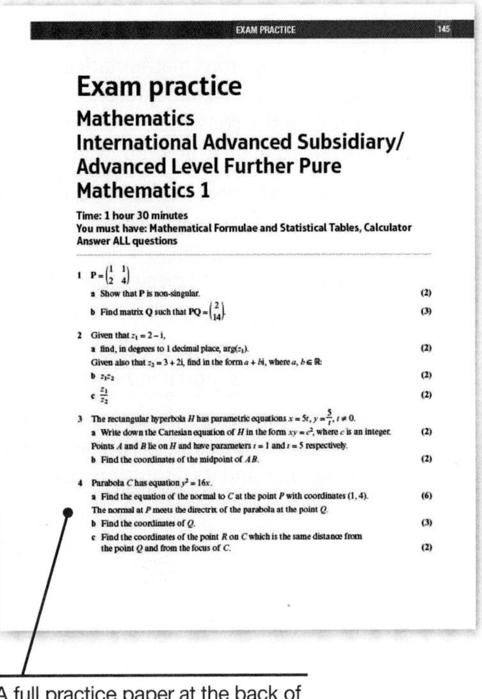

A full practice paper at the back of the book helps you prepare for the real thing

QUALIFICATION AND ASSESSMENT OVERVIEW

Qualification and content overview

Further Pure Mathematics 1 (FP1) is a **compulsory** unit in the following qualifications:

International Advanced Subsidiary in Further Mathematics

International Advanced Level in Further Mathematics

Assessment overview

The following table gives an overview of the assessment for this unit.

We recommend that you study this information closely to help ensure that you are fully prepared for this course and know exactly what to expect in the assessment.

Unit	Percentage	Mark	Time	Availability
FP1: Further Pure Mathematics 1 Paper code WFM01/01	$33\frac{1}{3}$ % of IAS $16\frac{2}{3}$ % of IAL	75	1 hour 30 mins	January and June First assessment June 2019

IAS: International Advanced Subsidiary, IAL: International Advanced A Level.

Assessment objectives and weightings

		Minimum weighting in IAS and IAL
AO1	Recall, select and use their knowledge of mathematical facts, concepts and techniques in a variety of contexts.	30%
AO2	Construct rigorous mathematical arguments and proofs through use of precise statements, logical deduction and inference and by the manipulation of mathematical expressions, including the construction of extended arguments for handling substantial problems presented in unstructured form.	30%
AO3	Recall, select and use their knowledge of standard mathematical models to represent situations in the real world; recognise and understand given representations involving standard models; present and interpret results from such models in terms of the original situation, including discussion of the assumptions made and refinement of such models.	10%
AO4	Comprehend translations of common realistic contexts into mathematics; use the results of calculations to make predictions, or comment on the context; and, where appropriate, read critically and comprehend longer mathematical arguments or examples of applications.	5%
AO5	Use contemporary calculator technology and other permitted resources (such as formulae booklets or statistical tables) accurately and efficiently; understand when not to use such technology, and its limitations. Give answers to appropriate accuracy.	5%

Relationship of assessment objectives to units

	Assessment objective				
FP1	AO1	AO2	AO3	AO4	AO5
Marks out of 75	25–30	25–30	0–5	5–10	5–10
%	$33\frac{1}{3}$–40	$33\frac{1}{3}$–40	0–$6\frac{2}{3}$	$6\frac{2}{3}$–$13\frac{1}{3}$	$6\frac{2}{3}$–$13\frac{1}{3}$

Calculators

Students may use a calculator in assessments for these qualifications. Centres are responsible for making sure that calculators used by their students meet the requirements given in the table below.

Students are expected to have available a calculator with at least the following keys: $+$, $-$, $\times$, $\div$, π, x^2, $\sqrt{x}$, $\frac{1}{x}$, x^y, $\ln x$, e^x, $x!$, sine, cosine and tangent and their inverses in degrees and decimals of a degree, and in radians; memory.

Prohibitions

Calculators with any of the following facilities are prohibited in all examinations:

- databanks
- retrieval of text or formulae
- built-in symbolic algebra manipulations
- symbolic differentiation and/or integration
- language translators
- communication with other machines or the internet

Extra online content

Whenever you see an *Online* box, it means that there is extra online content available to support you.

SolutionBank

SolutionBank provides worked solutions for questions in the book.
Download solutions as a PDF or quickly find the solution you need online.

Use of technology

Explore topics in more detail, visualise problems and consolidate your understanding. Use pre-made GeoGebra activities or Casio resources for a graphic calculator.

Online Find the point of intersection graphically using technology.

GeoGebra

GeoGebra-powered interactives

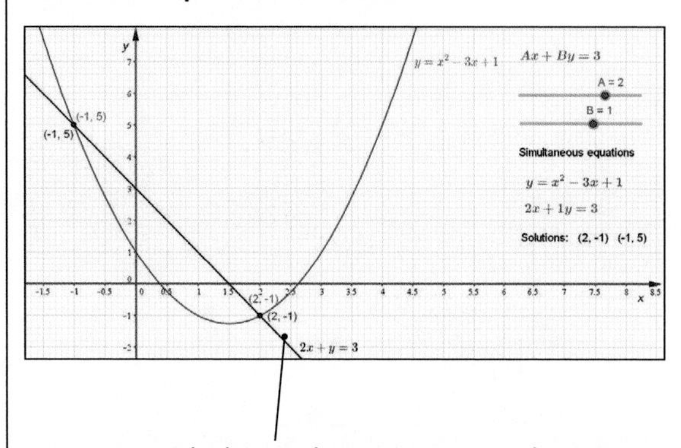

Interact with the mathematics you are learning using GeoGebra's easy-to-use tools

CASIO

Graphic calculator interactives

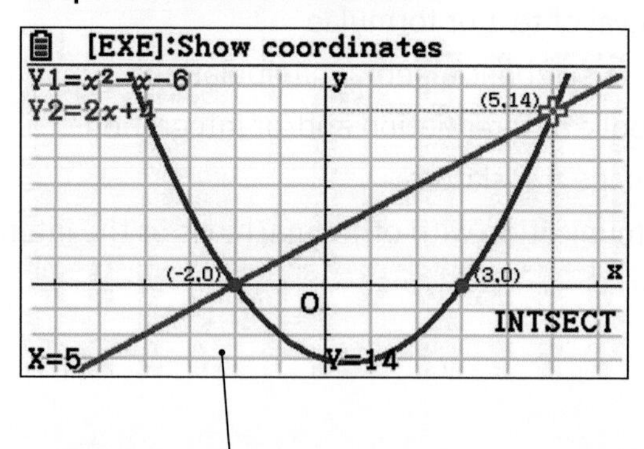

Explore the mathematics you are learning and gain confidence in using a graphic calculator

Calculator tutorials

Our helpful video tutorials will guide you through how to use your calculator in the exams. They cover both Casio's scientific and colour graphic calculators.

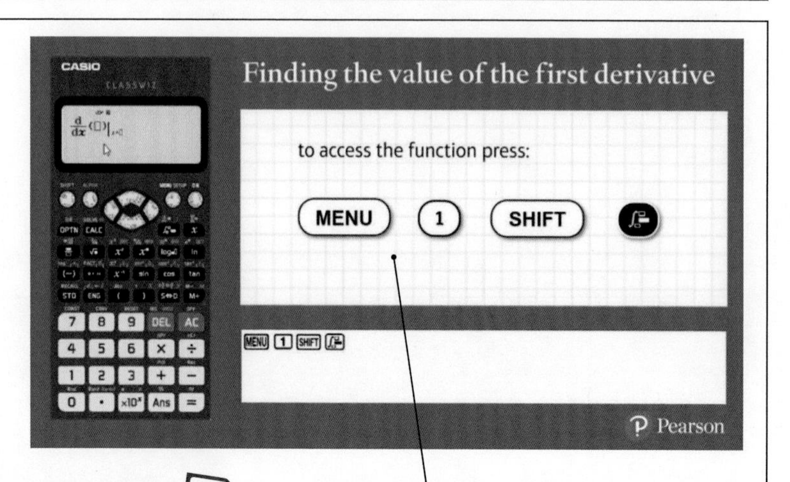

Step-by-step guide with audio instructions on exactly which buttons to press and what should appear on your calculator's screen

Online Work out each coefficient quickly using the ${}^nC_r$ and power functions on your calculator.

1 COMPLEX NUMBERS

Learning objectives

After completing this chapter you should be able to:

- Understand and use the definitions of imaginary and complex numbers → page 2
- Add and subtract complex numbers → pages 2–3
- Find solutions to any quadratic equation with real coefficients → pages 4–5
- Multiply complex numbers → pages 5–6
- Understand the definition of a complex conjugate → pages 7–8
- Divide complex numbers → pages 7–8
- Show complex numbers on an Argand diagram → pages 9–10
- Find the modulus and argument of a complex number → pages 11–14
- Write a complex number in modulus-argument form → pages 15–16
- Solve quadratic equations that have complex roots → pages 16–18
- Solve cubic or quartic equations that have complex roots → pages 18–22

Prior knowledge check

1 Simplify each of the following:

a $\sqrt{50}$ **b** $\sqrt{108}$ **c** $\sqrt{180}$ ← Pure 1 Section 1.5

2 In each case, determine the number of distinct real roots of the equation $f(x) = 0$.

a $f(x) = 3x^2 + 8x + 10$

b $f(x) = 2x^2 - 9x + 7$

c $f(x) = 4x^2 + 12x + 9$ ← Pure 1 Section 2.3

3 For the triangle shown, find the values of:

a x **b** θ

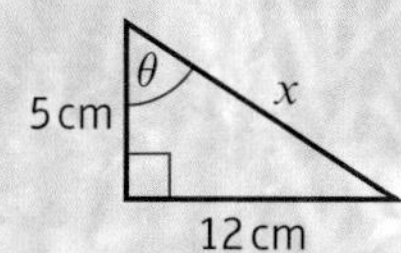

← International GCSE Mathematics

4 Find the solutions of $x^2 - 8x + 6 = 0$, giving your answers in the form $a \pm \sqrt{b}$ where a and b are integers. ← Pure 1 Section 2.1

5 Write $\frac{7}{4 - \sqrt{3}}$ in the form $p + q\sqrt{3}$ where p and q are rational numbers. ← Pure 1 Section 1.6

Complex numbers contain a real part and an imaginary part. Engineers and physicists often describe quantities with two components using a single complex number. This allows them to model complicated situations such as air flow over a cyclist.

1.1 Imaginary and complex numbers

The **quadratic equation** $ax^2 + bx + c = 0$ has solutions given by

$$x = \frac{-b \pm \sqrt{b^2 - 4ac}}{2a}$$

If the **expression** under the square root is negative, there are no real solutions.

Links For the equation $ax^2 + bx + c = 0$, the **discriminant** is $b^2 - 4ac$.
- If $b^2 - 4ac > 0$, there are two distinct real **roots**.
- If $b^2 - 4ac = 0$, there are two equal real roots.
- If $b^2 - 4ac < 0$, there are no real roots.

← Pure 1 Section 2.5

You can find solutions to the equation in all cases by extending the number system to include $\sqrt{-1}$. Since there is no **real number** that squares to produce −1, the number $\sqrt{-1}$ is called an **imaginary number**, and is represented using the letter **i**. **Complex numbers** have a real part and an imaginary part, for example 3 + 2i.

- **$i = \sqrt{-1}$**
- **An imaginary number is a number of the form bi, where $b \in \mathbb{R}$.**
- **A complex number is written in the form $a + b$i, where $a, b \in \mathbb{R}$.**

Notation The set of all complex numbers is written as $\mathbb{C}$.
For the complex number $z = a + bi$:
- $\text{Re}(z) = a$ is the real part
- $\text{Im}(z) = b$ is the imaginary part

Example 1 SKILLS INTERPRETATION

Write each of the following in terms of i.

a $\sqrt{-36}$ **b** $\sqrt{-28}$

a $\sqrt{-36} = \sqrt{36 \times (-1)} = \sqrt{36}\sqrt{-1} = 6i$

b $\sqrt{-28} = \sqrt{28 \times (-1)} = \sqrt{28}\sqrt{-1}$
$= \sqrt{4}\sqrt{7}\sqrt{-1} = (2\sqrt{7})i$

You can use the rules of **surds** to manipulate imaginary numbers.

Watch out An alternative way of writing $(2\sqrt{7})i$ is $2i\sqrt{7}$. Avoid writing $2\sqrt{7}i$ as this can easily be confused with $2\sqrt{7i}$.

In a complex number, the real part and the imaginary part cannot be combined to form a single term.

- **Complex numbers can be added or subtracted by adding or subtracting their real parts and adding or subtracting their imaginary parts.**
- **You can multiply a real number by a complex number by multiplying out the brackets in the usual way.**

Example 2

Simplify each of the following, giving your answers in the form $a + bi$, where $a, b \in \mathbb{R}$.

a $(2 + 5i) + (7 + 3i)$ **b** $(2 - 5i) - (5 - 11i)$ **c** $2(5 - 8i)$ **d** $\frac{10 + 6i}{2}$

a $(2 + 5i) + (7 + 3i) = (2 + 7) + (5 + 3)i$
$= 9 + 8i$

Add the real parts and add the imaginary parts.

b $(2 - 5i) - (5 - 11i) = (2 - 5) + (-5 -(-11))i$
$= -3 + 6i$

Subtract the real parts and subtract the imaginary parts.

c $2(5 - 8i) = (2 \times 5) - (2 \times 8)i = 10 - 16i$ — $2(5 - 8i)$ can also be written as $(5 - 8i) + (5 - 8i)$.

d $\frac{10 + 6i}{2} = \frac{10}{2} + \frac{6}{2}i = 5 + 3i$ — First separate into real and imaginary parts.

Exercise 1A SKILLS INTERPRETATION

Do not use your calculator in this exercise.

1 Write each of the following in the form bi, where b is a real number.

a $\sqrt{-9}$ **b** $\sqrt{-49}$ **c** $\sqrt{-121}$ **d** $\sqrt{-10000}$ **e** $\sqrt{-225}$

f $\sqrt{-5}$ **g** $\sqrt{-12}$ **h** $\sqrt{-45}$ **i** $\sqrt{-200}$ **j** $\sqrt{-147}$

2 Simplify, giving your answers in the form $a + bi$, where $a, b \in \mathbb{R}$.

a $(5 + 2i) + (8 + 9i)$ **b** $(4 + 10i) + (1 - 8i)$

c $(7 + 6i) + (-3 - 5i)$ **d** $\left(\frac{1}{2} + \frac{1}{3}i\right) + \left(\frac{5}{2} + \frac{5}{3}i\right)$

e $(20 + 12i) - (11 + 3i)$ **f** $(2 - i) - (-5 + 3i)$

g $(-4 - 6i) - (-8 - 8i)$ **h** $(3\sqrt{2} + i) - (\sqrt{2} - i)$

i $(-2 - 7i) + (1 + 3i) - (-12 + i)$ **j** $(18 + 5i) - (15 - 2i) - (3 + 7i)$

3 Simplify, giving your answers in the form $a + bi$, where $a, b \in \mathbb{R}$.

a $2(7 + 2i)$ **b** $3(8 - 4i)$

c $2(3 + i) + 3(2 + i)$ **d** $5(4 + 3i) - 4(-1 + 2i)$

e $\frac{6 - 4i}{2}$ **f** $\frac{15 + 25i}{5}$

g $\frac{9 + 11i}{3}$ **h** $\frac{-8 + 3i}{4} - \frac{7 - 2i}{2}$

(P) 4 Write in the form $a + bi$, where a and b are simplified surds.

a $\frac{4 - 2i}{\sqrt{2}}$ **b** $\frac{2 - 6i}{1 + \sqrt{3}}$

5 Given that $z = 7 - 6i$ and $w = 7 + 6i$, find, in the form $a + bi$, where $a, b \in \mathbb{R}$:

a $z - w$ **b** $w + z$

Notation Complex numbers are often represented by the letter z or the letter w.

(E) 6 Given that $z_1 = a + 9i$, $z_2 = -3 + bi$ and $z_2 - z_1 = 7 + 2i$, find a and b where $a, b \in \mathbb{R}$. **(2 marks)**

(P) 7 Given that $z_1 = 4 + i$ and $z_2 = 7 - 3i$, find, in the form $a + bi$, where $a, b \in \mathbb{R}$:

a $z_1 - z_2$ **b** $4z_2$ **c** $2z_1 + 5z_2$

(P) 8 Given that $z = a + bi$ and $w = a - bi$, where $a, b \in \mathbb{R}$, show that:

a $z + w$ is always real **b** $z - w$ is always imaginary

You can use complex numbers to find solutions to any quadratic equation with real **coefficients**.

- **If $b^2 - 4ac < 0$ then the quadratic equation $ax^2 + bx + c = 0$ has two distinct complex roots, neither of which are real.**

Example 3 SKILLS PROBLEM-SOLVING

Solve the equation $z^2 + 9 = 0$.

$z^2 = -9$

$z = \pm\sqrt{-9} = \pm\sqrt{9 \times (-1)} = \pm\sqrt{9}\sqrt{-1} = \pm 3i$

$z = +3i, z = -3i$

Note that just as $z^2 = 9$ has two roots +3 and −3, $z^2 = -9$ also has two roots +3i and −3i.

Example 4

Solve the equation $z^2 + 6z + 25 = 0$.

Method 1 (Completing the square)

$z^2 + 6z = (z + 3)^2 - 9$

Because $(z + 3)^2 = (z + 3)(z + 3) = z^2 + 6z + 9$

$z^2 + 6z + 25 = (z + 3)^2 - 9 + 25 = (z + 3)^2 + 16$

$(z + 3)^2 + 16 = 0$

$(z + 3)^2 = -16$

$z + 3 = \pm\sqrt{-16} = \pm 4i$

$\sqrt{-16} = \sqrt{16 \times (-1)} = \sqrt{16}\sqrt{-1} = \pm 4i$

$z = -3 \pm 4i$

$z = -3 + 4i, z = -3 - 4i$

You can use your calculator to find the complex roots of a quadratic equation like this one.

Method 2 (Quadratic formula)

$$z = \frac{-6 \pm \sqrt{6^2 - 4 \times 1 \times 25}}{2}$$

Using $z = \frac{-b \pm \sqrt{b^2 - 4ac}}{2a}$

$$= \frac{-6 \pm \sqrt{-64}}{2}$$

$$z = \frac{-6 \pm 8i}{2} = -3 \pm 4i$$

$\sqrt{-64} = \sqrt{64 \times (-1)} = \sqrt{64}\sqrt{-1} = \pm 8i$

$z = -3 + 4i, z = -3 - 4i$

Exercise 1B SKILLS PROBLEM-SOLVING

Do not use your calculator in this exercise.

1 Solve each of the following equations. Write your answers in the form $\pm bi$.

a $z^2 + 121 = 0$ **b** $z^2 + 40 = 0$ **c** $2z^2 + 120 = 0$

d $3z^2 + 150 = 38 - z^2$ **e** $z^2 + 30 = -3z^2 - 66$ **f** $6z^2 + 1 = 2z^2$

2 Solve each of the following equations.
Write your answers in the form $a \pm b\text{i}$.

> **Hint** The left-hand side of each equation is in completed square form already. Use **inverse operations** to find the values of z.

a $(z-3)^2 - 9 = -16$
b $2(z-7)^2 + 30 = 6$
c $16(z+1)^2 + 11 = 2$

3 Solve each of the following equations. Write your answers in the form $a \pm b\text{i}$.

a $z^2 + 2z + 5 = 0$ **b** $z^2 - 2z + 10 = 0$ **c** $z^2 + 4z + 29 = 0$
d $z^2 + 10z + 26 = 0$ **e** $z^2 + 5z + 25 = 0$ **f** $z^2 + 3z + 5 = 0$

4 Solve each of the following equations. Write your answers in the form $a \pm b\text{i}$.

a $2z^2 + 5z + 4 = 0$ **b** $7z^2 - 3z + 3 = 0$ **c** $5z^2 - z + 3 = 0$

5 The solutions to the quadratic equation $z^2 - 8z + 21 = 0$ are z_1 and z_2.
Find z_1 and z_2, giving each in the form $a \pm \text{i}\sqrt{b}$.

(E/P) **6** The equation $z^2 + bz + 11 = 0$, where $b \in \mathbb{R}$, has distinct non-real complex roots.
Find the range of possible values of b. **(3 marks)**

1.2 Multiplying complex numbers

You can multiply complex numbers using the same technique that you use for multiplying brackets in algebra. You can use the fact that $\text{i} = \sqrt{-1}$ to simplify powers of i.

- $\text{i}^2 = -1$

Example 5

Express each of the following in the form $a + b\text{i}$, where a and b are real numbers.

a $(2+3\text{i})(4+5\text{i})$ **b** $(7-4\text{i})^2$

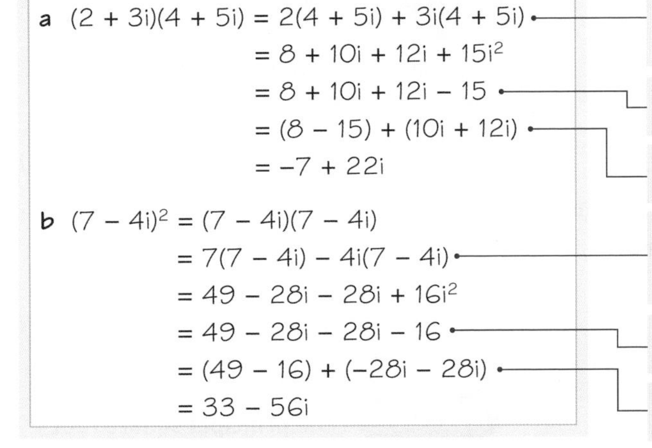

a $(2+3\text{i})(4+5\text{i}) = 2(4+5\text{i}) + 3\text{i}(4+5\text{i})$
$= 8 + 10\text{i} + 12\text{i} + 15\text{i}^2$
$= 8 + 10\text{i} + 12\text{i} - 15$
$= (8-15) + (10\text{i} + 12\text{i})$
$= -7 + 22\text{i}$

b $(7-4\text{i})^2 = (7-4\text{i})(7-4\text{i})$
$= 7(7-4\text{i}) - 4\text{i}(7-4\text{i})$
$= 49 - 28\text{i} - 28\text{i} + 16\text{i}^2$
$= 49 - 28\text{i} - 28\text{i} - 16$
$= (49-16) + (-28\text{i} - 28\text{i})$
$= 33 - 56\text{i}$

Multiply out the two brackets the same as you would with real numbers.

Use the fact that $\text{i}^2 = -1$.

Add real parts and add imaginary parts.

Multiply out the two brackets the same as you would with real numbers.

Use the fact that $\text{i}^2 = -1$.

Add real parts and add imaginary parts.

Example 6 SKILLS ANALYSIS

Simplify: **a** i^3 **b** i^4 **c** $(2i)^5$

a $i^3 = i \times i \times i = i^2 \times i = -i$

$i^2 = -1$

b $i^4 = i \times i \times i \times i = i^2 \times i^2 = (-1) \times (-1) = 1$

c $(2i)^5 = 2i \times 2i \times 2i \times 2i \times 2i$

$= 32(i \times i \times i \times i \times i) = 32(i^2 \times i^2 \times i)$

$= 32 \times (-1) \times (-1) \times i = 32i$

$(2i)^5 = 2^5 \times i^5$
First work out $2^5 = 32$.

Exercise 1C SKILLS EXECUTIVE FUNCTION

Do not use your calculator in this exercise.

1 Simplify each of the following, giving your answers in the form $a + bi$.

a $(5 + i)(3 + 4i)$ **b** $(6 + 3i)(7 + 2i)$ **c** $(5 - 2i)(1 + 5i)$
d $(13 - 3i)(2 - 8i)$ **e** $(-3 - i)(4 + 7i)$ **f** $(8 + 5i)^2$
g $(2 - 9i)^2$ **h** $(1 + i)(2 + i)(3 + i)$
i $(3 - 2i)(5 + i)(4 - 2i)$ **j** $(2 + 3i)^3$

Hint For part **h**, begin by multiplying the first pair of brackets.

(P) **2** **a** Simplify $(4 + 5i)(4 - 5i)$, giving your answer in the form $a + bi$.
b Simplify $(7 - 2i)(7 + 2i)$, giving your answer in the form $a + bi$.
c Comment on your answers to parts **a** and **b**.
d Show that $(a + bi)(a - bi)$ is a real number for any real numbers a and b.

(P) **3** Given that $(a + 3i)(1 + bi) = 25 - 39i$, find two possible pairs of values for a and b.

4 Write each of the following in its simplest form.
a i^6 **b** $(3i)^4$ **c** $i^5 + i$ **d** $(4i)^3 - 4i^3$

(P) **5** Express $(1 + i)^6$ in the form $a - bi$, where a and b are **integers** to be found.

(P) **6** Find the value of the real part of $(3 - 2i)^4$.

(P) **7** $f(z) = 2z^2 - z + 8$
Find: **a** $f(2i)$ **b** $f(3 - 6i)$

Problem-solving
You can use the **binomial** theorem to expand $(a + b)^n$. ← Pure 2 Section 4.3

(E/P) **8** $f(z) = z^2 - 2z + 17$
Show that $z = 1 - 4i$ is a solution to $f(z) = 0$. **(2 marks)**

9 **a** Given that $i^1 = i$ and $i^2 = -1$, write i^3 and i^4 in their simplest forms.
b Write i^5, i^6, i^7 and i^8 in their simplest forms.
c Write down the value of:
i i^{100} **ii** i^{253} **iii** i^{301}

Challenge

a Expand $(a + bi)^2$.
b Hence, or otherwise, find $\sqrt{40 - 42i}$, giving your answer in the form $a - bi$, where a and b are positive integers.

Notation The **principal square root** of a complex number, $\sqrt{z}$, has a positive real part.

1.3 Complex conjugation

- For any complex number $z = a + bi$, the **complex conjugate** of the number is defined as $z^* = a - bi$.

Notation Together, z and z^* are called a **complex conjugate pair**.

Example 7 SKILLS INTERPRETATION

Given that $z = 2 - 7i$,

a write down z^* **b** find the value of $z + z^*$ **c** find the value of zz^*.

a $z^* = 2 + 7i$

b $z + z^* = (2 - 7i) + (2 + 7i)$
$= (2 + 2) + (-7 + 7)i = 4$

c $zz^* = (2 - 7i)(2 + 7i)$
$= 2(2 + 7i) - 7i(2 + 7i)$
$= 4 + 14i - 14i - 49i^2$
$= 4 + 49 = 53$

Change the sign of the imaginary part from – to +.

Notation Notice that $z + z^*$ is real.

Remember $i^2 = -1$.

Notation Notice that zz^* is real.

For any complex number z, the **product** of z and z^* is a real number. You can use this property (i.e. characteristic) to **divide two complex numbers**. To do this, you multiply both the numerator and the denominator by the complex conjugate of the denominator and then simplify the result.

Links The method used to divide complex numbers is similar to the method used to rationalise a denominator when simplifying surds.
← Pure 1 Section 1.5

Example 8

Write $\frac{5 + 4i}{2 - 3i}$ in the form $a + bi$.

$$\frac{5 + 4i}{2 - 3i} = \frac{5 + 4i}{2 - 3i} \times \frac{2 + 3i}{2 + 3i}$$
$$= \frac{(5 + 4i)(2 + 3i)}{(2 - 3i)(2 + 3i)}$$
$$= \frac{10 + 8i + 15i + 12i^2}{4 + 6i - 6i - 9i^2}$$
$$= \frac{10 - 12 + 23i}{4 + 9}$$
$$= \frac{-2 + 23i}{13} = -\frac{2}{13} + \frac{23}{13}i$$

The complex conjugate of the denominator is $2 + 3i$. Multiply both the numerator and the denominator by the complex conjugate.

zz^* is real, so $(2 - 3i)(2 + 3i)$ will be a real number.

You can enter complex numbers directly into your calculator to multiply or divide them quickly.

Divide each term in the numerator by 13.

Exercise 1D SKILLS INTERPRETATION

Do not use your calculator in this exercise.

1 Write down the complex conjugate z^* for:

a $z = 8 + 2\text{i}$ b $z = 6 - 5\text{i}$ c $z = \frac{2}{3} - \frac{1}{2}\text{i}$ d $z = \sqrt{5} + \text{i}\sqrt{10}$

2 Find $z + z^*$ and zz^* for:

a $z = 6 - 3\text{i}$ b $z = 10 + 5\text{i}$ c $z = \frac{3}{4} + \frac{1}{4}\text{i}$ d $z = \sqrt{5} - 3\text{i}\sqrt{5}$

3 Write each of the following in the form $a + b\text{i}$.

a $\frac{3 - 5\text{i}}{1 + 3\text{i}}$ b $\frac{3 + 5\text{i}}{6 - 8\text{i}}$ c $\frac{28 - 3\text{i}}{1 - \text{i}}$ d $\frac{2 + \text{i}}{1 + 4\text{i}}$

4 Write $\frac{(3 - 4\text{i})^2}{1 + \text{i}}$ in the form $x + \text{i}y$, where $x, y \in \mathbb{R}$.

5 Given that $z_1 = 1 + \text{i}$, $z_2 = 2 + \text{i}$ and $z_3 = 3 + \text{i}$, write each of the following in the form $a + b\text{i}$.

a $\frac{z_1 z_2}{z_3}$ b $\frac{(z_2)^2}{z_1}$ c $\frac{2z_1 + 5z_3}{z_2}$

(E) 6 Given that $\frac{5 + 2\text{i}}{z} = 2 - \text{i}$, find z in the form $a + b\text{i}$. **(2 marks)**

7 Simplify $\frac{6 + 8\text{i}}{1 + \text{i}} + \frac{6 + 8\text{i}}{1 - \text{i}}$, giving your answer in the form $a + b\text{i}$.

8 $w = \frac{4}{8 - \text{i}\sqrt{2}}$

Express w in the form $a + b\text{i}\sqrt{2}$, where a and b are **rational numbers.**

9 $w = 1 - 9\text{i}$

Express $\frac{1}{w}$ in the form $a + b\text{i}$, where a and b are rational numbers.

10 $z = 4 - \text{i}\sqrt{2}$

Use algebra to express $\frac{z + 4}{z - 3}$ in the form $p + q\text{i}\sqrt{2}$, where p and q are rational numbers.

(E/P) 11 The complex number z satisfies the equation $(4 + 2\text{i})(z - 2\text{i}) = 6 - 4\text{i}$.
Find z, giving your answer in the form $a + b\text{i}$ where a and b are rational numbers. **(4 marks)**

(E/P) 12 The complex numbers z_1 and z_2 are given by $z_1 = p - 7\text{i}$ and $z_2 = 2 + 5\text{i}$, where p is an integer.
Find $\frac{z_1}{z_2}$ in the form $a + b\text{i}$, where a and b are rational, and are given in terms of p. **(4 marks)**

(E) 13 $z = \sqrt{5} + 4\text{i}$. z^* is the complex conjugate of z.
Show that $\frac{z}{z^*} = a + b\text{i}\sqrt{5}$, where a and b are rational numbers to be found. **(4 marks)**

(E/P) 14 The complex number z is defined by $z = \frac{p + 5\text{i}}{p - 2\text{i}}$, $p \in \mathbb{R}$, $p > 0$.
Given that the real part of z is $\frac{1}{2}$,

a find the value of p **(4 marks)**

b write z in the form $a + b\text{i}$, where a and b are real. **(1 mark)**

1.4 Argand diagrams

- You can represent complex numbers on an **Argand diagram**. The x-axis on an Argand diagram is called the real axis and the y-axis is called the imaginary axis. The complex number $z = x + iy$ is represented on the diagram by the point $P(x, y)$, where x and y are **Cartesian coordinates.**

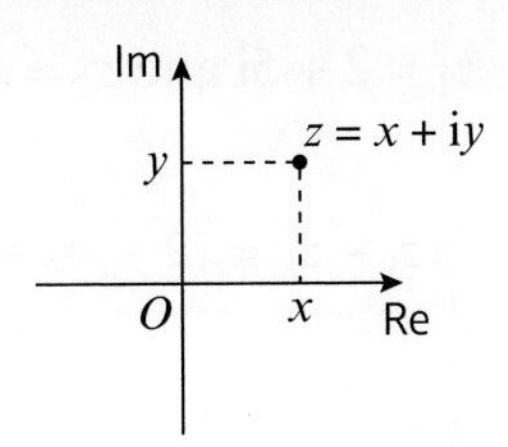

Example 9 SKILLS INTERPRETATION

Show the complex numbers $z_1 = -4 + i$, $z_2 = 2 + 3i$ and $z_3 = 2 - 3i$ on an Argand diagram.

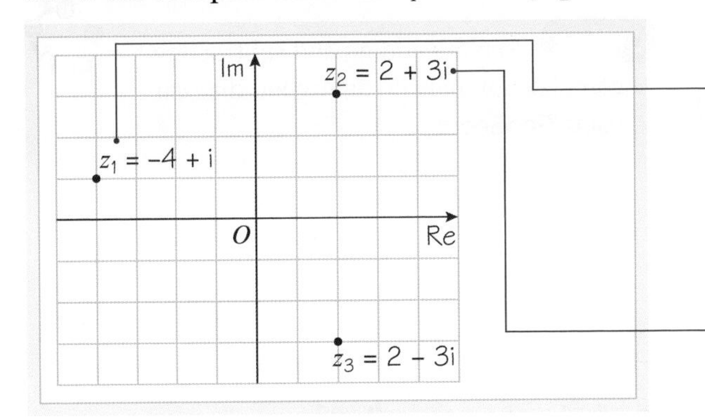

The real part of each number describes its horizontal position, and the imaginary part describes its vertical position. For example, $z_1 = -4 + i$ has real part -4 and imaginary part 1.

Note that z_2 and z_3 are complex conjugates. On an Argand diagram, complex conjugate pairs are symmetrical about the real axis.

Complex numbers can also be represented as vectors on an Argand diagram.

- **The complex number $z = x + iy$ can be represented as the vector $\begin{pmatrix} x \\ y \end{pmatrix}$ on an Argand diagram.**

You can add or subtract complex numbers on an Argand diagram by adding or subtracting their **corresponding** (i.e. equivalent) vectors.

Example 10

$z_1 = 4 + i$ and $z_2 = 3 + 3i$. Show z_1, z_2 and $z_1 + z_2$ on an Argand diagram.

$z_1 + z_2 = (4 + 3) + (1 + 3)i = 7 + 4i$

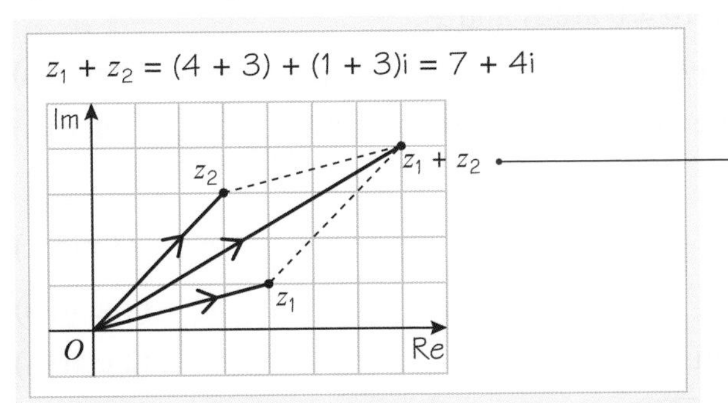

The **vector** representing $z_1 + z_2$ is the diagonal of the parallelogram with vertices at O, z_1 and z_2. You can use vector addition to find $z_1 + z_2$:

$$\begin{pmatrix} 4 \\ 1 \end{pmatrix} + \begin{pmatrix} 3 \\ 3 \end{pmatrix} = \begin{pmatrix} 7 \\ 4 \end{pmatrix}$$

Example 11

$z_1 = 2 + 5i$ and $z_2 = 4 + 2i$. Show z_1, z_2 and $z_1 - z_2$ on an Argand diagram.

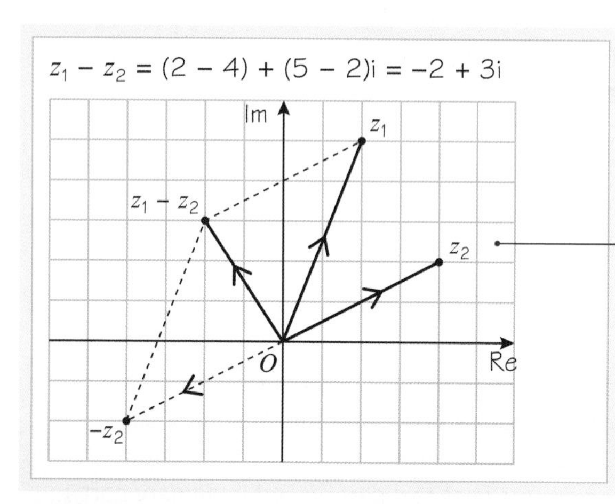

The vector corresponding to z_2 is $\begin{pmatrix}4\\2\end{pmatrix}$, so the vector corresponding to $-z_2$ is $\begin{pmatrix}-4\\-2\end{pmatrix}$.

The vector representing $z_1 - z_2$ is the diagonal of the parallelogram with vertices at O, z_1 and $-z_2$.

Online Explore adding and subtracting complex numbers on an Argand diagram using GeoGebra.

Exercise 1E

SKILLS INTERPRETATION

1 Show these numbers on an Argand diagram.

a $7 + 2i$ **b** $5 - 4i$ **c** $-6 - i$ **d** $-2 + 5i$

e $3i$ **f** $\sqrt{2} + 2i$ **g** $-\frac{1}{2} + \frac{5}{2}i$ **h** -4

2 $z_1 = 11 + 2i$ and $z_2 = 2 + 4i$. Show z_1, z_2 and $z_1 + z_2$ on an Argand diagram.

3 $z_1 = -3 + 6i$ and $z_2 = 8 - i$. Show z_1, z_2 and $z_1 + z_2$ on an Argand diagram.

4 $z_1 = 8 + 4i$ and $z_2 = 6 + 7i$. Show z_1, z_2 and $z_1 - z_2$ on an Argand diagram.

5 $z_1 = -6 - 5i$ and $z_2 = -4 + 4i$. Show z_1, z_2 and $z_1 - z_2$ on an Argand diagram.

(P) **6** $z_1 = 7 - 5i$, $z_2 = a + bi$ and $z_3 = -3 + 2i$, where $a, b \in \mathbb{Z}$. Given that $z_3 = z_1 + z_2$,

a find the values of a and b **b** show z_1, z_2 and z_3 on an Argand diagram.

(P) **7** $z_1 = p + qi$, $z_2 = 9 - 5i$ and $z_3 = -8 + 5i$, where $p, q \in \mathbb{Z}$. Given that $z_3 = z_1 + z_2$,

a find the values of p and q **b** show z_1, z_2 and z_3 on an Argand diagram.

(E) **8** The solutions to the quadratic equation $z^2 - 6z + 10 = 0$ are z_1 and z_2.

a Find z_1 and z_2, giving your answers in the form $p \pm qi$, where p and q are integers. **(3 marks)**

b Show, on an Argand diagram, the points representing the complex numbers z_1 and z_2. **(2 marks)**

(E/P) **9** $f(z) = 2z^3 - 19z^2 + 64z - 60$

a Show that $f\left(\frac{3}{2}\right) = 0$. **(1 mark)**

b Use algebra to solve $f(z) = 0$ completely. **(4 marks)**

c Show all three solutions on an Argand diagram. **(2 marks)**

Challenge

SKILLS CREATIVITY

a Find all the solutions to the equation $z^6 = 1$.

b Show each solution on an Argand diagram.

c Show that each solution lies on a circle with centre (0, 0) and radius 1.

Hint There will be six distinct roots in total. Write $z^6 = 1$ as $(z^3 - 1)(z^3 + 1) = 0$, then find three distinct roots of $z^3 - 1 = 0$ and three distinct roots of $z^3 + 1 = 0$.

1.5 Modulus and argument

The **modulus** or absolute value of a complex number is the magnitude (i.e. size) of its corresponding vector.

- **The modulus of a complex number, $|z|$, is the distance from the origin to that number on an Argand diagram. For a complex number $z = x + \mathrm{i}y$, the modulus is given by $|z| = \sqrt{x^2 + y^2}$.**

Notation The modulus of the complex number z is written as r, $|z|$ or $|x + \mathrm{i}y|$.

The **argument** of a complex number is the angle its corresponding vector makes with the positive real axis.

- **The argument of a complex number, arg z, is the angle between the positive real axis and the line joining that number to the origin on an Argand diagram, measured in an anticlockwise direction (i.e. moving in the opposite direction to the hands of a clock). For a complex number $z = x + \mathrm{i}y$, the argument, θ, satisfies $\tan\theta = \frac{y}{x}$.**

Notation The argument of the complex number z is written as arg z. It is usually given in radians, where
- 2π radians = 360°
- π radians = 180°

← Pure 1 Section 7.1

The argument θ of any complex number is usually given in the range $-\pi < \theta \leqslant \pi$. This is sometimes referred to as the **principal argument**.

Example 12 SKILLS PROBLEM-SOLVING

Given the complex number $z = 2 + 7\mathrm{i}$, find:

a the modulus of z **b** the argument of z, giving your answer in **radians** to 2 d.p.

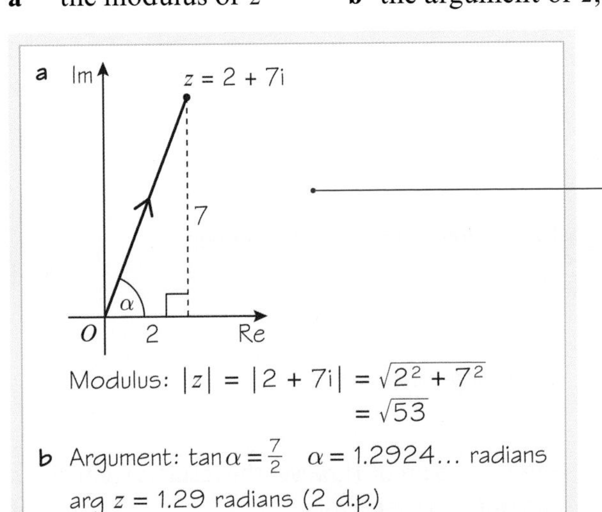

Sketch the Argand diagram, showing the position of the number.

If z does not lie in the first **quadrant**, you can use an Argand diagram to help you find its argument.

- Let α be the positive **acute** angle made with the real axis by the line joining the origin and z.
  - If z lies in the first quadrant, then $\arg z = \alpha$.
  - If z lies in the second quadrant, then $\arg z = \pi - \alpha$.
  - If z lies in the third quadrant, then $\arg z = -(\pi - \alpha)$.
  - If z lies in the fourth quadrant, then $\arg z = -\alpha$.

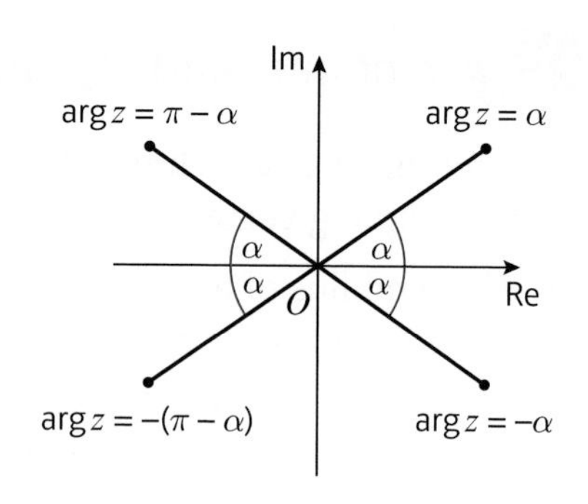

Example 13

Given the complex number $z = -4 - i$, find:

a the modulus of z **b** the argument of z, giving your answer in radians to 2 d.p.

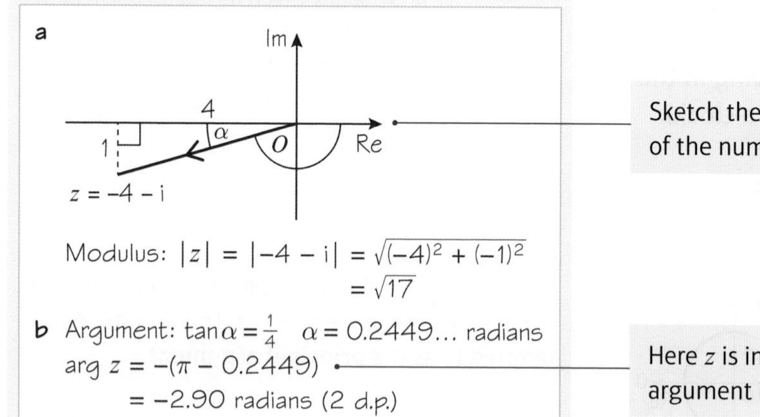

Sketch the Argand diagram, showing the position of the number.

Here z is in the third quadrant, so the required argument is $-(\pi - \alpha)$.

You can use the following rule to multiply the moduli of complex numbers quickly.

For any two complex numbers z_1 and z_2,

$$|z_1||z_2| = |z_1 z_2|.$$

The proof of this result is beyond the scope of this book.

Example 14

$z_1 = 3 + 4i$ and $z_2 = 5 - 12i$

Find:

a the modulus of z_1 and the modulus of z_2

b $z_1 z_2$

c hence, find $|z_1 z_2|$ and verify that $|z_1||z_2| = |z_1 z_2|$

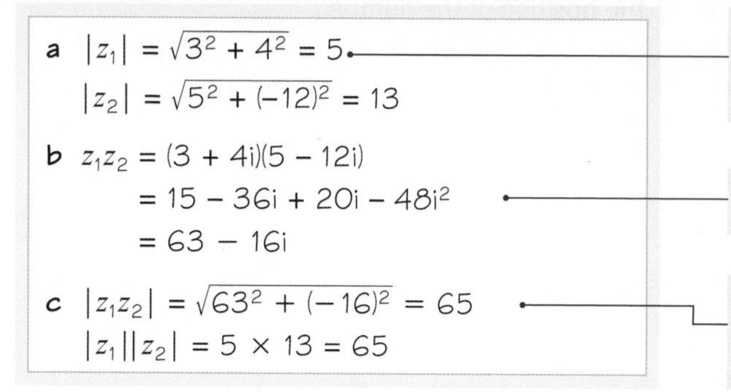

The modulus of a complex number $|a + bi| = \sqrt{a^2 + b^2}$

Don't forget that $i^2 = -1$.

This is not a proof. However the result is verified and works in every case.

Exercise 1F SKILLS PROBLEM-SOLVING

1 For each of the following complex numbers,

i find the modulus, writing your answer in surd form if necessary

ii find the argument, writing your answer in radians to 2 decimal places.

a $z = 12 + 5\text{i}$ **b** $z = \sqrt{3} + \text{i}$ **c** $z = -3 + 6\text{i}$

d $z = 2 - 2\text{i}$ **e** $z = -8 - 7\text{i}$ **f** $z = -4 + 11\text{i}$

g $z = 2\sqrt{3} - \text{i}\sqrt{3}$ **h** $z = -8 - 15\text{i}$

Hint In part **c**, the complex number is in the second quadrant, so the argument will be $\pi - \alpha$. In part **d**, the complex number is in the fourth quadrant, so the argument will be $-\alpha$.

2 For each of the following complex numbers,

i find the modulus, writing your answer in surd form

ii find the argument, writing your answer in terms of π.

a $2 + 2\text{i}$ **b** $5 + 5\text{i}$ **c** $-6 + 6\text{i}$ **d** $-a - a\text{i}, a \in \mathbb{R}$

3 The complex number z_1 is such that $z_1 = 3 + 5\text{i}$.

a Find $|z_1|$. **b** Find $(z_1)^2$. **c** Hence verify that $|(z_1)^2| = |z_1|^2$.

4 The complex number z_1 is such that $z_1 = \dfrac{26}{3 + 2\text{i}}$.

a Write z_1 in the form $a + b\text{i}$, where a and b are integers.

b Find $|z_1|$.

c Given that $|z_1 z_2| = 26\sqrt{13}$, find $|z_2|$.

d Given also that $z_2 = 5 + p\text{i}$, find the possible values of p.

(E) **5** $z = -40 - 9\text{i}$

a Show z on an Argand diagram. **(1 mark)**

b Calculate arg z, giving your answer in radians to 2 decimal places. **(2 marks)**

(E) **6** $z = 3 + 4\text{i}$

a Show that $z^2 = -7 + 24\text{i}$. **(2 marks)**

Find, showing your working:

b $|z^2|$ **(2 marks)**

c $\arg(z^2)$, giving your answer in radians to 2 decimal places. **(2 marks)**

d Show z and z^2 on an Argand diagram. **(1 mark)**

(E) **7** The complex numbers z_1 and z_2 are given by $z_1 = 4 + 6\text{i}$ and $z_2 = 1 + \text{i}$.

Find, showing your working:

a $\dfrac{z_1}{z_2}$ in the form $a + b\text{i}$, where a and b are real **(3 marks)**

b $\left|\dfrac{z_1}{z_2}\right|$ **(2 marks)**

c $\arg \dfrac{z_1}{z_2}$, giving your answer in radians to 2 decimal places. **(2 marks)**

(E/P) **8** The complex numbers z_1 and z_2 are such that $z_1 = 3 + 2p\text{i}$ and $\frac{z_1}{z_2} = 1 - \text{i}$, where p is a real constant.

a Find z_2 in the form $a + b\text{i}$, giving the real numbers a and b in terms of p. **(3 marks)**

Given that $\arg z_2 = \tan^{-1} 5$,

b find the value of p **(2 marks)**

c find the value of $|z_2|$ **(2 marks)**

d show z_1, z_2 and $\frac{z_1}{z_2}$ on a single Argand diagram. **(2 marks)**

(E) **9** Given the complex number $z = \frac{26}{2 - 3\text{i}}$, find:

a z in the form $a + \text{i}b$, where $a, b \in \mathbb{R}$ **(2 marks)**

b z^2 in the form $a + \text{i}b$, where $a, b \in \mathbb{R}$ **(2 marks)**

c $|z|$ **(2 marks)**

d $\arg (z^2)$, giving your answer in radians to 2 decimal places. **(2 marks)**

(E/P) **10** Given that $z_1 = 4 + 2\text{i}$, $z_2 = 2 + 4\text{i}$, $z_3 = a + b\text{i}$, where $a, b \in \mathbb{R}$,

a find the exact value of $|z_1 + z_2|$. **(2 marks)**

Given that $w = \frac{z_1 z_3}{z_2}$,

b find w in terms of a and b, giving your answer in the form $x + \text{i}y$, where $x, y \in \mathbb{R}$. **(4 marks)**

Given also that $w = \frac{21}{5} - \frac{22}{5}\text{i}$, find:

c the values of a and b **(3 marks)**

d $\arg w$, giving your answer in radians to 2 decimal places. **(2 marks)**

(E/P) **11** The complex number w is given by $w = 6 + 3\text{i}$. Find:

a $|w|$ **(1 mark)**

b $\arg w$, giving your answer in radians to 2 decimal places. **(2 marks)**

Given that $\arg(\lambda + 5\text{i} + w) = \frac{\pi}{4}$ where λ is a real constant,

c find the value of λ. **(2 marks)**

(E) **12** Given the complex number $z = -1 - \text{i}\sqrt{3}$, find:

a $|z|$ **(1 mark)**

b $\left|\frac{z}{z^*}\right|$ **(4 marks)**

c $\arg z$, $\arg (z^*)$ and $\arg \frac{z}{z^*}$, giving your answers in terms of π. **(3 marks)**

(E/P) **13** The complex numbers w and z are given by $w = k + \text{i}$ and $z = -4 + 5k\text{i}$, where k is a real constant. Given that $\arg(w + z) = \frac{2\pi}{3}$, find the exact value of k. **(6 marks)**

(E/P) **14** The complex numbers w and z are defined such that $\arg w = \frac{\pi}{10}$, $|w| = 5$ and $\arg z = \frac{2\pi}{5}$.
Given that $\arg(w + z) = \frac{\pi}{5}$, find the value of $|z|$. **(4 marks)**

1.6 Modulus–argument form of complex numbers

You can write any complex number in terms of its modulus and argument.

- **For a complex number z with $|z| = r$ and $\arg z = \theta$, the modulus–argument form of z is $z = r(\cos\theta + \mathrm{i}\sin\theta)$.**

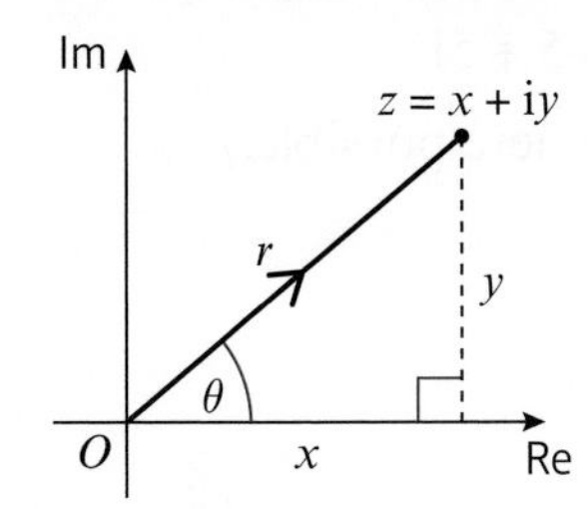

From the right-angled triangle, $x = r\cos\theta$ and $y = r\sin\theta$.

$z = x + \mathrm{i}y = r\cos\theta + \mathrm{i}r\sin\theta = r(\cos\theta + \mathrm{i}\sin\theta)$

This formula works for a complex number in any quadrant of the Argand diagram. The argument, θ, is usually given in the range $-\pi < \theta \leqslant \pi$, although the formula works for any value of θ measured anticlockwise from the positive real axis.

Example 15 SKILLS INTERPRETATION

Express $z = -\sqrt{3} + \mathrm{i}$ in the form $r(\cos\theta + \mathrm{i}\sin\theta)$, where $-\pi < \theta \leqslant \pi$.

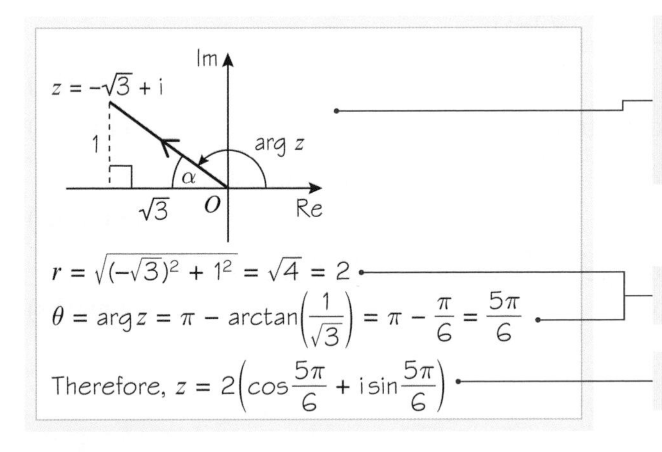

$r = \sqrt{(-\sqrt{3})^2 + 1^2} = \sqrt{4} = 2$

$\theta = \arg z = \pi - \arctan\left(\frac{1}{\sqrt{3}}\right) = \pi - \frac{\pi}{6} = \frac{5\pi}{6}$

Therefore, $z = 2\left(\cos\frac{5\pi}{6} + \mathrm{i}\sin\frac{5\pi}{6}\right)$

Sketch the Argand diagram, showing the position of the number.

Here z is in the second quadrant, so the required argument is $\pi - \alpha$.

Find r and θ.

Apply $z = r(\cos\theta + \mathrm{i}\sin\theta)$.

Example 16

Express $z = -1 - \mathrm{i}$ in the form $r(\cos\theta + \mathrm{i}\sin\theta)$, where $-\pi < \theta \leqslant \pi$.

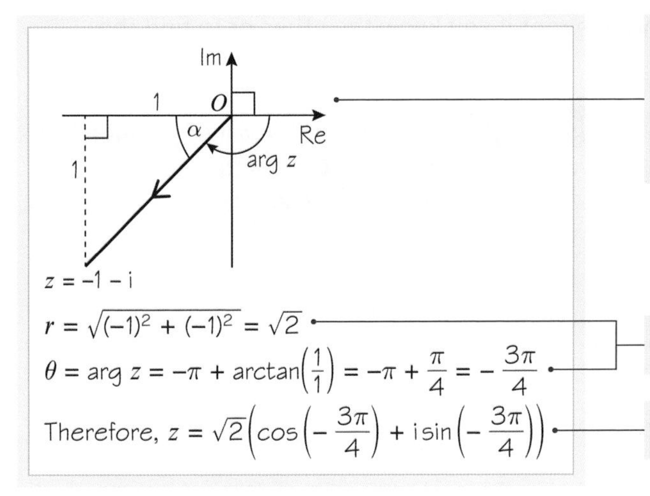

$r = \sqrt{(-1)^2 + (-1)^2} = \sqrt{2}$

$\theta = \arg z = -\pi + \arctan\left(\frac{1}{1}\right) = -\pi + \frac{\pi}{4} = -\frac{3\pi}{4}$

Therefore, $z = \sqrt{2}\left(\cos\left(-\frac{3\pi}{4}\right) + \mathrm{i}\sin\left(-\frac{3\pi}{4}\right)\right)$

Sketch the Argand diagram, showing the position of the number.

Here z is in the third quadrant, so the required argument is $-(\pi - \alpha)$.

Find r and θ.

Apply $z = r(\cos\theta + \mathrm{i}\sin\theta)$.

Exercise 1G SKILLS INTERPRETATION

1 Express the following in the form $r(\cos\theta + \text{i}\sin\theta)$, where $-\pi < \theta \leqslant \pi$.
Give the exact values of r and θ where possible, or values to 2 d.p. otherwise.

a $2 + 2\text{i}$ **b** 3i **c** $-3 + 4\text{i}$ **d** $1 - \text{i}\sqrt{3}$

e $-2 - 5\text{i}$ **f** -20 **g** $7 - 24\text{i}$ **h** $-5 + 5\text{i}$

2 Express these in the form $r(\cos\theta + \text{i}\sin\theta)$, giving exact values of r and θ where possible, or values to 2 d.p. otherwise.

a $\frac{3}{1 + \text{i}\sqrt{3}}$ **b** $\frac{1}{2 - \text{i}}$ **c** $\frac{1 + \text{i}}{1 - \text{i}}$

3 Express the following in the form $x + \text{i}y$, where $x, y \in \mathbb{R}$.

a $5\left(\cos\frac{\pi}{2} + \text{i}\sin\frac{\pi}{2}\right)$ **b** $\frac{1}{2}\left(\cos\frac{\pi}{6} + \text{i}\sin\frac{\pi}{6}\right)$ **c** $6\left(\cos\frac{5\pi}{6} + \text{i}\sin\frac{5\pi}{6}\right)$

d $3\left(\cos\left(-\frac{2\pi}{3}\right) + \text{i}\sin\left(-\frac{2\pi}{3}\right)\right)$ **e** $2\sqrt{2}\left(\cos\left(-\frac{\pi}{4}\right) + \text{i}\sin\left(-\frac{\pi}{4}\right)\right)$ **f** $-4\left(\cos\frac{7\pi}{6} + \text{i}\sin\frac{7\pi}{6}\right)$

(E) 4 **a** Express the complex number $z = 4\left(\cos\left(\frac{2\pi}{3}\right) + \text{i}\sin\left(\frac{2\pi}{3}\right)\right)$ in the form $x + \text{i}y$, where $x, y \in \mathbb{R}$. **(2 marks)**

b Show the complex number z on an Argand diagram. **(1 mark)**

(E) 5 The complex number z is such that $|z| = 7$ and $\arg z = \frac{11\pi}{6}$. Find z in the form $p + q\text{i}$, where p and q are exact real numbers to be found. **(3 marks)**

(E) 6 The complex number z is such that $|z| = 5$ and $\arg z = -\frac{4\pi}{3}$. Find z in the form $a + b\text{i}$, where a and b are exact real numbers to be found. **(3 marks)**

1.7 Roots of quadratic equations

- **For real numbers a, b and c, if the roots of the quadratic equation $az^2 + bz + c = 0$ are non-real complex numbers, then they occur as a conjugate pair.**

Another way of stating this is that for a real-valued quadratic **function** $f(z)$, if z_1 is a root of $f(z) = 0$ then z_1^* is also a root. You can use this fact to find one root if you know the other, or to find the original equation.

- **If the roots of a quadratic equation are α and β, then you can write the equation as $(z - \alpha)(z - \beta) = 0$**

or $z^2 - (\alpha + \beta)z + \alpha\beta = 0$

Notation Roots of complex-valued **polynomials** are often written using Greek letters such as α (alpha), β (beta) and γ (gamma).

Example 17 SKILLS EXECUTIVE FUNCTION

Given that $\alpha = 7 + 2i$ is one of the roots of a quadratic equation with real coefficients,

a state the value of the other root, β

b find the quadratic equation

c find the values of $\alpha + \beta$ and $\alpha\beta$ and interpret the results.

a $\beta = 7 - 2i$

α and β will always be a complex conjugate pair.

b
$$(z - \alpha)(z - \beta) = 0$$
$$(z - (7 + 2i))(z - (7 - 2i)) = 0$$
$$z^2 - z(7 - 2i) - z(7 + 2i) + (7 + 2i)(7 - 2i) = 0$$
$$z^2 - 7z + 2iz - 7z - 2iz + 49 - 14i + 14i - 4i^2 = 0$$
$$z^2 - 14z + 49 + 4 = 0$$
$$z^2 - 14z + 53 = 0$$

The quadratic equation with roots α and β is $(z - \alpha)(z - \beta) = 0$.

Collect like terms. Use the fact that $i^2 = -1$.

c $\alpha + \beta = (7 + 2i) + (7 - 2i)$
$= (7 + 7) + (2 + (-2))i = 14$

The coefficient of z in $z^2 - 14z + 53$ is $-(\alpha + \beta)$.

$\alpha\beta = (7 + 2i)(7 - 2i) = 49 - 14i + 14i - 4i^2$
$= 49 + 4 = 53$

The constant term in $z^2 - 14z + 53$ is $\alpha\beta$.

Hint For $z = a + bi$, you should learn the results:

$$z + z^* = 2a$$
$$zz^* = a^2 + b^2$$

You can use these to find the quadratic equation quickly.

Exercise 1H SKILLS EXECUTIVE FUNCTION

1 The roots of the quadratic equation $z^2 + 2z + 26 = 0$ are α and β.

Find: **a** α and β **b** $\alpha + \beta$ **c** $\alpha\beta$

2 The roots of the quadratic equation $z^2 - 8z + 25 = 0$ are α and β.

Find: **a** α and β **b** $\alpha + \beta$ **c** $\alpha\beta$

(E) **3** Given that $2 + 3i$ is one of the roots of a quadratic equation with real coefficients,

a write down the other root of the equation **(1 mark)**

b find the quadratic equation, giving your answer in the form $z^2 + bz + c = 0$ where b and c are real constants. **(3 marks)**

(E) **4** Given that $5 - i$ is a root of the equation $z^2 + pz + q = 0$, where p and q are real constants,

a write down the other root of the equation **(1 mark)**

b find the value of p and the value of q. **(3 marks)**

(E/P) **5** Given that $z_1 = -5 + 4i$ is one of the roots of the quadratic equation $z^2 + bz + c = 0$, where b and c are real constants, find the values of b and c. **(4 marks)**

(E/P) **6** Given that $1 + 2i$ is one of the roots of a quadratic equation with real coefficients, find the equation, giving your answer in the form $z^2 + bz + c = 0$, where b and c are integers to be found. **(4 marks)**

(E/P) **7** Given that $3 - 5\mathrm{i}$ is one of the roots of a quadratic equation with real coefficients, find the equation, giving your answer in the form $z^2 + bz + c = 0$, where b and c are real constants. **(4 marks)**

(E/P) **8** $z = \dfrac{5}{3 - \mathrm{i}}$

a Find z in the form $a + b\mathrm{i}$, where a and b are real constants. **(1 mark)**

Given that z is a complex root of the quadratic equation $z^2 + pz + q = 0$, where p and q are rational numbers,

b find the value of p and the value of q. **(4 marks)**

(E/P) **9** Given that $z = 5 + q\mathrm{i}$ is a root of the equation $z^2 - 4pz + 34 = 0$, where p and q are positive real constants, find the value of p and the value of q. **(4 marks)**

1.8 Solving cubic and quartic equations

You can generalise the rule for the roots of quadratic equations to any polynomial with real coefficients.

- **If f(z) is a polynomial with real coefficients, and z_1 is a root of f(z) = 0, then z_1^* is also a root of f(z) = 0.**

Notation If z_1 is real, then $z_1^* = z_1$.

You can use this property (i.e. characteristic) to find roots of **cubic** and **quartic** equations with real coefficients.

- **An equation of the form $az^3 + bz^2 + cz + d = 0$ is called a cubic equation, and has three roots.**
- **For a cubic equation with real coefficients, either:**
  - **all three roots are real, or**
  - **one root is real and the other two roots form a complex conjugate pair.**

Watch out A real-valued cubic equation might have two or three repeated real roots.

Example 18 SKILLS EXECUTIVE FUNCTION

Given that -1 is a root of the equation $z^3 - z^2 + 3z + k = 0$,

a find the value of k **b** find the other two roots of the equation.

a If -1 is a root,

$(-1)^3 - (-1)^2 + 3(-1) + k = 0$

$-1 - 1 - 3 + k = 0$

$k = 5$

b -1 is a root of the equation, so $z + 1$ is a factor of $z^3 - z^2 + 3z + 5$.

$$
\begin{array}{r}
z^2 - 2z + 5 \\
z + 1 \overline{) z^3 - z^2 + 3z + 5} \\
\underline{z^3 + z^2} \\
-2z^2 + 3z \\
\underline{-2z^2 - 2z} \\
5z + 5 \\
\underline{5z + 5} \\
0
\end{array}
$$

Problem-solving Use the factor theorem to help: if f(α) = 0, then α is a root of the polynomial and $z - \alpha$ is a factor of the polynomial.

Use long division (or inspection) to find the quadratic factor.

$z^3 - z^2 + 3z + 5 = (z + 1)(z^2 - 2z + 5) = 0$

Solving $z^2 - 2z + 5 = 0$,

$z^2 - 2z = (z - 1)^2 - 1$

$z^2 - 2z + 5 = (z - 1)^2 - 1 + 5 = (z - 1)^2 + 4$

$(z - 1)^2 + 4 = 0$

$(z - 1)^2 = -4$

$z - 1 = \pm\sqrt{-4} = \pm 2\text{i}$

$z = 1 \pm 2\text{i}$

$z = 1 + 2\text{i}, z = 1 - 2\text{i}$

So the other two roots of the equation are $1 + 2\text{i}$ and $1 - 2\text{i}$.

The other two roots are found by solving the quadratic equation.

Solve by completing the square. Alternatively, you could use the quadratic formula.

The quadratic equation has complex roots which must be a conjugate pair.

You could write the equation as $(z + 1)[z - (1 + 2\text{i})][z - (1 - 2\text{i})] = 0$

- **An equation of the form $az^4 + bz^3 + cz^2 + dz + e = 0$ is called a quartic equation, and has four roots.**
- **For a quartic equation with real coefficients, either:**
  - **all four roots are real, or**
  - **two roots are real and the other two roots form a complex conjugate pair, or**
  - **two roots form a complex conjugate pair and the other two roots also form a complex conjugate pair.**

Watch out A real-valued quartic equation might have repeated real roots or repeated complex roots.

Example 19

Given that $3 + \text{i}$ is a root of the quartic equation $2z^4 - 3z^3 - 39z^2 + 120z - 50 = 0$, solve the equation completely.

Another root is $3 - \text{i}$.

So $(z - (3 + \text{i}))(z - (3 - \text{i}))$ is a factor of $2z^4 - 3z^3 - 39z^2 + 120z - 50$

$(z - (3 + \text{i}))(z - (3 - \text{i})) = z^2 - z(3 - \text{i}) - z(3 + \text{i}) + (3 + \text{i})(3 - \text{i})$
$= z^2 - 6z + 10$

So $z^2 - 6z + 10$ is a factor of $2z^4 - 3z^3 - 39z^2 + 120z - 50$.

$(z^2 - 6z + 10)(az^2 + bz + c) = 2z^4 - 3z^3 - 39z^2 + 120z - 50$

Consider $2z^4$:

The only z^4 term in the expansion is $z^2 \times az^2$, so $a = 2$.

$(z^2 - 6z + 10)(2z^2 + bz + c) = 2z^4 - 3z^3 - 39z^2 + 120z - 50$

Consider $-3z^3$:

The z^3 terms in the expansion are $z^2 \times bz$ and $-6z \times 2z^2$,

so $bz^3 - 12z^3 = -3z^3$

$b - 12 = -3$

$b = 9$

so $(z^2 - 6z + 10)(2z^2 + 9z + c) = 2z^4 - 3z^3 - 39z^2 + 120z - 50$

Complex roots occur in conjugate pairs.

If α and β are roots of $f(z) = 0$, then $(z - \alpha)(z - \beta)$ is a factor of $f(z)$.

You can work this out quickly by noting that
$[z - (a + b\text{i})][z - (a - b\text{i})]$
$= z^2 - 2az + a^2 + b^2$

Problem-solving

It is possible to **factorise** a polynomial without using a formal **algebraic** method. Here, the polynomial is factorised by 'inspection' (i.e. looking carefully). By considering each term of the quartic separately, it is possible to work out the missing coefficients.

Consider −50:

The only constant term in the expansion is $10 \times c$, so $c = -5$.

$2z^4 - 3z^3 - 39z^2 + 120z - 50 = (z^2 - 6z + 10)(2z^2 + 9z - 5)$

Solving $2z^2 + 9z - 5 = 0$:

$(2z - 1)(z + 5) = 0$

$z = \frac{1}{2}, z = -5$

So the roots of $2z^4 - 3z^3 - 39z^2 + 120z - 50 = 0$ are:

$\frac{1}{2}, -5, 3 + \mathrm{i}$ and $3 - \mathrm{i}$

You can check this by considering the z and z^2 terms in the expansion.

Example 20 SKILLS EXECUTIVE FUNCTION

Show that $z^2 + 4$ is a factor of $z^4 - 2z^3 + 21z^2 - 8z + 68$.

Hence solve the equation $z^4 - 2z^3 + 21z^2 - 8z + 68 = 0$.

Using long division:

$$\begin{array}{r} z^2 - 2z + 17 \\ z^2 + 4 \overline{) z^4 - 2z^3 + 21z^2 - 8z + 68} \\ \underline{z^4 \quad\quad + 4z^2} \quad\quad\quad\quad \\ -2z^3 + 17z^2 - 8z \quad\quad \\ \underline{-2z^3 \quad\quad\quad - 8z} \quad\quad \\ 17z^2 \quad\quad + 68 \\ \underline{17z^2 \quad\quad + 68} \\ 0 \end{array}$$

There is no remainder and hence $z^2 + 4$ is a factor of $z^4 - 2z^3 + 21z^2 - 8z + 68$.

Alternatively, the quartic can be factorised by inspection:

$z^4 - 2z^3 + 21z^2 - 8z + 68$

$= (z^2 + 4)(az^2 + bz + c)$

$a = 1$, as the leading coefficient is 1.

The only z^3 term is formed by $z^2 \times bz$ so $b = -2$.

The constant term is formed by $4 \times c$, so $4c = 68$, and $c = 17$.

So $z^4 - 2z^3 + 21z^2 - 8z + 68 = (z^2 + 4)(z^2 - 2z + 17) = 0$

Either $z^2 + 4 = 0$ or $z^2 - 2z + 17 = 0$

Solving $z^2 + 4 = 0$:

$z^2 = -4$

$z = \pm 2\mathrm{i}$

Solving $z^2 - 2z + 17 = 0$:

$(z - 1)^2 + 16 = 0$

$(z - 1)^2 = -16$

$z - 1 = \pm 4\mathrm{i}$

$z = 1 \pm 4\mathrm{i}$

So the roots of $z^4 - 2z^3 + 21z^2 - 8z + 68 = 0$ are:

$2\mathrm{i}, -2\mathrm{i}, 1 + 4\mathrm{i}$ and $1 - 4\mathrm{i}$

Solve by completing the square. Alternatively, you could use the quadratic formula.

Watch out You could use your calculator to solve $z^2 - 2z + 17 = 0$. However, you should still write down the equation you are solving, and both roots.

Exercise 1I SKILLS EXECUTIVE FUNCTION

(E) **1** $f(z) = z^3 - 6z^2 + 21z - 26$

a Show that $f(2) = 0$. **(1 mark)**

b Hence solve $f(z) = 0$ completely. **(3 marks)**

(E) **2** $f(z) = 2z^3 + 5z^2 + 9z - 6$

a Show that $f\left(\frac{1}{2}\right) = 0$. **(1 mark)**

b Hence write $f(z)$ in the form $(2z - 1)(z^2 + bz + c)$, where b and c are real constants to be found. **(2 marks)**

c Use algebra to solve $f(z) = 0$ completely. **(2 marks)**

(E/P) **3** $g(z) = 2z^3 - 4z^2 - 5z - 3$

Given that $z = 3$ is a root of the equation $g(z) = 0$, solve $g(z) = 0$ completely. **(4 marks)**

(E) **4** $p(z) = z^3 + 4z^2 - 15z - 68$

Given that $z = -4 + i$ is a solution to the equation $p(z) = 0$,

a show that $z^2 + 8z + 17$ is a factor of $p(z)$. **(2 marks)**

b Hence solve $p(z) = 0$ completely. **(2 marks)**

(E) **5** $f(z) = z^3 + 9z^2 + 33z + 25$

Given that $f(z) = (z + 1)(z^2 + az + b)$, where a and b are real constants,

a find the value of a and the value of b **(2 marks)**

b find the three roots of $f(z) = 0$ **(4 marks)**

c find the **sum** of the three roots of $f(z) = 0$. **(1 mark)**

(E/P) **6** $g(z) = z^3 - 12z^2 + cz + d = 0$, where $c, d \in \mathbb{R}$.

Given that 6 and $3 + i$ are roots of the equation $g(z) = 0$,

a write down the other complex root of the equation **(1 mark)**

b find the value of c and the value of d. **(4 marks)**

(E/P) **7** $h(z) = 2z^3 + 3z^2 + 3z + 1$

Given that $2z + 1$ is a factor of $h(z)$, find the three roots of $h(z) = 0$. **(4 marks)**

(E/P) **8** $f(z) = z^3 - 6z^2 + 28z + k$

Given that $f(2) = 0$,

a find the value of k **(1 mark)**

b find the other two roots of the equation. **(4 marks)**

9 Find the four roots of the equation $z^4 - 16 = 0$.

(E) **10** $f(z) = z^4 - 12z^3 + 31z^2 + 108z - 360$

a Write $f(z)$ in the form $(z^2 - 9)(z^2 + bz + c)$, where b and c are real constants to be found. **(2 marks)**

b Hence find all the solutions to $f(z) = 0$. **(3 marks)**

(P) **11** $g(z) = z^4 + 2z^3 - z^2 + 38z + 130$

Given that $g(2 + 3i) = 0$, find all the roots of $g(z) = 0$.

(E/P) **12** $f(z) = z^4 - 10z^3 + 71z^2 + Qz + 442$, where Q is a real constant.

Given that $z = 2 - 3i$ is a root of the equation $f(z) = 0$,

a show that $z^2 - 6z + 34$ is a factor of $f(z)$ **(4 marks)**

b find the value of Q **(1 mark)**

c solve completely the equation $f(z) = 0$. **(2 marks)**

Challenge

Three of the roots of the equation $z^5 + bz^4 + cz^3 + dz^2 + ez + f = 0$, where $b, c, d, e, f \in \mathbb{R}$, are -2, $2i$ and $1 + i$. Find the values of b, c, d, e and f.

Chapter review 1

1 Given that $z_1 = 8 - 3i$ and $z_2 = -2 + 4i$, find, in the form $a + bi$, where $a, b \in \mathbb{R}$:

a $z_1 + z_2$

b $3z_2$

c $6z_1 - z_2$

(E/P) **2** The equation $z^2 + bz + 14 = 0$, where $b \in \mathbb{R}$, has no real roots.

Find the range of possible values of b. **(3 marks)**

3 The solutions to the quadratic equation $z^2 - 6z + 12 = 0$ are z_1 and z_2.

Find z_1 and z_2, giving each answer in the form $a \pm i\sqrt{b}$.

(E/P) **4** By using the binomial expansion, or otherwise, show that $(1 + 2i)^5 = 41 - 38i$. **(3 marks)**

(E) **5** $f(z) = z^2 - 6z + 10$

Show that $z = 3 + i$ is a solution to $f(z) = 0$. **(2 marks)**

6 You are given the complex numbers $z_1 = 4 + 2i$ and $z_2 = -3 + i$.

Express, in the form $a + bi$, where $a, b \in \mathbb{R}$:

a z_1^* **b** $z_1 z_2$ **c** $\dfrac{z_1}{z_2}$

7 Write $\dfrac{(7 - 2i)^2}{1 + i\sqrt{3}}$ in the form $x + iy$, where $x, y \in \mathbb{R}$.

(E/P) **8** Given that $\dfrac{4 - 7i}{z} = 3 + i$, find z in the form $a + bi$, where $a, b \in \mathbb{R}$. **(2 marks)**

9 You are given the complex number $z = \frac{1}{2 + \text{i}}$.

Express in the form $a + b\text{i}$, where $a, b \in \mathbb{R}$:

a z^2 **b** $z - \frac{1}{z}$

E/P **10** Given that $z = a + b\text{i}$, show that $\frac{z}{z^*} = \left(\frac{a^2 - b^2}{a^2 + b^2}\right) + \left(\frac{2ab}{a^2 + b^2}\right)\text{i}$ **(4 marks)**

E/P **11** The complex number z is defined by $z = \frac{3 + q\text{i}}{q - 5\text{i}}$, where $q \in \mathbb{R}$.

Given that the real part of z is $\frac{1}{13}$,

a find the possible values of q **(4 marks)**

b write the possible values of z in the form $a + b\text{i}$, where a and b are real constants. **(1 mark)**

E/P **12** Given that $z = x + \text{i}y$, find the value of x and the value of y such that $z + 4\text{i}z^* = -3 + 18\text{i}$, where z^* is the complex conjugate of z. **(5 marks)**

13 $z = 9 + 6\text{i}$, $w = 2 - 3\text{i}$

Express $\frac{z}{w}$ in the form $a + b\text{i}$, where a and b are real constants.

E/P **14** The complex number z is given by $z = \frac{q + 3\text{i}}{4 + q\text{i}}$ where q is an integer.

Express z in the form $a + b\text{i}$ where a and b are rational and are given in terms of q. **(4 marks)**

E **15** $\text{f}(z) = z^2 + 5z + 10$

a Find the roots of the equation $\text{f}(z) = 0$, giving your answers in the form $a \pm \text{i}b$, where a and b are real numbers. **(3 marks)**

b Show these roots on an Argand diagram. **(1 mark)**

E **16** Given that $6 - 2\text{i}$ is one of the roots of a quadratic equation with real coefficients,

a write down the other root of the equation **(1 mark)**

b find the quadratic equation, giving your answer in the form $z^2 + bz + c = 0$ where b and c are real constants. **(2 marks)**

E/P **17** Given that $z = 4 - k\text{i}$ is a root of the equation $z^2 - 2mz + 52 = 0$, where k and m are positive real constants, find the value of k and the value of m. **(4 marks)**

E/P **18** $\text{h}(z) = z^3 - 11z + 20$

Given that $2 + \text{i}$ is a root of the equation $\text{h}(z) = 0$, solve $\text{h}(z) = 0$ completely. **(4 marks)**

E/P **19** $\text{f}(z) = z^3 + 6z + 20$

Given that $\text{f}(1 + 3\text{i}) = 0$, solve $\text{f}(z) = 0$ completely. **(4 marks)**

E/P **20** $\text{f}(z) = z^3 + 3z^2 + kz + 48$, $k \in \mathbb{R}$

Given that $\text{f}(4\text{i}) = 0$,

a find the value of k **(2 marks)**

b find the other two roots of the equation. **(3 marks)**

(E/P) **21** $f(z) = z^3 + z^2 + 3z - 5$

Given that $f(-1 + 2i) = 0$,

a find all the solutions to the equation $f(z) = 0$ **(4 marks)**

b show all the roots of $f(z) = 0$ on a single Argand diagram **(2 marks)**

c prove that these three points are the vertices of a right-angled triangle. **(2 marks)**

(E) **22** $f(z) = z^4 - z^3 - 16z^2 - 74z - 60$

a Write $f(z)$ in the form $(z^2 - 5z - 6)(z^2 + bz + c)$, where b and c are real constants to be found. **(2 marks)**

b Hence find all the solutions to $f(z) = 0$. **(3 marks)**

(E/P) **23** $g(z) = z^4 - 6z^3 + 19z^2 - 36z + 78$

Given that $g(3 - 2i) = 0$, find all the roots of $g(z) = 0$. **(4 marks)**

(E/P) **24** $f(z) = z^4 - 2z^3 - 5z^2 + pz + 24$

Given that $f(4) = 0$,

a find the value of p **(1 mark)**

b solve completely the equation $f(z) = 0$. **(5 marks)**

(E/P) **25** $f(z) = z^4 - z^3 + 13z^2 - 47z + 34$

Given that $z = -1 + 4i$ is a solution to the equation,

a find all the solutions to the equation $f(z) = 0$ **(4 marks)**

b show all the roots on a single Argand diagram. **(2 marks)**

(E) **26** The real and imaginary parts of the complex number $z = x + iy$ satisfy the equation $(4 - 3i)x - (1 + 6i)y - 3 = 0$.

a Find the value of x and the value of y. **(3 marks)**

b Show z on an Argand diagram. **(1 mark)**

Find the values of:

c $|z|$ **(2 marks)**

d $\arg z$ **(2 marks)**

(E) **27** A complex number z is given by $z = a + 4i$ where a is a non-zero real number.

a Find $z^2 + 2z$ in the form $x + iy$, where x and y are real expressions in terms of a. **(4 marks)**

Given that $z^2 + 2z$ is real,

b find the value of a. **(1 mark)**

Using this value for a,

c find the values of the modulus and argument of z, giving the argument in radians and giving your answers correct to 3 significant figures. **(3 marks)**

d Show the complex numbers z, z^2 and $z^2 + 2z$ on a single Argand diagram. **(3 marks)**

(E) **28** The complex number z is defined by $z = \frac{3 + 5i}{2 - i}$.

Find:

a $|z|$ **(4 marks)**

b $\arg z$ **(2 marks)**

(E) **29** You are given the complex number $z = 1 + 2i$.

a Show that $|z^2 - z| = 2\sqrt{5}$. **(4 marks)**

b Find $\arg(z^2 - z)$, giving your answer in radians to 2 decimal places. **(2 marks)**

c Show z and $z^2 - z$ on a single Argand diagram. **(2 marks)**

(E) **30** You are given the complex number $z = \frac{1}{2 + i}$.

a Express in the form $a + bi$, where $a, b \in \mathbb{R}$:

i z^2 **ii** $z - \frac{1}{z}$ **(4 marks)**

b Find $|z^2|$. **(2 marks)**

c Find $\arg\left(z - \frac{1}{z}\right)$, giving your answer in radians to 2 decimal places. **(2 marks)**

(E/P) **31** Given is the complex number $z = \frac{a + 3i}{2 + ai}$, where $a \in \mathbb{R}$.

a Given that $a = 4$, find $|z|$. **(3 marks)**

b Show that there is only one value of a for which $\arg z = \frac{\pi}{4}$, and find this value. **(3 marks)**

(E) **32** Express $4 - 4i$ in the form $r(\cos\theta + i\sin\theta)$, where $r > 0$, $-\pi < \theta \leqslant \pi$, giving r and θ as exact values. **(3 marks)**

Challenge

a Explain why a cubic equation with real coefficients cannot have a repeated non-real root.

b By means of an example, show that a quartic equation with real coefficients can have a repeated non-real root.

Summary of key points

1 $i = \sqrt{-1}$ and $i^2 = -1$

2 An **imaginary number** is a number of the form bi, where $b \in \mathbb{R}$.

3 A **complex number** is written in the form $a + bi$, where $a, b \in \mathbb{R}$.

4 Complex numbers can be added or subtracted by adding or subtracting their real parts and adding or subtracting their imaginary parts.

5 You can multiply a real number by a complex number by multiplying out the brackets in the usual way.

6 If $b^2 - 4ac < 0$ then the quadratic equation $ax^2 + bx + c = 0$ has two distinct complex roots, neither of which is real.

7 For any complex number $z = a + bi$, the **complex conjugate** of the number is defined as $z^* = a - bi$.

8 For real numbers a, b and c, if the roots of the quadratic equation $az^2 + bz + c = 0$ are non-real complex numbers, then they occur as a conjugate pair.

9 If the roots of a quadratic equation are α and β, then you can write the equation as

$(z - \alpha)(z - \beta) = 0$ or $z^2 - (\alpha + \beta)z + \alpha\beta = 0$.

10 If $f(z)$ is a polynomial with real coefficients, and z_1 is a root of $f(z) = 0$, then z_1^* is also a root of $f(z) = 0$.

11 An equation of the form $az^3 + bz^2 + cz + d = 0$ is called a cubic equation, and has three roots. For a cubic equation with real coefficients, either:

- all three roots are real, or
- one root is real and the other two roots form a complex conjugate pair.

12 An equation of the form $az^4 + bz^3 + cz^2 + dz + e = 0$ is called a quartic equation, and has four roots.

For a quartic equation with real coefficients, either:

- all four roots are real, or
- two roots are real and the other two roots form a complex conjugate pair, or
- two roots form a complex conjugate pair and the other two roots also form a complex conjugate pair.

13 You can represent complex numbers on an **Argand diagram**. The x-axis on an Argand diagram is called the **real axis** and the y-axis is called the **imaginary axis**. The complex number $z = x + iy$ is represented on the diagram by the point $P(x, y)$, where x and y are Cartesian coordinates.

14 The complex number $z = x + iy$ can be represented as the vector $\begin{pmatrix} x \\ y \end{pmatrix}$ on an Argand diagram.

15 The **modulus** of a complex number, $|z|$, is the distance from the origin to that number on an Argand diagram. For a complex number $z = x + iy$, the modulus is given by $|z| = \sqrt{x^2 + y^2}$.

16 The **argument** of a complex number, $\arg z$, is the angle between the positive real axis and the line joining that number to the origin on an Argand diagram. For a complex number $z = x + iy$, the argument, θ, satisfies $\tan\theta = \frac{y}{x}$.

17 Let α be the positive acute angle made with the real axis by the line joining the origin and z.

- If z lies in the first quadrant, then $\arg z = \alpha$.
- If z lies in the second quadrant, then $\arg z = \pi - \alpha$.
- If z lies in the third quadrant, then $\arg z = -(\pi - \alpha)$.
- If z lies in the fourth quadrant, then $\arg z = -\alpha$.

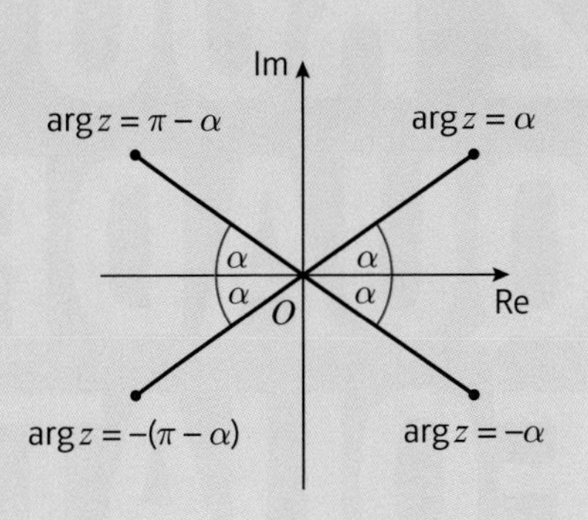

18 For a complex number z with $|z| = r$ and $\arg z = \theta$, the modulus–argument form of z is $z = r(\cos\theta + i\sin\theta)$.

19 For any two complex numbers z_1 and z_2, $|z_1 z_2| = |z_1||z_2|$

2 ROOTS OF QUADRATIC EQUATIONS

Learning objectives

After completing this chapter you should be able to:

- Find the sum of roots and the product of roots of a quadratic equation → pages 29–30
- Manipulate expressions involving the sum of roots and the product of roots → pages 29–30
- Form quadratic equations with new roots → pages 31–33

Prior knowledge check

1 Solve the following quadratic equations.

a $x^2 + 4x + 5 = 0$

b $2x^2 - 7x + 8 = 0$ ← Pure 1 Section 2.1

2 $f(x) = x^2 - 2x - 3$. Find the roots of:

a $f(x) = 0$

b $f(x - 5) = 0$

c $f(2x) = 0$ ← International GCSE Mathematics

Planets orbit the sun in shapes called ellipses whilst some other celestial objects move in hyperbolic orbits. Quadratic equations are used to describe these shapes mathematically.

2.1 Roots of a quadratic equation

A quadratic equation of the form $ax^2 + bx + c = 0$, $x \in \mathbb{C}$, where a, b and c are real constants, can have two real roots, one repeated (real) root or two complex roots.

Links If the roots of a quadratic equation with real coefficients are complex, then they occur as a conjugate pair.
← Further Pure 1 Section 1.7

If the roots of this equation are α and β, you can determine the relationship between the **coefficients** of the terms in the quadratic equation and the values of α and β:

$$ax^2 + bx + c = a(x - \alpha)(x - \beta)$$
$$= a(x^2 - \alpha x - \beta x + \alpha\beta)$$
$$= ax^2 - a(\alpha + \beta)x + a\alpha\beta$$

Write the quadratic expression in factorised form, then rearrange into the form $ax^2 + bx + c$.

So $b = -a(\alpha + \beta)$ and $c = a\alpha\beta$.

Notation The sum of the roots is $-\frac{b}{a}$ and the product of the roots is $\frac{c}{a}$. Note that these values are real even if the roots are complex, because the sum or product of a conjugate pair is real.

If α and β are roots of the equation $ax^2 + bx + c = 0$, then:

- $\alpha + \beta = -\frac{b}{a}$
- $\alpha\beta = \frac{c}{a}$

Example 1

SKILLS PROBLEM-SOLVING

The roots of the quadratic equation $2x^2 - 5x - 4 = 0$ are α and β.
Without solving the equation, find the values of:

a $\alpha + \beta$ **b** $\alpha\beta$ **c** $\frac{1}{\alpha} + \frac{1}{\beta}$ **d** $\alpha^2 + \beta^2$

a $\alpha + \beta = \frac{5}{2}$

Use the result $\alpha + \beta = -\frac{b}{a}$

b $\alpha\beta = -2$

Use the result $\alpha\beta = \frac{c}{a}$

c $\frac{1}{\alpha} + \frac{1}{\beta} = \frac{\alpha + \beta}{\alpha\beta} = \frac{\frac{5}{2}}{-2} = -\frac{5}{4}$

d $\alpha^2 + \beta^2 = (\alpha + \beta)^2 - 2\alpha\beta$

$= \left(\frac{5}{2}\right)^2 - 2(-2) = \frac{41}{4}$

Problem-solving
Write each expression in terms of $\alpha + \beta$ and $\alpha\beta$:
$(\alpha + \beta)^2 = \alpha^2 + \beta^2 + 2\alpha\beta \Rightarrow \alpha^2 + \beta^2 = (\alpha + \beta)^2 - 2\alpha\beta$

Example 2

The roots of the quadratic equation $ax^2 + bx + c = 0$ are $\alpha = -\frac{3}{2}$ and $\beta = \frac{5}{4}$.
Find integer values for a, b and c.

$\alpha + \beta = -\frac{3}{2} + \frac{5}{4} = -\frac{1}{4}$, so $-\frac{1}{4} = -\frac{b}{a}$

$\alpha\beta = -\frac{3}{2} \times \frac{5}{4} = -\frac{15}{8}$, so $-\frac{15}{8} = \frac{c}{a}$

$ax^2 + bx + c = 0$ can be written as $x^2 + \frac{b}{a}x + \frac{c}{a} = 0$
Use the values of $\alpha + \beta$ and $\alpha\beta$ to write down a quadratic equation with roots α and β.

$x^2 + \frac{1}{4}x - \frac{15}{8} = 0$

$8x^2 + 2x - 15 = 0$

$a = 8, b = 2, c = -15$

Any constant multiple of this equation will have roots α and β.

You could also set $a = 8$ to find integer solutions to the equations $-\frac{1}{4} = -\frac{b}{a}$ and $-\frac{15}{8} = \frac{c}{a}$

Exercise 2A

SKILLS PROBLEM-SOLVING

1 α and β are the roots of the quadratic equation $3x^2 + 7x - 4 = 0$.
Without solving the equation, find the values of:

a $\alpha + \beta$
b $\alpha\beta$
c $\frac{1}{\alpha} + \frac{1}{\beta}$
d $\alpha^2 + \beta^2$

2 α and β are the roots of the quadratic equation $7x^2 - 3x + 1 = 0$.
Without solving the equation, find the values of:

a $\alpha + \beta$
b $\alpha\beta$
c $\frac{1}{\alpha} + \frac{1}{\beta}$
d $\alpha^2 + \beta^2$

3 α and β are the roots of the quadratic equation $6x^2 - 9x + 2 = 0$.
Without solving the equation, find the values of:

a $\alpha + \beta$
b $\alpha^2 \times \beta^2$
c $\frac{1}{\alpha} + \frac{1}{\beta}$
d $\alpha^3 + \beta^3$

Hint Try expanding $(\alpha + \beta)^3$.

4 The roots of the quadratic equation $ax^2 + bx + c = 0$ are $\alpha = 2$ and $\beta = -3$.
Find integer values for a, b and c.

5 The roots of the quadratic equation $ax^2 + bx + c = 0$ are $\alpha = -\frac{1}{2}$ and $\beta = -\frac{1}{3}$.
Find integer values for a, b and c.

6 The roots of the quadratic equation $ax^2 + bx + c = 0$ are $\alpha = \frac{-1 + i}{2}$ and $\beta = \frac{-1 - i}{2}$.
Find integer values for a, b and c.

7 One of the roots of the quadratic equation $ax^2 + bx + c = 0$ is $\alpha = -1 - 4i$.

a Write down the other root, β.

b Given that $a = 1$, find the values of b and c.

(P) **8** Given that $kx^2 + (k - 3)x - 2 = 0$, find the value of k if the sum of the roots is 4.

(P) **9** The equation $nx^2 - (16 + n)x + 256 = 0$ has real roots α and $-\alpha$. Find the value of n.

(P) **10** The roots of the equation $6x^2 + 36x + k = 0$ are **reciprocals** of each other. Find the value of k.

(P) **11** The equation $mx^2 + 4x + 4m = 0$ has roots of the form k and $2k$. Find the values of m and k.

2.2 Forming quadratic equations with new roots

You can use the sum of roots and product of roots of a quadratic equation to form a new quadratic equation with roots that are related to the roots of the original equation.

You can rewrite the quadratic equation $ax^2 + bx + c = 0$ in the form $x^2 + \frac{b}{a}x + \frac{c}{a} = 0$ where $\frac{b}{a} = -$ sum of the roots and $\frac{c}{a} =$ product of the roots.

Example 3 SKILLS PROBLEM-SOLVING

The roots of the quadratic equation $3x^2 + x - 6 = 0$ are α and β.

Form a quadratic equation with integer coefficients which has roots:

a α^2 and β^2 **b** $\frac{1}{\alpha}$ and $\frac{1}{\beta}$

Problem-solving

Find $\alpha + \beta$ and $\alpha\beta$ first.

a α^2 and β^2

$3x^2 + x - 6 = 0 \Rightarrow a = 3, b = 1, c = -6$

$\Rightarrow \alpha + \beta = -\frac{1}{3}$

$\Rightarrow \alpha\beta = -\frac{6}{3} = -2$

Find the value of $\alpha + \beta$ and $\alpha\beta$.

For the sum of roots we need $\alpha^2 + \beta^2$

$(\alpha + \beta)^2 = \alpha^2 + 2\alpha\beta + \beta^2$

$\Rightarrow \alpha^2 + \beta^2 = (\alpha + \beta)^2 - 2\alpha\beta$

To find the sum of roots you must expand $(\alpha + \beta)^2$ and rearrange.

$= \left(-\frac{1}{3}\right)^2 - 2(-2) = \frac{37}{9}$

For the product of roots we need $\alpha^2 \times \beta^2$

$\alpha^2 \times \beta^2 = (\alpha\beta)^2 = (-2)^2 = 4$

When you have expressions for the required sum and product in terms of $(\alpha + \beta)$ and $\alpha\beta$ only, substitute these values in to your expressions.

Sum $\alpha^2 + \beta^2 = \frac{37}{9} = -\frac{b}{a}$

Product $\alpha^2 \times \beta^2 = 4 = \frac{c}{a}$

Find the values of the required sum and product.

The new equation with roots α^2 and β^2 can be written as:

$x^2 - \frac{37}{9}x + 4 = 0$

$\Rightarrow 9x^2 - 37x + 36 = 0$

Form a new quadratic equation and multiply through to achieve integer coefficients.

b $\frac{1}{\alpha}$ and $\frac{1}{\beta}$

Sum $\frac{1}{\alpha} + \frac{1}{\beta} = \frac{\alpha + \beta}{\alpha\beta}$

To find the sum of roots you must add $\frac{1}{\alpha}$ and $\frac{1}{\beta}$ and find an expression into which you can substitute $\alpha + \beta$ and $\alpha\beta$.

$\Rightarrow \frac{\alpha + \beta}{\alpha\beta} = \frac{-\frac{1}{3}}{-2} = \frac{1}{6}$

Product $\frac{1}{\alpha} \times \frac{1}{\beta} = \frac{1}{\alpha\beta} = \frac{1}{-2} = -\frac{1}{2}$

Substitute in the values for $\alpha + \beta$ and $\alpha\beta$ into the Sum and Product.

The new equation with roots $\frac{1}{\alpha}$ and $\frac{1}{\beta}$ can be written as:

$x^2 - \frac{1}{6}x - \frac{1}{2} = 0$

$\Rightarrow 6x^2 - x - 3 = 0$

Form a new quadratic equation and multiply through to achieve integer coefficients.

You need to know an identity for $\alpha^3 + \beta^3$.

$(\alpha + \beta)^3 = \alpha^3 + 3\alpha^2\beta + 3\alpha\beta^2 + \beta^3$

So, $\alpha^3 + \beta^3 = (\alpha + \beta)^3 - 3\alpha\beta(\alpha + \beta)$.

Notice that you are able to substitute $(\alpha + \beta)$ and $\alpha\beta$ directly into this identity for $\alpha^3 + \beta^3$.

Example  4

SKILLS PROBLEM-SOLVING

The roots of the quadratic equation $2x^2 + 5x - 4 = 0$ are α and β.
Form a quadratic equation with integer coefficients which has roots:

a α^3 and β^3 **b** $\frac{\alpha + \beta}{\alpha^2}$ and $\frac{\alpha + \beta}{\beta^2}$

Problem-solving

Find $\alpha + \beta$ and $\alpha\beta$ for the equation first.

a α^3 and β^3

$2x^2 + 5x - 4 = 0 \Rightarrow a = 2, b = 5, c = -4$

$\alpha + \beta = -\frac{5}{2}, \alpha\beta = -\frac{4}{2} = -2$

Find the value of $\alpha + \beta$ and $\alpha\beta$.

Sum of roots

$\alpha^3 + \beta^3 = (\alpha + \beta)^3 - 3\alpha\beta(\alpha + \beta)$

$\alpha^3 + \beta^3 = \left(-\frac{5}{2}\right)^3 - 3(-2)\left(-\frac{5}{2}\right) = -\frac{245}{8}$

Use the identity for $\alpha^3 + \beta^3$ to find an expression and value of the sum of roots.

Product of roots

$\alpha^3 \times \beta^3 = (\alpha\beta)^3 = (-2)^3 = -8$

Use the identity for $\alpha^3\beta^3$ to find an expression and value of the product of roots.

Equation

$x^2 - \left(-\frac{245}{8}\right)x - 8 = 0$

$\Rightarrow 8x^2 + 245x - 64 = 0$

Form a new quadratic equation and multiply through to achieve integer coefficients.

b $\frac{\alpha + \beta}{\alpha^2}$ and $\frac{\alpha + \beta}{\beta^2}$

Sum of roots

$$\left(\frac{\alpha + \beta}{\alpha^2}\right) + \left(\frac{\alpha + \beta}{\beta^2}\right) = \frac{\alpha\beta^2 + \beta^3 + \alpha^3 + \alpha^2\beta}{\alpha^2\beta^2}$$

$$= \frac{(\alpha + \beta)^3 - 3\alpha\beta(\alpha + \beta) + \alpha\beta(\alpha + \beta)}{\alpha^2\beta^2}$$

$$= \frac{(\alpha + \beta)^3 - 2\alpha\beta(\alpha + \beta)}{\alpha^2\beta^2}$$

$$= \frac{\left(-\frac{5}{2}\right)^3 - 2(-2)\left(-\frac{5}{2}\right)}{(-2)^2} = -\frac{205}{32}$$

Manipulate the algebra to form expressions for the sum and product into which $\alpha + \beta$ and $\alpha\beta$ can be substituted directly.

Product of roots

$$\left(\frac{\alpha + \beta}{\alpha^2}\right) \times \left(\frac{\alpha + \beta}{\beta^2}\right) = \frac{(\alpha + \beta)^2}{\alpha^2\beta^2}$$

$$\frac{\left(-\frac{5}{2}\right)^2}{(-2)^2} = \frac{25}{16}$$

$$x^2 - \left(-\frac{205}{32}\right)x + \frac{25}{16} = 0$$

$$\Rightarrow 32x^2 + 205x + 50 = 0$$

Form a new quadratic equation and multiply through to achieve integer coefficients.

Exercise 2B

SKILLS PROBLEM-SOLVING

(P) **1** The roots of the equation $x^2 + 5x + 2 = 0$ are α and β.
Find an equation with integer coefficients which has roots:

a $2\alpha + 1$ and $2\beta + 1$

b $\alpha\beta$ and $\alpha^2\beta^2$.

(P) **2** The roots of the equation $3x^2 - 2x + 3 = 0$ are α and β.
Find an equation with integer coefficients which has roots:

a $\frac{1}{\alpha^2}$ and $\frac{1}{\beta^2}$

b $\frac{\beta}{\alpha^2}$ and $\frac{\alpha}{\beta^2}$.

(P) **3** The roots of the equation $3x^2 + 7x + 6 = 0$ are α and β.
Find an equation with integer coefficients which has roots:

a $\alpha^2 + \beta$ and $\alpha + \beta^2$

b α^3 and β^3.

(P) **4** The roots of the equation $6x^2 - 3x + 4 = 0$ are α and β.
Find an equation with integer coefficients which has roots:

a $\frac{1}{\alpha^3}$ and $\frac{1}{\beta^3}$

b $\alpha^2\beta$ and $\alpha\beta^2$.

Chapter review 2

SKILLS EXECUTIVE FUNCTION

E/P **1** The equation $3x^2 - 4x + 6 = 0$ has roots α and β.

a Without solving the equation, write down:

i the value of $\alpha + \beta$

ii the value of $\alpha\beta$. **(2 marks)**

b Without solving the equation, show that the value of $\alpha^3 + \beta^3 = -\frac{152}{27}$. **(3 marks)**

c Form a quadratic equation, with integer coefficients, that has roots $\frac{\alpha}{\beta^2}$ and $\frac{\beta}{\alpha^2}$. **(4 marks)**

E/P **2** The roots α and β of a quadratic equation are such that $\alpha + \beta = -\frac{5}{2}$ and $\alpha\beta = -6$.

a Form a quadratic equation with integer coefficients that has roots α and β. **(4 marks)**

b Find the value of:

i $\alpha^2 + \beta^2$

ii $\alpha^3 + \beta^3$. **(3 marks)**

c Hence form a quadratic equation with integer coefficients that has roots $\left(\alpha - \frac{1}{\alpha^2}\right)$ and $\left(\beta - \frac{1}{\beta^2}\right)$. **(4 marks)**

E/P **3** The roots of a quadratic equation are α and β where $\alpha + \beta = -\frac{7}{3}$ and $\alpha\beta = -2$.

a Find a quadratic equation, with integer coefficients, which has roots α and β. **(4 marks)**

Given that $\alpha > \beta$, and without solving the equation:

b show that $\alpha - \beta = \frac{11}{3}$ **(4 marks)**

c form a quadratic equation, with integer coefficients, which has roots $\frac{\alpha + \beta}{\alpha}$ and $\frac{\alpha - \beta}{\beta}$. **(4 marks)**

Challenge

SKILLS
CRITICAL THINKING

1 The equation $2x^2 + px + q = 0$ has roots α and β.

The equation $16x^2 + 57x + 16 = 0$ has roots $\frac{\alpha}{\beta}$ and $\frac{\beta}{\alpha}$.

a Given that $p + q = -3$ and p is a positive integer, find the value of:

i p

ii q

b Given also that $\alpha > \beta$, find the exact value of $\alpha - \beta$.

2 The equation $4x^2 + px + q = 0$, where p and q are real constants, has roots α and α^*.

a Given that $\text{Re}(\alpha) = -3$, find the value of p.

b Given that $\text{Im}(\alpha) \neq 0$, find the possible range of values of q.

Summary of key points

1 If α and β are roots of the equation $ax^2 + bx + c = 0$, then:

- $\alpha + \beta = -\frac{b}{a}$
- $\alpha\beta = \frac{c}{a}$

2 The identity for $\alpha^2 + \beta^2$ is: $\alpha^2 + \beta^2 = (\alpha + \beta)^2 - 2\alpha\beta$

3 The identity for $\alpha^3 + \beta^3$ is: $\alpha^3 + \beta^3 = (\alpha + \beta)^3 - 3\alpha\beta(\alpha + \beta)$

4 To form an equation with new roots, always rearrange the required roots into a form so you can substitute $\alpha + \beta$ and $\alpha\beta$.

3 NUMERICAL SOLUTIONS OF EQUATIONS

3.1

Learning objectives

After completing this chapter you should be able to:

- Locate roots of $f(x) = 0$ by considering changes of sign → pages 37–39
- Use interval bisection to find approximations to the solutions of equations of the form $f(x) = 0$ → pages 39–41
- Use linear interpolation to find approximations to the solutions of equations of the form $f(x) = 0$ → pages 41–44
- Use the Newton–Raphson method to find approximations to the solutions of equations of the form $f(x) = 0$ → pages 44–46

Prior knowledge check

1 $f(x) = x^2 - 6x + 10$. Evaluate:

a $f(1.5)$ **b** $f(-0.2)$ ← Pure 1 Section 2.3

2 Find $f'(x)$ given that:

a $f(x) = 7x^3 - 2x^2 + 8$

b $f(x) = 3\sqrt{x} + 4x^2 - \dfrac{5}{x^3}$ ← Pure 1 Section 8.4

3 Given that $u_{n+1} = u_n + \dfrac{1}{u_n}$ and that $u_0 = 1$, find the values of u_1, u_2 and u_3. ← Pure 2 Section 5.7

You can use numerical methods to find solutions to equations that are hard or impossible to solve exactly. The Newton–Raphson method was developed 400 years ago to describe the positions of planets as they orbit the sun.

3.1 Locating roots

A root of a function is a value of x for which $f(x) = 0$. The graph of $y = f(x)$ will cross the x-axis at points corresponding to the roots of the function.

Links The following two things are identical:
- the roots of the function $f(x)$
- the roots of the equation $f(x) = 0$ ← Pure 1 Section 2.3

You can sometimes show that a root exists within a given **interval** by showing that the function changes sign (from positive to negative, or vice versa) within the interval.

- **If the function $f(x)$ is continuous on the interval $[a, b]$ and $f(a)$ and $f(b)$ have opposite signs, then $f(x)$ has at least one root, x, which satisfies $a < x < b$.**

Notation **Continuous** means that the function does not 'jump' from one value to another. If the graph of the function has a vertical **asymptote** between a and b then the function is not continuous on $[a, b]$.

Example 1 SKILLS ANALYSIS

The diagram shows a sketch of the curve $y = f(x)$, where $f(x) = x^3 - 4x^2 + 3x + 1$.

a Explain how the graph shows that $f(x)$ has a root between $x = 2$ and $x = 3$.

b Show that $f(x)$ has a root between $x = 1.4$ and $x = 1.5$.

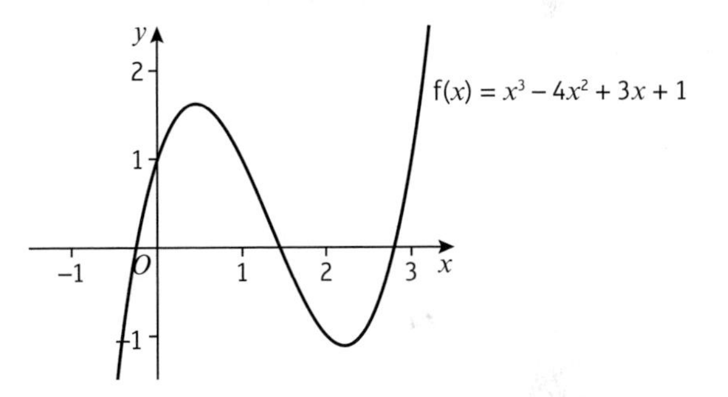

a The graph crosses the x-axis between $x = 2$ and $x = 3$. This means that a root of $f(x)$ lies between $x = 2$ and $x = 3$.

The graph of $y = f(x)$ crosses the x-axis whenever $f(x) = 0$.

b $f(1.4) = (1.4)^3 - 4(1.4)^2 + 3(1.4) + 1 = 0.104$
$f(1.5) = (1.5)^3 - 4(1.5)^2 + 3(1.5) + 1 = -0.125$
There is a change of sign for $f(x)$ between 1.4 and 1.5, so there is at least one root between $x = 1.4$ and $x = 1.5$.

$f(1.4) > 0$ and $f(1.5) < 0$, so there is a change of sign.

$f(x)$ changes sign in the interval $[1.4, 1.5]$, so $f(x)$ must equal zero within this interval.

There are three situations you need to watch out for when using the change of sign rule to locate roots. A change of sign does not necessarily mean there is exactly one root, and the absence of a sign change does not necessarily mean that a root does not exist in the interval.

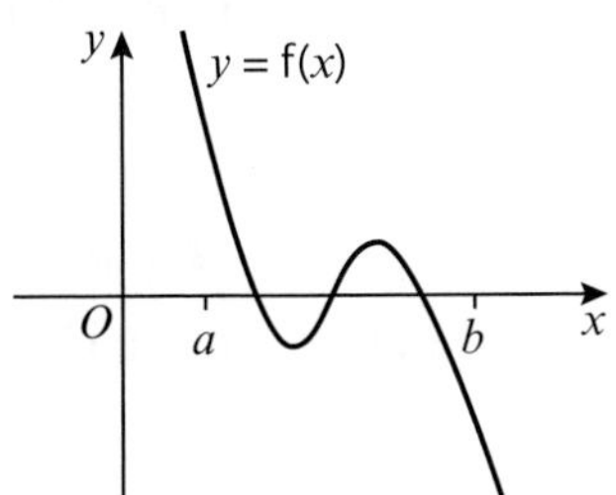

There are multiple roots within the interval $[a, b]$. In this case there is an **odd number** of roots.

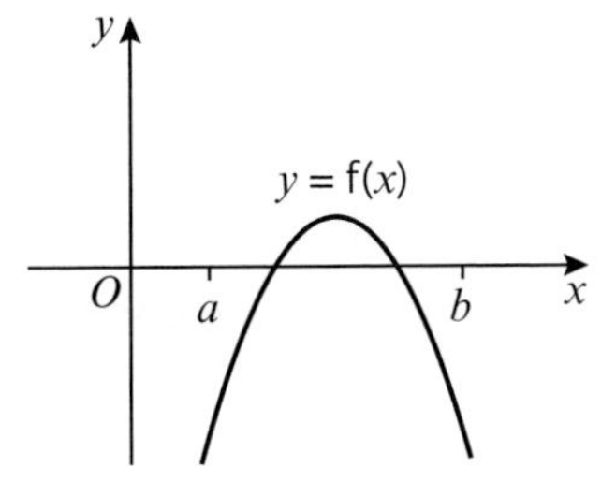

There are multiple roots within the interval $[a, b]$, but a sign change does not occur. In this case there is an **even number** of roots.

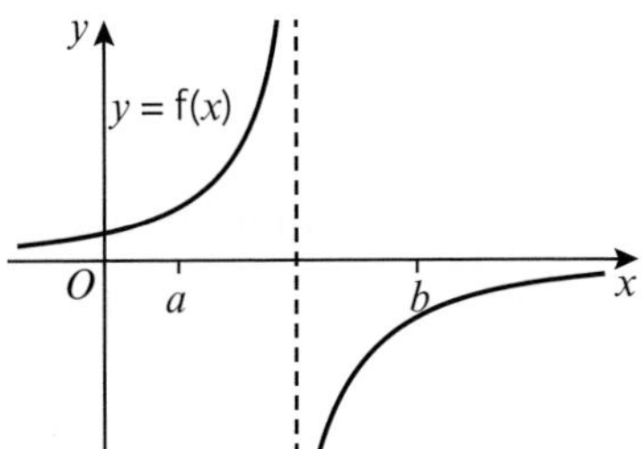

There is a vertical asymptote within interval $[a, b]$. A sign change does occur, but there is **no** root.

Example 2 SKILLS EXECUTIVE FUNCTION

The diagram shows the graph of the function $f(x) = 54x^3 - 225x^2 + 309x - 140$.

A student observes that f(1.1) and f(1.6) are both negative, and states that f(x) has no roots in the interval [1.1, 1.6].

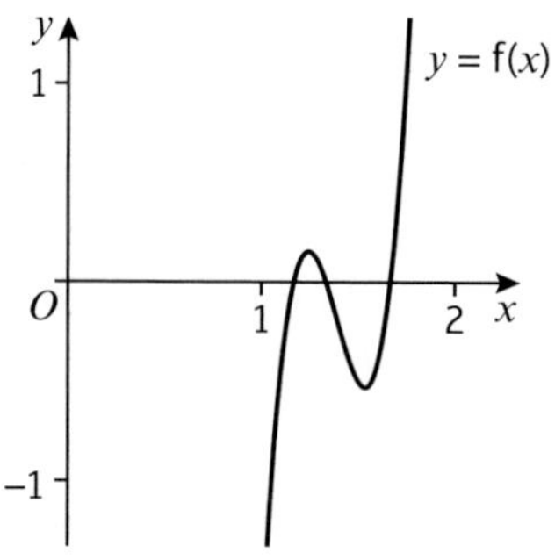

a Explain by reference to the diagram why the student is incorrect.

b Calculate f(1.3) and f(1.5) and use your answer to explain why there are at least three roots in the interval $1.1 < x < 1.7$.

a The diagram shows that there could be two roots in the interval [1.1, 1.6].

b $f(1.1) = -0.476 < 0$
$f(1.3) = 0.088 > 0$
$f(1.5) = -0.5 < 0$
$f(1.7) = 0.352 > 0$
There is a change of sign between 1.1 and 1.3, between 1.3 and 1.5, and between 1.5 and 1.7, so there are at least three roots in the interval $1.1 < x < 1.7$.

Notation The interval [1.1, 1.6] is the set of all real numbers, x, that satisfy $1.1 < x < 1.6$.

Calculate the values of f(1.1), f(1.3), f(1.5) and f(1.7). Comment on the sign of each answer.

f(x) changes sign at least three times in the interval $1.1 < x < 1.7$ so f(x) must equal zero at least three times within this interval.

Exercise 3A SKILLS EXECUTIVE FUNCTION

1 Show that each of these functions has at least one root in the given interval.

a $f(x) = x^3 - x + 5$, $-2 < x < -1$

b $f(x) = x^2 - \sqrt{x} - 10$, $3 < x < 4$

c $f(x) = x^3 - \frac{1}{x} - 2$, $-0.5 < x < -0.2$

d $f(x) = e^x - \ln x - 5$, $1.65 < x < 1.75$

(E) **2** $f(x) = 3 + x^2 - x^3$

a Show that the equation $f(x) = 0$ has a root, α, in the interval [1.8, 1.9]. **(2 marks)**

b By considering a change of sign of f(x) in a suitable interval, verify that $\alpha = 1.864$, correct to 3 decimal places. **(3 marks)**

(E) **3** $h(x) = \sqrt[3]{x} - \cos x - 1$, where x is in radians.

a Show that the equation $h(x) = 0$ has a root, α, between $x = 1.4$ and $x = 1.5$. **(2 marks)**

b By choosing a suitable interval, show that $\alpha = 1.441$ is correct to 3 decimal places. **(3 marks)**

(P) **4** $f(x) = 2 + \tan x$, $0 < x < \pi$, where x is in radians.

a Show that f(x) changes sign in the interval [1.5, 1.6].

b State with a reason whether or not f(x) has a root in the interval [1.5, 1.6].

(P) **5** A student observes that the function $f(x) = \frac{1}{x} + 2, x \neq 0$, has a change of sign on the interval $[-1, 1]$. The student writes:

> $y = f(x)$ has a vertical asymptote within this interval, so even though there is a change of sign, $f(x)$ has no roots in this interval.

By means of a sketch, or otherwise, explain why the student is incorrect.

(E/P) **6** **a** On the same **axes**, sketch the graphs of $y = \sqrt{x}$ and $y = \frac{2}{x}$. **(2 marks)**

b With reference to your sketch, explain why the equation $\sqrt{x} = \frac{2}{x}$ has exactly one real root. **(1 mark)**

c Given that $f(x) = \sqrt{x} - \frac{2}{x}$, show that the equation $f(x) = 0$ has a root r, where $1 < r < 2$. **(2 marks)**

d Show that the equation $\sqrt{x} = \frac{2}{x}$ may be written in the form $x^p = q$, where p and q are integers to be found. **(2 marks)**

e Hence write down the exact value of the root of the equation $\sqrt{x} - \frac{2}{x} = 0$. **(1 mark)**

(E/P) **7** $f(x) = x^4 - 21x - 18$

a Show that there is a root of the equation $f(x) = 0$ in the interval $[-0.9, -0.8]$. **(3 marks)**

b Find the **coordinates** of any **stationary points** on the graph $y = f(x)$. **(3 marks)**

c Given that $f(x) = (x - 3)(x^3 + ax^2 + bx + c)$, find the values of the constants a, b and c. **(3 marks)**

d Sketch the graph of $y = f(x)$. **(3 marks)**

3.2 Interval bisection

- If you find an interval in which $f(x)$ changes sign, then the interval must contain a root of the equation $f(x) = 0$.
- You then take the **midpoint** as the first **approximation** and repeat this process until you get the required accuracy. This method is known as **interval bisection.**

Example 3 SKILLS PROBLEM-SOLVING

Use interval bisection to find the positive root of $\sqrt{11}$ to 1 decimal place.

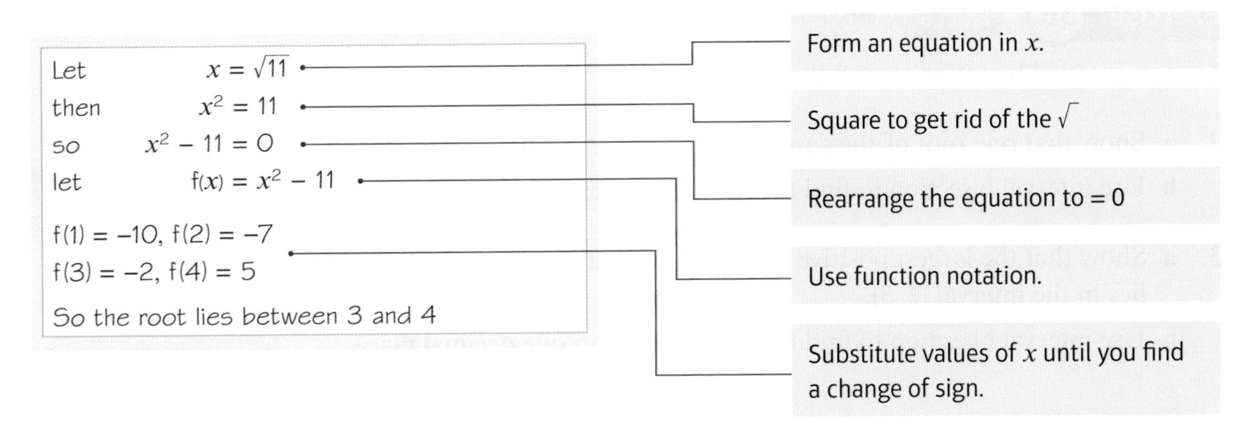

a	f(a)	b	f(b)	$\frac{a+b}{2}$	$f\left(\frac{a+b}{2}\right)$
3	−2	4	5	3.5	1.25
3	−2	3.5	1.25	3.25	−0.4375
3.25	−0.4375	3.5	1.25	3.375	0.390625
3.25	−0.4375	3.375	0.390635	3.3125	−0.0273437
3.3125	−0.0273437	3.375	0.390635	3.34375	0.180664

Hence $\sqrt{11} = 3.3$ to 1 d.p.

Make a table of values.

Let the interval $[a, b]$ be the interval in which the root lies. $\frac{a+b}{2}$ is the midpoint.

The sign changes between f(3) and f(3.5) so the root lies between them.

The sign changes between f(3.25) and f(3.5) so the root lies between them.

The sign changes between f(3.25) and f(3.375) so the root lies between them.

The sign changes between f(3.3125) and f(3.34375) so the root lies between them.

Both 3.3125 and 3.34375 are 3.3 when corrected to 1 d.p.

Example 4

$f(x) = 2^x - x - 3$

The equation $f(x) = 0$ has a root in the interval [2, 3]. Using the end points of this interval, find by interval bisection, a first and second approximation to x.

Let $a = 2$, $b = 3$

a	f(a)	b	f(b)	$\frac{a+b}{2}$	$f\left(\frac{a+b}{2}\right)$
2	−1	3	2	2.5	0.1569
2	−1	2.5	0.1569	2.25	−0.493

1st approximation = 2.5

2nd approximation = 2.25

Make a table of values. Use the change of sign rule to find the second approximation.

Exercise 3B SKILLS INTERPRETATION

1 Use interval bisection to find the positive root of $x^2 - 7 = 0$, correct to one decimal place.

2 a Show that one root of the equation $x^3 - 7x + 2 = 0$ lies in the interval [2, 3].

b Use interval bisection to find the root correct to two decimal places.

3 a Show that the largest positive root of the equation $x^3 + 2x^2 - 8x + 3 = 0$ lies in the interval [2, 3].

b Use interval bisection to find this root correct to one decimal place.

4 **a** Show that the equation $\frac{x}{2} - \frac{1}{x} = 0$, $x > 0$, has a root in the interval [1, 2].

b Obtain the root, using interval bisection three times. Give your answer to two significant figures.

(P) **5** $f(x) = 6x - x^3$

The equation $f(x) = 0$ has a root between $x = 2$ and $x = 3$. Starting with the interval [2, 3] use interval bisection three times to give an approximation to this root.

3.3 Linear interpolation

- **In linear interpolation, draw a sketch of the function f(x) for a given interval [a, b].**
- **You can call the first approximation to the root of the function that lies in this interval x_1.**
- **Use similar triangles to find x_1.**
- **Repeat the process using an interval involving the first approximation and one of the initial limits, where there is a change of sign to find a second approximation.**
- **Repeat until an approximation to the required degree of accuracy is found.**

Example 5

a Show that the equation $x^3 + 4x - 9 = 0$ has a root in the interval [1, 2].

b Use linear interpolation to find this root to one decimal place.

a Let $f(x) = x^3 + 4x - 9$

$f(1) = 1^3 + 4 \times 1 - 9 = -4$

$f(2) = 2^3 + 4 \times 2 - 9 = 7$

Use the change of rule sign to show that there is a root in the interval [1, 2].

Since there is a change of sign between f(1) and f(2), the equation $x^3 + 4x - 9 = 0$ has a root in the interval [1, 2]

Explain why there is a root between $x = 1$ and $x = 2$

b

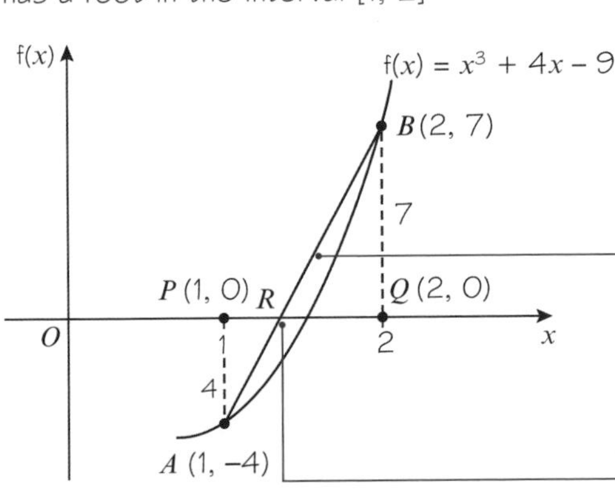

Draw a graph of the function between $x = 1$ and $x = 2$

Join $A(1, -4)$ to $B(2, 7)$ with a straight line.

The point where the straight line crosses the x-axis is the first approximation of the root. Call this x.

Then using similar triangles: $\frac{QR}{RP} = \frac{QB}{PA}$

So, $\frac{2 - x_1}{x_1 - 1} = \frac{7}{4}$

Use similar triangles to work out x_1

So, $8 - 4x_1 = 7x_1 - 7$

$x_1 = \frac{15}{11} = 1.3636...$

f(1.3636...) = −1.0097671 using the interval $[1.\dot{3}\dot{6}, 2]$

The root lies in the interval $[1.\dot{3}\dot{6}, 2]$ as the sign changes between these values.

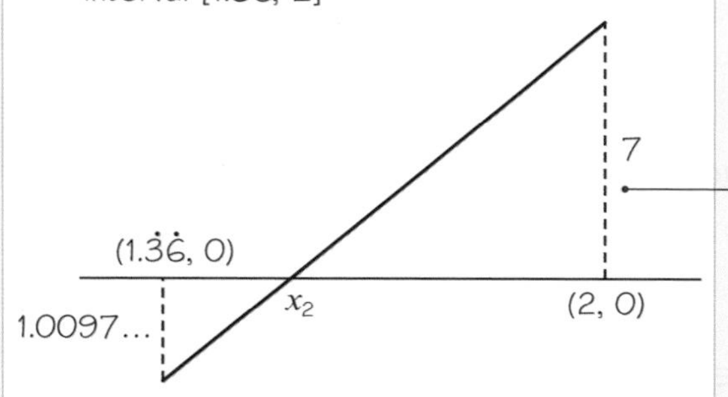

Draw another diagram for the interval $[1.\dot{3}\dot{6}, 2]$. You do not need to draw the graph of f(x). Call the next approximation x_2.

Using similar triangles

$\frac{2 - x_2}{x_2 - 1.3636...} = \frac{7}{1.0097...}$

$x_2 = 1.4438607...$

Use your calculator and **do not** clear the values as you complete the calculation.

f(1.4438607...) = −0.2144918 using the interval [1.443..., 2]

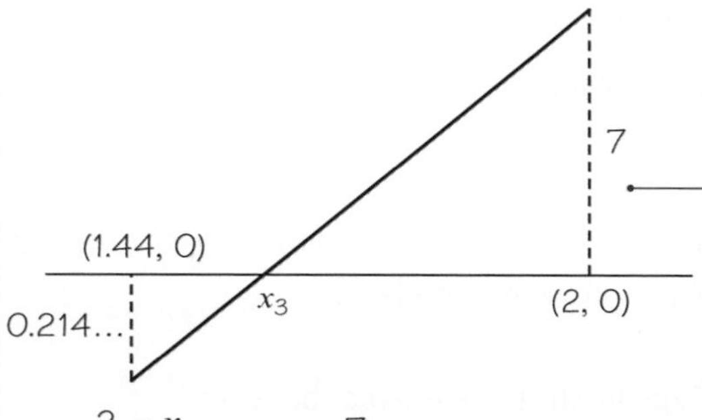

Repeat the process as above. You may be able to do this without drawing a diagram.

$\frac{2 - x_3}{x_3 - 1.443...} = \frac{7}{0.214...}$

$x_3 = 1.4603952...$

f(1.4603952...) = −0.0437552 using the interval [1.460..., 2]

Repeat the process again.

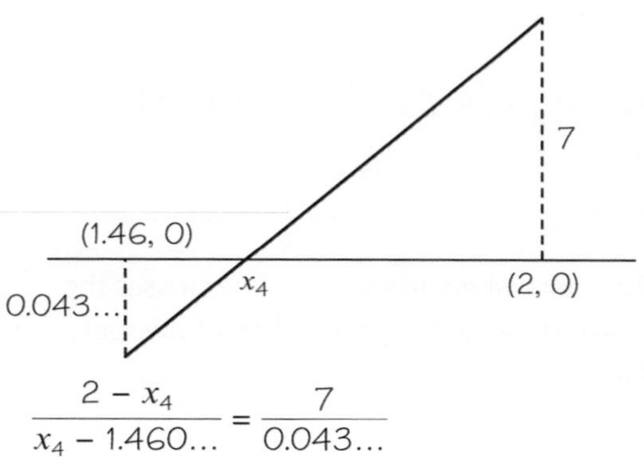

$\frac{2 - x_4}{x_4 - 1.460...} = \frac{7}{0.043...}$

$x_4 = 1.4637472...$

Hence, the root is 1.5, accurate to one d.p.

Two **successive** approximations (x_3 and x_4) give the root as 1.5, accurate to one d.p.

Example 6

$f(x) = 3^x - 5x$

The equation $f(x) = 0$ has a root α in the interval [2, 3].

Using the end points of this interval, find, by linear interpolation, an approximation to α.

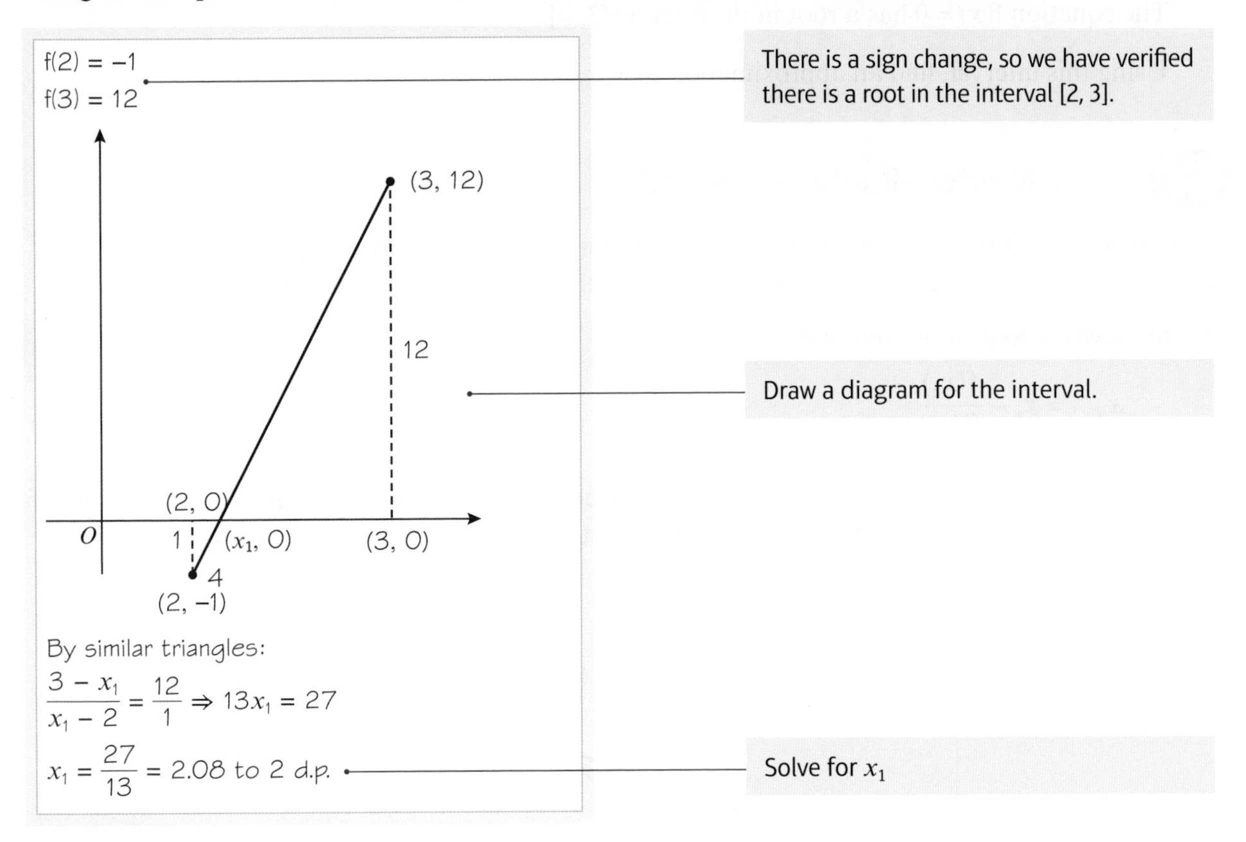

Exercise 3C SKILLS INTERPRETATION

1 a Show that a root of the equation $x^3 - 3x - 5 = 0$ lies in the interval [2, 3].
  b Find this root using linear interpolation, correct to one decimal place.

2 a Show that a root of the equation $5x^3 - 8x^2 + 1 = 0$ lies in the interval [1, 2].
  b Find this root using linear interpolation, correct to one decimal place.

3 a Show that a root of the equation $\frac{3}{x} + 3 = x$ lies in the interval [3, 4].
  b Use linear interpolation to find this root, correct to one decimal place.

4 a Show that a root of the equation $2x \cos x - 1 = 0$ lies in the interval [1, 1.5].
  b Find this root using linear interpolation, correct to one decimal place.

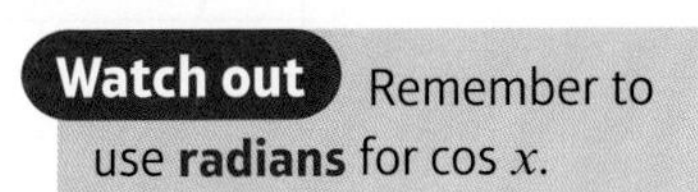

5 **a** Show that the largest possible root of the equation $x^3 - 2x^2 - 3 = 0$ lies in the interval [2, 3].

b Find this root correct to one decimal place using linear interpolation.

(P) **6** $f(x) = 2^x - 3x - 1$

The equation $f(x) = 0$ has a root in the interval [3, 4].

Using this interval, find an approximation to x.

3.4 The Newton–Raphson method

The Newton–Raphson method can be used to find **numerical** solutions to equations of the form $f(x) = 0$. You need to be able to **differentiate** $f(x)$ to use this method.

- The Newton–Raphson formula is:

$$x_{n+1} = x_n - \frac{f(x_n)}{f'(x_n)}$$

The method uses **tangent** lines to find increasingly accurate approximations of a root. The value of x_{n+1} is the point at which the tangent to the graph at $(x_n, f(x_n))$ **intersects** the x-axis.

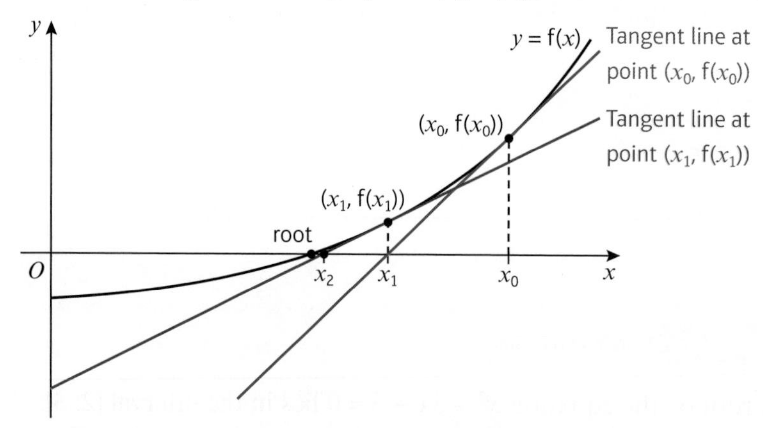

If the starting value is not chosen carefully, the Newton–Raphson method might **converge** on a root very slowly, or can fail completely. If the initial value, x_0, is near a turning point or the **derivative** at this point, $f'(x_0)$, is close to zero, then the tangent at $(x_0, f(x_0))$ will intercept the x-axis a long way from x_0.

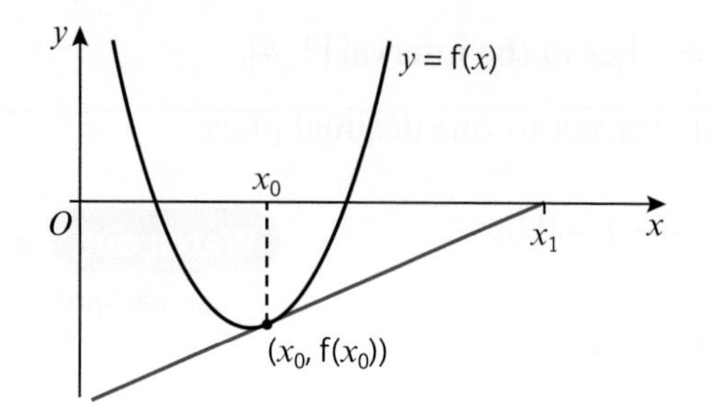

Because x_0 is close to a turning point, the **gradient** of the tangent at $(x_0, f(x_0))$ is small, so it intercepts the x-axis a long way from x_0.

If any value, x_i, in the Newton–Raphson method is at a turning point, the method will fail because $f'(x_i) = 0$ and the formula would result in division by zero, which is not valid. Graphically, the tangent line will run parallel to the x-axis, therefore never intersecting the x-axis.

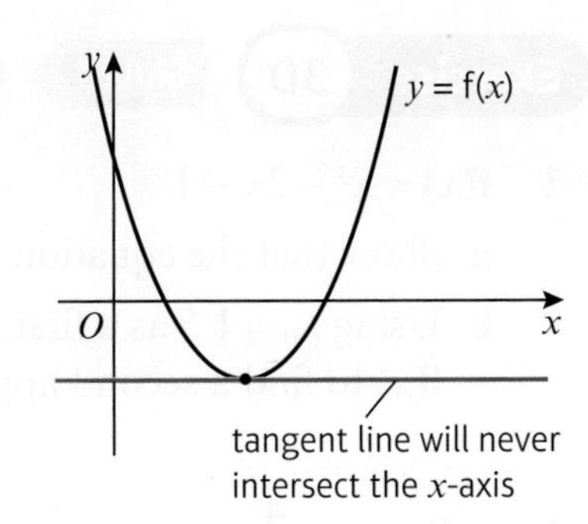

Example 7

SKILLS EXECUTIVE FUNCTION

The diagram shows part of the curve with equation $y = f(x)$, where $f(x) = x^3 + 2x^2 - 5x - 4$.

The point A, with x-coordinate p, is a stationary point on the curve.

The equation $f(x) = 0$ has a root, α, in the interval $1.8 < \alpha < 1.9$.

a Explain why $x_0 = p$ is not suitable to use as a first approximation to α when applying the Newton–Raphson method to $f(x)$.

b Using $x_0 = 2$ as a first approximation to α, apply the Newton–Raphson method twice to $f(x)$ to find a second approximation to α, giving your answer to 3 decimal places.

c By considering the change of sign in $f(x)$ over an appropriate interval, show that your answer to part **b** is accurate to 3 decimal places.

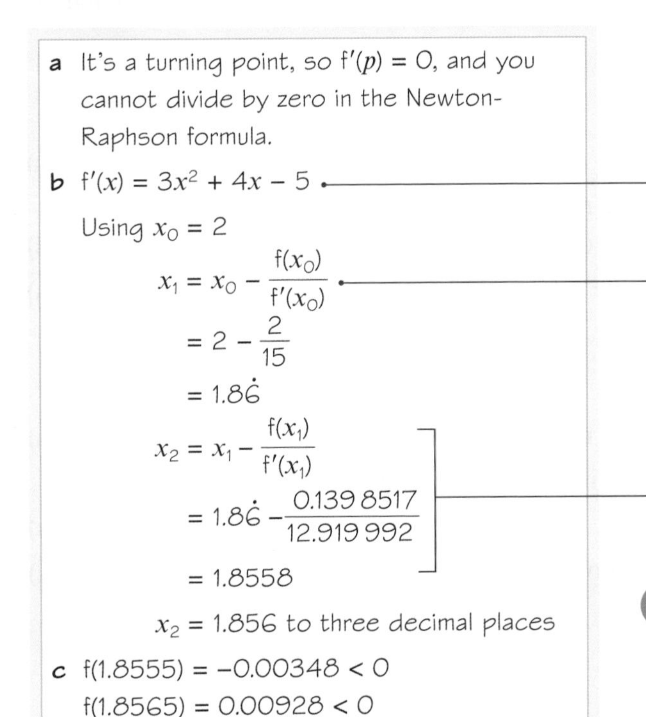

a It's a turning point, so $f'(p) = 0$, and you cannot divide by zero in the Newton-Raphson formula.

b $f'(x) = 3x^2 + 4x - 5$

Using $x_0 = 2$

$$x_1 = x_0 - \frac{f(x_0)}{f'(x_0)}$$

$$= 2 - \frac{2}{15}$$

$$= 1.8\dot{6}$$

$$x_2 = x_1 - \frac{f(x_1)}{f'(x_1)}$$

$$= 1.8\dot{6} - \frac{0.139\,8517}{12.919\,992}$$

$$= 1.8558$$

$x_2 = 1.856$ to three decimal places

c $f(1.8555) = -0.00348 < 0$

$f(1.8565) = 0.00928 < 0$

Sign change in interval [1.8555, 1.8565] therefore $x = 1.856$ is accurate to 3 decimal places.

Use $\frac{d}{dx}(ax^n) = anx^{n-1}$

Use the Newton–Raphson method twice.

Substitute $x_1 = 1.8\dot{6}$ into the Newton–Raphson formula.

Hint Use a spreadsheet package to find successive Newton–Raphson approximations.

Online Explore how the Newton–Raphson method works graphically and algebraically using technology.

Exercise 3D

SKILLS EXECUTIVE FUNCTION

1 $f(x) = x^3 - 2x - 1$

a Show that the equation $f(x) = 0$ has a root, α, in the interval $1 < \alpha < 2$.

b Using $x_0 = 1.5$ as a first approximation to α, apply the Newton–Raphson method once to $f(x)$ to find a second approximation to α, giving your answer to 3 decimal places.

E **2** $f(x) = x^2 - \frac{4}{x} + 6x - 10, x \neq 0.$

a Use **differentiation** to find $f'(x)$. **(2 marks)**

The root, α, of the equation $f(x) = 0$ lies in the interval $[-0.4, -0.3]$.

b Taking -0.4 as a first approximation to α, apply the Newton–Raphson method once to $f(x)$ to obtain a second approximation to α.
Give your answer to 3 decimal places. **(4 marks)**

E **3** $f(x) = x^2 - \frac{3}{x^2}, x > 0$

a Show that a root α of the equation $f(x) = 0$ lies in the interval $[1.3, 1.4]$. **(1 mark)**

b Differentiate $f(x)$ to find $f'(x)$. **(2 marks)**

c By taking 1.3 as a first approximation to α, apply the Newton–Raphson method once to $f(x)$ to obtain a second approximation to α.
Give your answer to 3 decimal places. **(3 marks)**

E/P **4** $f(x) = \frac{x^3}{3} - x + 2$

a Sketch the curve $y = f(x)$. **(3 marks)**

b Use the Newton-Raphson method to find the root of the equation $f(x) = 0$, starting with $x_0 = -2$ and giving your answer to 3 decimal places. **(4 marks)**

c Explain what happens when the starting value is taken to be $x_0 = 1$. **(2 marks)**

E/P **5** $f(x) = x^4 - 7x^3 + 1$

a Show that the equation $x^4 - 7x^3 + 1 = 0$ has a root in the interval $[0, 1]$. **(2 marks)**

b Use the Newton-Raphson method, starting with $x_0 = 0.5$ to find this root.
Give your answer to 4 decimal places. **(4 marks)**

c Explain why $x_0 = 0$ is not a suitable starting point. **(2 marks)**

Chapter review 3

SKILLS EXECUTIVE FUNCTION

E/P **1** $f(x) = \frac{1}{4-x} + 3$

a Calculate f(3.9) and f(4.1). **(2 marks)**

b Explain why the equation $f(x) = 0$ does not have a root in the interval $3.9 < x < 4.1$. **(2 marks)**

The equation $f(x) = 0$ has a single root, α.

c Use algebra to find the exact value of α. **(2 marks)**

E **2** Given that the equation $x^3 - 2x + 2 = 0$ has a root in the interval [−1, −2], use interval bisection on the interval [−1, −2] to obtain the root, correct to one decimal place. **(4 marks)**

E/P **3** Show that the equation $x^3 - 12x - 7.2 = 0$ has one positive root and two negative roots. Obtain the positive root correct to three significant figures using the Newton-Raphson method. **(6 marks)**

E/P **4** **a** On the same axes, sketch the graphs of $y = \frac{1}{x}$ and $y = x + 3$. **(2 marks)**

b Write down the number of roots of the equation $\frac{1}{x} = x + 3$. **(1 mark)**

c Show that the positive root of the equation $\frac{1}{x} = x + 3$ lies in the interval [0.30, 0.31]. **(2 marks)**

d Show that the equation $\frac{1}{x} = x + 3$ may be written in the form $x^2 + 3x - 1 = 0$. **(2 marks)**

e Use the quadratic formula to find the positive root of the equation $x^2 + 3x - 1 = 0$ to 3 decimal places. **(2 marks)**

E/P **5** $g(x) = x^3 - 7x^2 + 2x + 4$

a Find $g'(x)$. **(2 marks)**

A root α of the equation $g(x) = 0$ lies in the interval [6.5, 6.7].

b Taking 6.6 as a first approximation to α, apply the Newton–Raphson method once to $g(x)$ to obtain a second approximation to α. Give your answer to 3 decimal places. **(4 marks)**

c Given that $g(1) = 0$, find the exact value of the other two roots of $g(x)$. **(3 marks)**

d Calculate the percentage error of your answer in part **b**. **(2 marks)**

P **6** The equation $\cos x = \frac{1}{4}x$ has a root in the interval [1, 2]. Use linear interpolation once in the interval [1.0, 1.4] to find an estimate of the root, giving your answer correct to two decimal places.

P **7** $f(x) = x^3 - 3x^2 + 1$

a Show that the equation $f(x) = 0$ has exactly three roots.

b Use the Newton-Raphson method to find each of the roots. Give your answers to 3 decimal places.

Challenge

$$f(x) = \begin{cases} 3 + \sqrt{x+2} & f(x) \geqslant 3 \\ 3 - \sqrt{x+2} & f(x) < 3 \end{cases} \quad x \geqslant -2$$

a Show that f(x) has two intersections with the y-axis in the intervals [4, 5] and [1, 2].

b Using the Newton-Raphson method with $y_0 = 5$, find the intersection with the y-axis in the interval [4, 5] giving your answer to five significant figures. Explain why your answer is correct to five significant figures.

Hint Rearrange f(x) to make x the subject.

Summary of key points

1. You can locate roots of an equation $f(x) = 0$, if $f(x)$ is continuous in the interval $[a, b]$, and $f(a)$ and $f(b)$ have opposite signs. The change of sign indicates an **intersection** on the x-axis, and hence there is a root of the equation.
2. You can solve equations of the form $f(x) = 0$ using interval bisection. If you find an interval in which $f(x)$ changes sign, then the interval must contain a root of the equation $f(x) = 0$. You can then take the midpoint as the first approximation and repeat this process until you get the required accuracy.
3. You can solve equations of the form $f(x) = 0$ using linear interpolation.
4. You can solve equations of the form $f(x) = 0$ using the Newton-Raphson method.
5. The Newton-Raphson formula is $x_{n+1} = x_n - \dfrac{f(x_n)}{f'(x_n)}$
6. The Newton-Raphson method may not always give you a better approximation and may take you further away from the root.

4 COORDINATE SYSTEMS

3.1

Learning objectives

After completing this chapter, you should be able to:

- Plot and sketch a curve parametrically → pages 50–53
- Work with Cartesian equations and parametric equations of
  - a parabola → pages 53–59
  - a rectangular hyperbola → pages 60–67
- Understand the focus – directrix property of a parabola → pages 53–59
- Find the equation of the tangent and the equation of a normal to a point on
  - a parabola → pages 53–59
  - a rectangular hyperbola → pages 60–67

Prior knowledge check

1 Points A and B are at (4, 3) and (–2, 7) respectively. Find:

a the midpoint of line segment AB

b the length of line segment AB.

← International GCSE Mathematics

2 Find the equation of the line that passes through the point (1, 3) and is parallel to the line with equation $y = 3x - 1$. ← International GCSE Mathematics

3 Find the equation of the normal to the curve with equation $y = 2x^2 + 5$ at the point where $x = 1$.

Give your answer in the form $ax + by = 0$, where a and b are integers. ← Pure 1 Section 8.6

The shapes of car headlights, television and radio antennae, and reflecting telescopes are all examples of parabolas.

The shape of a hyperbola (hour glass shape) can be found in gear systems, and examples in buildings include the cooling towers in power stations, the Eiffel tower and the Kobe Port Tower in Japan.

 4.1 Parametric equations

Parametric equations are where the x and y coordinates of each point on a curve are expressed in the form of an independent variable, say t, which is called a **parameter**. The **parametric** equation of a curve is written in the form:

$$x = \mathrm{f}(t),\ y = \mathrm{g}(t)$$

You can define the coordinates of any point on a curve by using parametric equations.

Example 1

Sketch the curve given by the parametric equations $x = at^2$, $y = 2at$, $t \in \mathbb{R}$, where a is a positive constant.

To get an idea of the shape of the curve, choose some values for t.
Let's say $t = -3, -2, -1, 0, 1, 2, 3$

t	−3	−2	−1	0	1	2	3
$x = at^2$	$9a$	$4a$	a	0	a	$4a$	$9a$
$y = 2at$	$-6a$	$-4a$	$-2a$	0	$2a$	$4a$	$6a$

Draw a table showing the values of t, x and y.

Work out the value of x and the value of y by substituting each value of t into the parametric equations $x = at^2$ and $y = at$.

e.g. for $t = -3$:

$x = at^2 = a(-3)^2 = 9a$

and

$y = 2at = 2a(-3) = -6a$

So, when $t = -3$ the curve passes through the point $(9a, -6a)$.

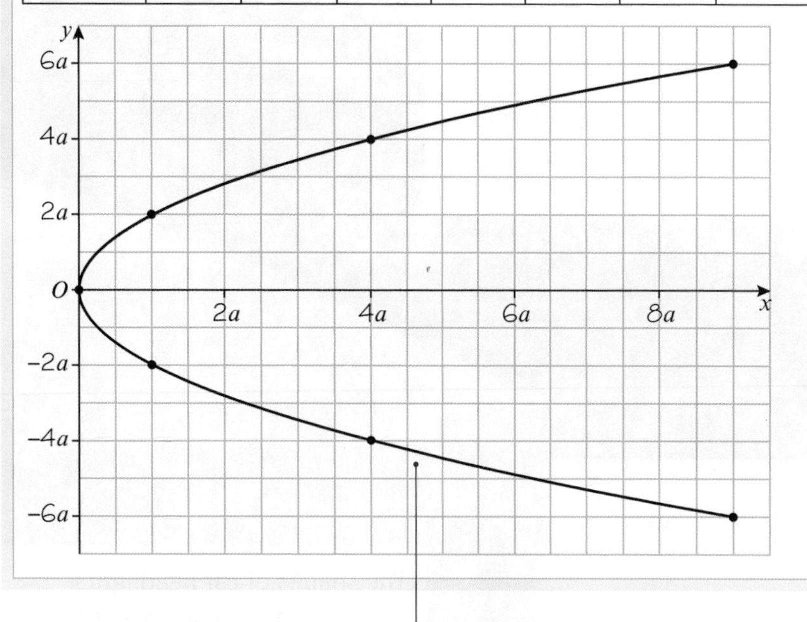

Don't worry about a. It's just a constant (or any positive real number).

Join points up in a smooth curve.

Example 2

A curve has parametric equations $x = at^2$, $y = 2at$, $t \in \mathbb{R}$, where a is a positive constant. Find the Cartesian equation of the curve.

$y = 2at$

So $t = \frac{y}{2a}$ **(1)**

$x = at^2$ **(2)**

Substitute **(1)** into **(2)**:

$x = a\left(\frac{y}{2a}\right)^2$

So $x = \frac{ay^2}{4a^2}$ which simplifies to

$x = \frac{y^2}{4a}$

Hence, the Cartesian equation is

$y^2 = 4ax$

A Cartesian equation is an equation in terms of x and y only. To obtain the Cartesian equation, eliminate t from the given parametric equations.

Rearrange $y = 2at$ for t.

Substitute $t = \frac{y}{2a}$ into $x = at^2$.

This equation now involves x and y. Note that a is a constant.

Example 3

A curve has parametric equations $x = ct$, $y = \frac{c}{t}$, $t \in \mathbb{R}$, $t \neq 0$, where c is a positive constant.

a Find the Cartesian equation of the curve.

b Hence sketch this curve.

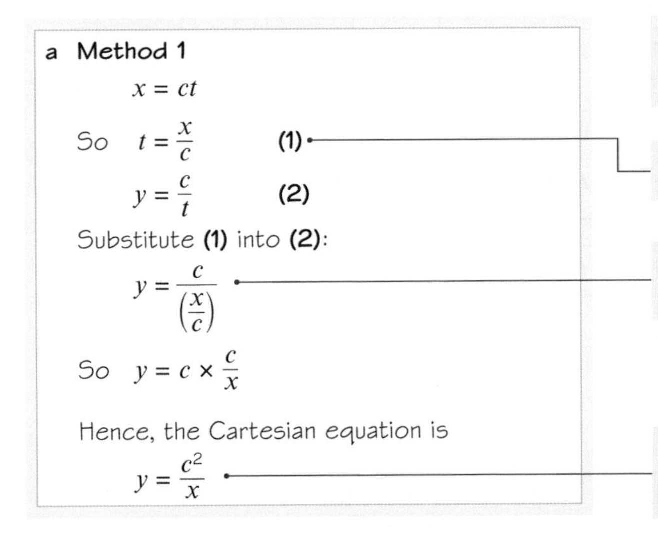

a Method 1

$x = ct$

So $t = \frac{x}{c}$ **(1)**

$y = \frac{c}{t}$ **(2)**

Substitute **(1)** into **(2)**:

$y = \frac{c}{\left(\frac{x}{c}\right)}$

So $y = c \times \frac{c}{x}$

Hence, the Cartesian equation is

$y = \frac{c^2}{x}$

To obtain the Cartesian equation, eliminate t from the given parametric equations.

Rearrange $x = ct$ for t.

Substitute $t = \frac{x}{c}$ into $y = \frac{c}{t}$.

This equation now involves x and y. Note that c is a constant.

Method 2

$xy = ct \times \left(\frac{c}{t}\right)$

Alternatively, you can multiply x by y on this occasion to eliminate t. The t's will now cancel out.

$xy = \frac{c^2t}{t}$

Hence, the Cartesian equation is

$xy = c^2$

This may also be expressed as

$y = \frac{c^2}{x}$

This equation now involves x and y. Note that c is a constant.

b

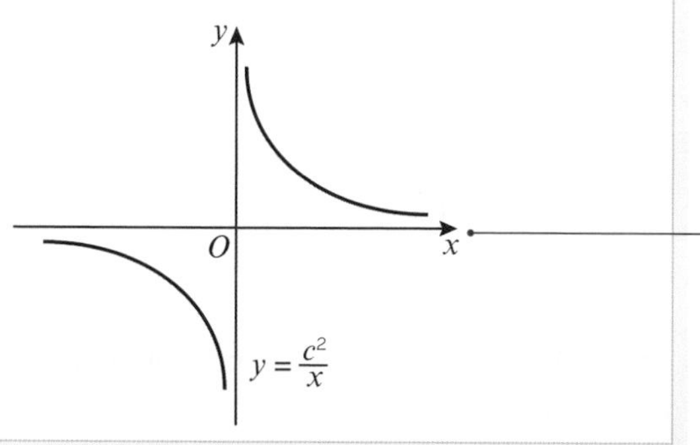

- As c is a positive constant, then c^2 is also a positive constant, which may be denoted by another constant, k.
- Hence the Cartesian equation represents a curve of the form $y = \frac{k}{x}$, $k > 0$.
- This is the reciprocal curve that you have seen in Pure 1.

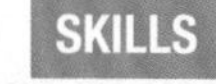

Exercise 4A SKILLS INTERPRETATION

1 A curve is given by the parametric equations $x = 2t^2$, $y = 4t$, $t \in \mathbb{R}$. Copy and complete the following table and draw a graph of the curve for $-4 \leqslant t \leqslant 4$.

t	−4	−3	−2	−1	−0.5	0	0.5	1	2	3	4
$x = 2t^2$	32					0	0.5				32
$y = 4t$	−16						2				16

2 A curve is given by the parametric equations $x = 3t^2$, $y = 6t$, $t \in \mathbb{R}$. Copy and complete the following table and draw a graph of the curve for $-3 \leqslant t \leqslant 3$.

t	−3	−2	−1	−0.5	0	0.5	1	2	3
$x = 3t^2$					0				
$y = 6t$					0				

3 A curve is given by the parametric equations $x = 4t$, $y = \frac{4}{t}$, $t \in \mathbb{R}$.

Copy and complete the following table and draw a graph of the curve for $-4 \leqslant t \leqslant 4$.

t	−4	−3	−2	−1	−0.5	0.5	1	2	3	4
$x = 4t$	−16				−2					
$y = \frac{4}{t}$	−1				−8					

4 Find the Cartesian equation of the curves given by these parametric equations.

a $x = 5t^2$, $y = 10t$ **b** $x = \frac{1}{2}t^2$, $y = t$ **c** $x = 50t^2$, $y = 100t$

d $x = \frac{1}{5}t^2$, $y = \frac{2}{5}t$ **e** $x = \frac{5}{2}t^2$, $y = 5t$ **f** $x = \sqrt{3}t^2$, $y = 2\sqrt{3}t$

g $x = 4t$, $y = 2t^2$ **h** $x = 6t$, $y = 3t^2$

5 Find the Cartesian equation of the curves given by these parametric equations.

a $x = t$, $y = \frac{1}{t}$, $t \neq 0$ **b** $x = 7t$, $y = \frac{7}{t}$, $t \neq 0$

c $x = 3\sqrt{5}t$, $y = \frac{3\sqrt{5}}{t}$, $t \neq 0$ **d** $x = \frac{t}{5}$, $y = \frac{1}{5t}$, $t \neq 0$

6 A curve has parametric equations $x = 3t$, $y = \frac{3}{t}$, $t \in \mathbb{R}$, $t \neq 0$.

a Find the Cartesian equation of the curve.

b Hence sketch this curve.

7 A curve has parametric equations $x = \sqrt{2}t$, $y = \frac{\sqrt{2}}{t}$, $t \in \mathbb{R}$, $t \neq 0$.

a Find the Cartesian equation of the curve.

b Hence sketch this curve.

4.2 The general equation of a parabola

- The curve opposite is an example of a **parabola** which has parametric equations:

  $x = at^2$, $y = 2at$, $t \in \mathbb{R}$,

  where a is a positive constant.

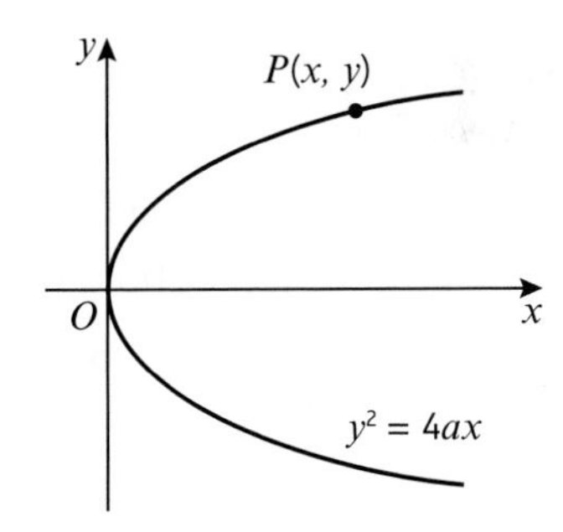

- The Cartesian equation of this curve is $y^2 = 4ax$ where a is a positive constant.
- This curve is **symmetrical** about the x-axis.
- A general point P on this curve has coordinates $P(x, y)$ or $P(at^2, 2at)$.

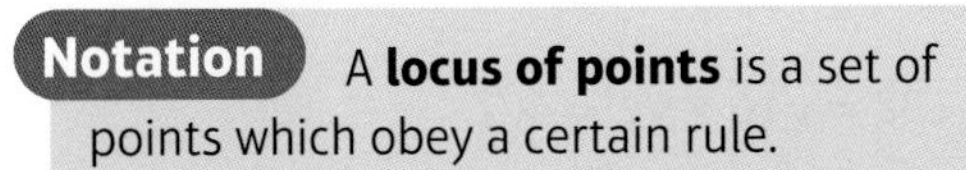
Notation A **locus of points** is a set of points which obey a certain rule.

- A parabola is the **locus** of points where every point $P(x, y)$ on the parabola is the same distance from a fixed point S, called the **focus**, and a fixed straight line called the **directrix**.
- The parabola is the set of points where $SP = PX$.

  The focus, S, has coordinates $(a, 0)$.

  The directrix has equation $x + a = 0$.

  The **vertex** is at the point $(0, 0)$.

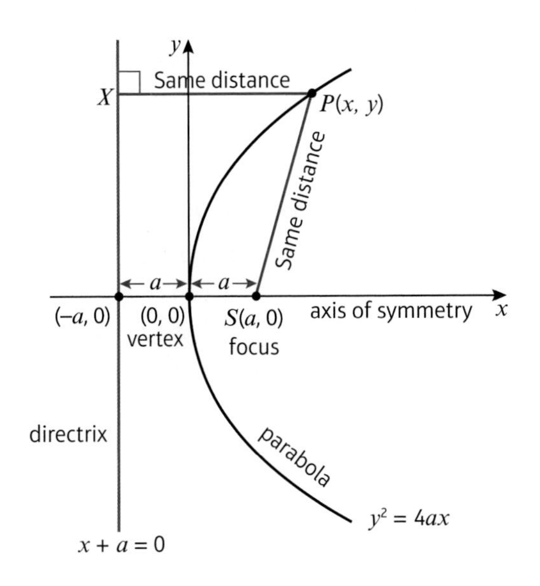

Example 4

Find an equation of the parabola with:

a focus (7, 0) and directrix $x + 7 = 0$

b focus $\left(\frac{\sqrt{3}}{4}, 0\right)$ and directrix $x = -\frac{\sqrt{3}}{4}$.

a focus (7, 0) and directrix

$x + 7 = 0$ — The focus and directrix are in the form $(a, 0)$ and $x + a = 0$, so $a = 7$

So the parabola has equation

$y^2 = 28x$ — Write the equation in the form $y^2 = 4ax$ with $a = 7$

b focus $\left(\frac{\sqrt{3}}{4}, 0\right)$ and directrix

$x = -\frac{\sqrt{3}}{4}$

$x + \frac{\sqrt{3}}{4} = 0$ — Write the focus and directrix in the form $(a, 0)$ and $x + a = 0$

So $a = \frac{\sqrt{3}}{4}$

So the parabola has equation

$y^2 = \sqrt{3}x$ — With $a = \frac{\sqrt{3}}{4}$, $y^2 = 4\left(\frac{\sqrt{3}}{4}\right)x$

Example  5

Find the coordinates of the focus and an equation for the directrix of a parabola with equation:

a $y^2 = 24x$ **b** $y^2 = \sqrt{32}x$

a $y^2 = 24x$ — This is in the form $y^2 = 4ax$. So $4a = 24$, gives $a = 6$

So the focus has coordinates (6, 0) — Focus has coordinates $(a, 0)$

and the directrix has equation $x + 6 = 0$. — Directrix has equation $x + a = 0$

b $y^2 = \sqrt{32}x$ — In surds, $\sqrt{32} = \sqrt{16} \times \sqrt{2} = 4\sqrt{2}$. So $4a = 4\sqrt{2}$, gives $a = \frac{4\sqrt{2}}{4} = \sqrt{2}$

So the focus has coordinates $(\sqrt{2}, 0)$ — Focus has coordinates $(a, 0)$

and the directrix has equation $x + \sqrt{2} = 0$. — Directrix has equation $x + a = 0$

Hint

To find the distance d between two points (x_1, y_1) and (x_2, y_2), you can use the formula: $d = \sqrt{(x_2 - x_1)^2 + (y_2 - y_1)^2}$.

This formula can also be written in the form $d^2 = (x_2 - x_1)^2 + (y_2 - y_1)^2$.

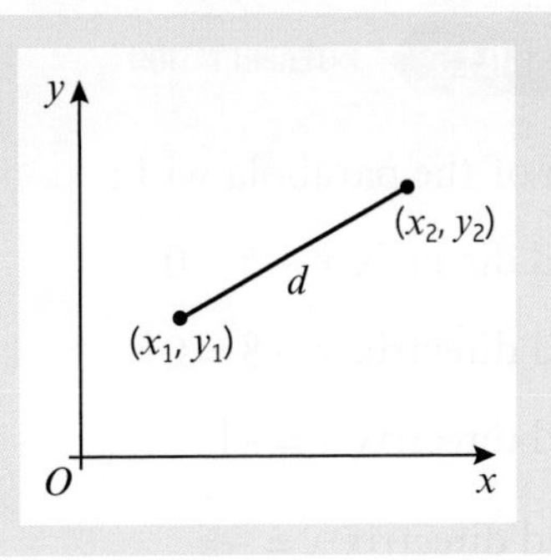

Example 6

A point $P(x, y)$ obeys a rule such that the distance of P to the point (6, 0) is the same as the distance of P to the straight line $x + 6 = 0$. Prove that the locus of P has an equation of the form $y^2 = 4ax$, stating the value of the constant a.

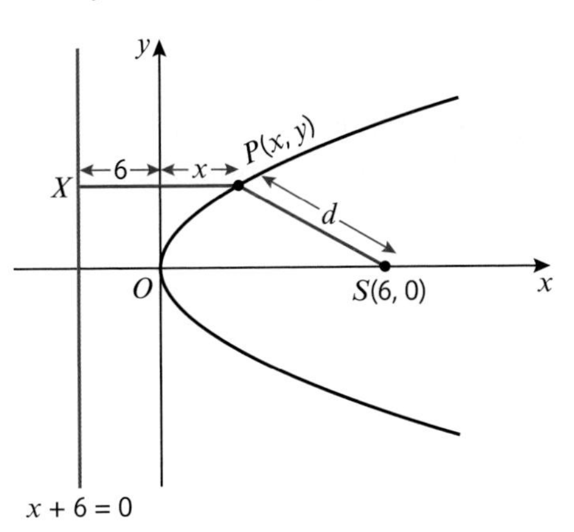

The (shortest) distance of P to the line $x + 6 = 0$ is the distance XP.

The distance SP is the same as the distance XP.

The line XP is horizontal and has distance $XP = x + 6$.

The locus of P is the curve shown.

Hint Always start by drawing a sketch if one is not provided in the question.

From the sketch, the locus satisfies
$SP = XP$ — This means the distance SP is the same as the distance XP.

Therefore, $SP^2 = XP^2$ — Square both sides.

So, $(x - 6)^2 + (y - 0)^2 = (x - (-6))^2$

$x^2 - 12x + 36 + y^2 = x^2 + 12x + 36$

$-12x + y^2 = 12x$

— Use $d^2 = (x_2 - x_1)^2 + (y_2 - y_1)^2$ on $SP^2 = XP^2$ with $S(6, 0)$, $P(x, y)$ and $X(-6, y)$. Remember, $(x - 6)^2 = (x - 6)(x - 6)$

which simplifies to $y^2 = 24x$.

So, the locus of P has an equation of the form $y^2 = 4ax$ where $a = 6$.

— This is in the form $y^2 = 4ax$
So $4a = 24$, gives $a = \frac{24}{4} = 6$

Exercise 4B SKILLS INTERPRETATION

1 Find an equation of the parabola with:

a focus (5, 0) and directrix $x + 5 = 0$

b focus (8, 0) and directrix $x + 8 = 0$

c focus (1, 0) and directrix $x = -1$

d focus $\left(\frac{3}{2}, 0\right)$ and directrix $x = -\frac{3}{2}$

e focus $\left(\frac{\sqrt{3}}{2}, 0\right)$ and directrix $x + \frac{\sqrt{3}}{2} = 0$.

2 Find the coordinates of the focus, and an equation for the directrix of a parabola with these equations.

a $y^2 = 12x$ **b** $y^2 = 20x$

c $y^2 = 10x$ **d** $y^2 = 4\sqrt{3}x$

e $y^2 = \sqrt{2}x$ **f** $y^2 = 5\sqrt{2}x$

(P) **3** A point $P(x, y)$ obeys a rule such that the distance of P to the point (3, 0) is the same as the distance of P to the straight line $x + 3 = 0$. Show that the locus of P has an equation of the form $y^2 = 4ax$, stating the value of the constant a.

(P) **4** A point $P(x, y)$ obeys a rule such that the distance of P to the point $(2\sqrt{5}, 0)$ is the same as the distance of P to the straight line $x = -2\sqrt{5}$. Show that the locus of P has an equation of the form $y^2 = 4ax$, stating the value of the constant a.

(P) **5** A point $P(x, y)$ obeys a rule such that the distance of P to the point (0, 2) is the same as the distance of P to the straight line $y = -2$.

a Show that the locus of P has an equation of the form $y = kx^2$, stating the value of the constant k.

Given that the locus of P is a parabola,

b state the coordinates of the focus of P, and an equation of the directrix of P

c sketch the locus of P with its focus and its directrix.

Example 7 SKILLS EXECUTIVE FUNCTION

The point $P(8, -8)$ lies on the parabola C with equation $y^2 = 8x$.
The point S is the focus of the parabola. The line l passes through S and P.

a Find the coordinates of S.

b Find an equation for l, giving your answer in the form $ax + by + c = 0$, where a, b and c are integers.

The line l meets the parabola C again at the point Q. The point M is the midpoint of PQ.

c Find the coordinates of Q.

d Find the coordinates of M.

e Draw a sketch showing parabola C, the line l and the points P, Q, S and M.

a $y^2 = 8x \Rightarrow a = 2$ — This is in the form $y^2 = 4ax$. So $4a = 8$, gives $a = \frac{8}{4} = 2$.

The focus, S has coordinates $(2, 0)$ — Focus has coordinates $(a, 0)$.

b $m = \frac{-8 - 0}{8 - 2} = \frac{-8}{6}$ — Use $m = \frac{y_2 - y_1}{x_2 - x_1}$, where $(x_1, y_1) = (2, 0)$ and $(x_2, y_2) = (8, -8)$.

So $m = -\frac{4}{3}$

l: $y - 0 = -\frac{4}{3}(x - 2)$ — Use $y - y_1 = m(x - x_1)$. Here $m = -\frac{4}{3}$ and $(x_1, y_1) = (2, 0)$.

l: $3y = -4(x - 2)$ — Multiply both sides by 3.

l: $3y = -4x + 8$ — Multiply out brackets.

l: $4x + 3y - 8 = 0$ — Simplify into the form $ax + by + c = 0$.

The line l has equation $4x + 3y - 8 = 0$.

c l: $4x + 3y - 8 = 0$ **(1)**
C: $y^2 = 8x$ **(2)** — As the line l meets the curve C, we solve these equations simultaneously.

$8x + 6y - 16 = 0$ **(3)** — Multiply **(1)** by 2 to get **(3)**.

$y^2 + 6y - 16 = 0$ — Substitute **(2)** into **(3)**.

$(y + 8)(y - 2) = 0$ — Factorise.

$y = -8, 2$ — $y = -8$ is at P and $y = 2$ is at Q.

$4 = 8x$ — Use $y^2 = 8x$ and $y = 2$ to find the x-coordinate of Q.

$x = \frac{4}{8} = \frac{1}{2}$

The point Q has coordinates $(\frac{1}{2}, 2)$.

d The midpoint is $\left(\frac{8 + \frac{1}{2}}{2}, \frac{-8 + 2}{2}\right)$ — Use $\left(\frac{x_1 + x_2}{2}, \frac{y_1 + y_2}{2}\right)$, where $P = (x_1, y_1) = (8, -8)$ and $Q = (x_2, y_2) = (\frac{1}{2}, 2)$.

The point M has coordinates $\left(\frac{17}{4}, -3\right)$. — Simplify.

e

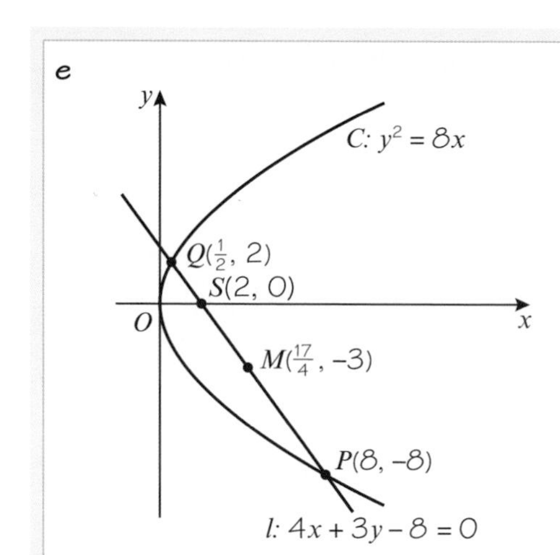

The parabola C has equation

$$y^2 = 8x.$$

The line l has equation

$$4x + 3y - 8 = 0.$$

The line l cuts the parabola at the points $P(8, -8)$ and $Q(\frac{1}{2}, 2)$.

The points $S(2, 0)$ and $M(\frac{17}{4}, -3)$ also lie on the line l.

Exercise 4C

1 The line $y = 2x - 3$ meets the parabola $y^2 = 3x$ at the points P and Q.
Find the coordinates of P and Q.

2 The line $y = x + 6$ meets the parabola $y^2 = 32x$ at the points A and B.
Find the exact length of AB, giving your answer as a surd in its simplest form.

3 The line $y = x - 20$ meets the parabola $y^2 = 10x$ at the points A and B.
The midpoint of AB is the point M. Find the coordinates of A, B and M.

(P) **4** The parabola C has parametric equations $x = 6t^2$, $y = 12t$. The focus of C is at the point S.

a Find a Cartesian equation of C.

b State the coordinates of S and the equation of the directrix of C.

c Sketch the graph of C.

The points P and Q on the parabola are both at a distance 9 units away from the directrix of the parabola.

d State the distance PS.

e Find the exact length PQ, giving your answer as a surd in its simplest form.

f Find the area of the triangle PQS, giving your answer in the form $k\sqrt{2}$, where k is an integer.

(P) **5** The parabola C has equation $y^2 = 4ax$, where a is a constant. The point $(\frac{5}{4}t^2, \frac{5}{2}t)$ is a general point on C.

a Find a Cartesian equation of C.

The point P lies on C with y-coordinate 5.

b Find the x-coordinate of P.

The point Q lies on the directrix of C where $y = 3$. The line l passes through the points P and Q.

c Find the coordinates of Q.

d Find an equation for l, giving your answer in the form $ax + by + c = 0$, where a, b and c are integers.

Ⓟ **6** A parabola C has equation $y^2 = 4x$. The point S is the focus to C.

a Find the coordinates of S.

The point P with y-coordinate 4 lies on C.

b Find the x-coordinate of P.

The line l passes through S and P.

c Find an equation for l, giving your answer in the form $ax + by + c = 0$, where a, b and c are integers.

The line l meets C again at the point Q.

d Find the coordinates of Q.

e Find the distance of the directrix of C to the point Q.

Ⓟ **7** The diagram shows the point P which lies on the parabola C with equation $y^2 = 12x$.

The point S is the focus of C. The points Q and R lie on the directrix of C. The line segment QP is parallel to the line segment RS as shown in the diagram. The distance of PS is 12 units.

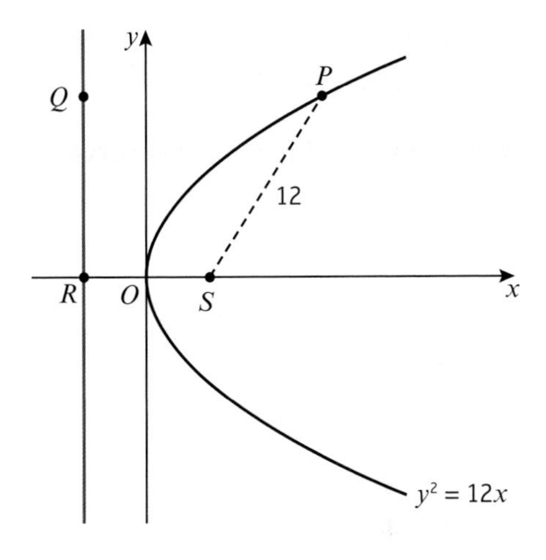

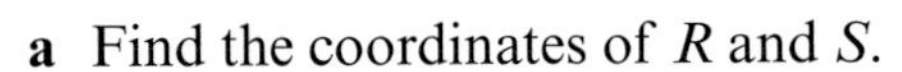

a Find the coordinates of R and S.

b Hence find the exact coordinates of P and Q.

c Find the area of the quadrilateral $PQRS$, giving your answer in the form $k\sqrt{3}$, where k is an integer.

Ⓟ **8** The points $P(16, 8)$ and $Q(4, b)$, where $b < 0$, lie on the parabola C with equation $y^2 = 4ax$.

a Find the values of a and b.

P and Q also lie on the line l. The midpoint of PQ is the point R.

b Find an equation of l, giving your answer in the form $y = mx + c$, where m and c are constants to be determined.

c Find the coordinates of R.

The line n is **perpendicular** to l and passes through R.

d Find an equation of n, giving your answer in the form $y = mx + c$, where m and c are constants to be determined.

The line n meets the parabola C at two points.

e Show that the x-coordinates of these two points can be written in the form $x = \lambda \pm \mu\sqrt{13}$, where λ and μ are integers to be determined.

4.3 The equation for a rectangular hyperbola
The equation of the tangent and the equation of the normal

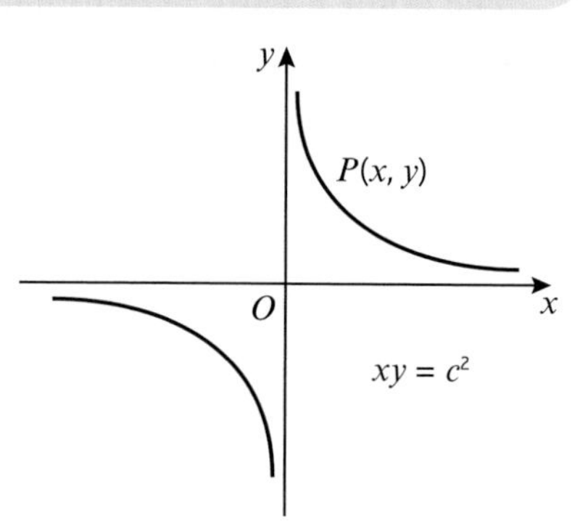

- The curve opposite is an example of a **rectangular hyperbola** which has parametric equations:

$$x = ct, y = \frac{c}{t}, t \in \mathbb{R}, t \neq 0$$

where c is a positive constant.

- The Cartesian equation of this curve is $xy = c^2$, where c is a positive constant.

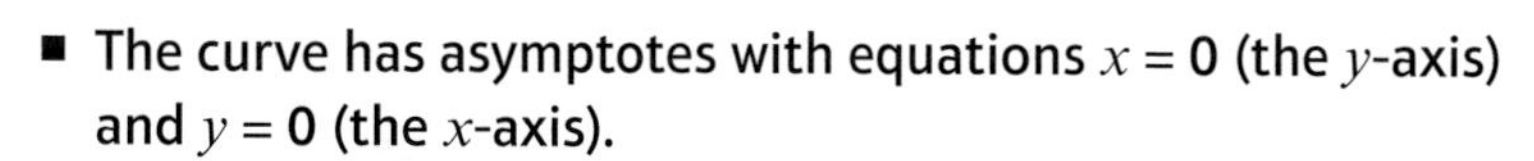

- The curve has asymptotes with equations $x = 0$ (the y-axis) and $y = 0$ (the x-axis).
- A general point P on this curve has coordinates $P(x, y)$ or $P\left(ct, \frac{c}{t}\right)$.

Example 8 SKILLS PROBLEM-SOLVING

The point P, where $x = 2$, lies on the rectangular hyperbola H with equation $xy = 8$.

Find:

a the equation of the tangent T

b the equation of the **normal** N, to H at the point P, giving your answers in the form $ax + by + c = 0$, where a, b and c are integers.

a H: $xy = 8$

$y = \frac{8}{x} \Rightarrow y = 8x^{-1}$ — Rearrange the equation for H in the form $y = x^n$ to find the derivative easily.

$\frac{dy}{dx} = -8x^{-2} = -\frac{8}{x^2}$ — Differentiate to determine the gradient of H and therefore the gradient of the tangent to H.

When $x = 2$, $m_T = \frac{dy}{dx} = -\frac{8}{2^2} = -2$ — Substitute $x = 2$ to calculate the gradient of the tangent to H.

When $x = 2$, $y = \frac{8}{2} = 4$ — Find the y-coordinate when $x = 2$. Hence P has coordinates (2, 4).

T: $y - 4 = -2(x - 2)$ — Use $y - y_1 = m_T(x - x_1)$ to find the equation of the tangent, T. Here $m_T = -2$ and $(x_1, y_1) = (2, 4)$

T: $2x + y - 8 = 0$

Therefore, the equation of the tangent to H at P is

$2x + y - 8 = 0$ — Then rearrange into the required form.

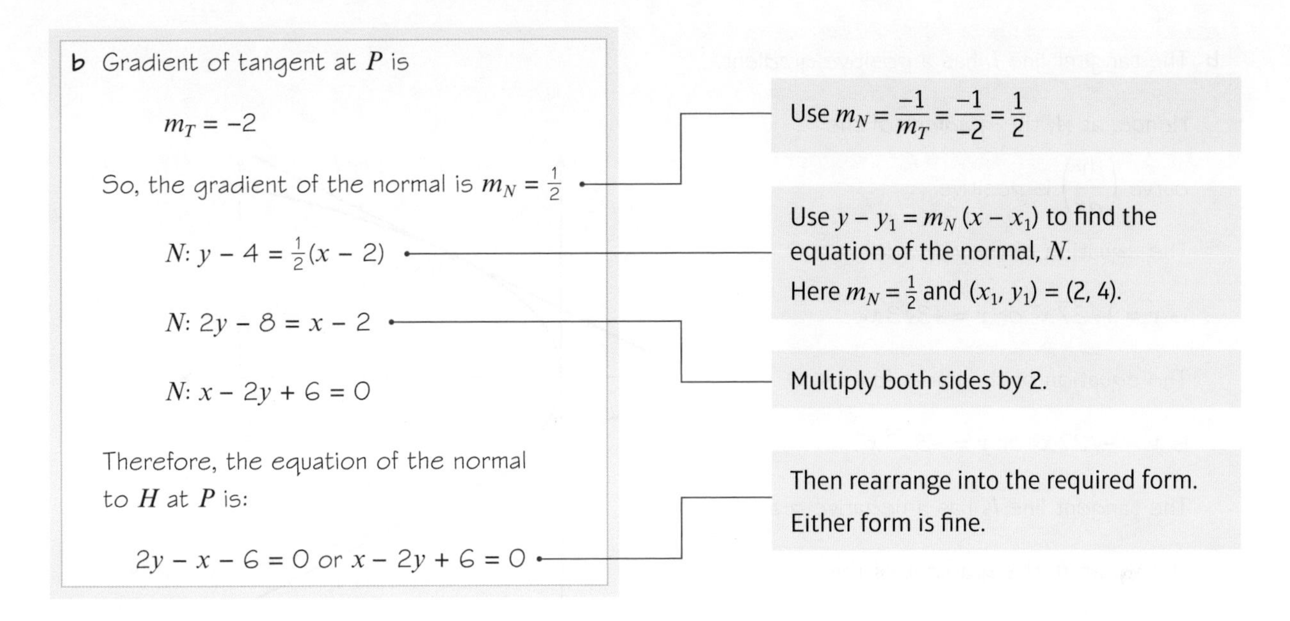

Example 9 SKILLS PROBLEM-SOLVING

The distinct points A and B lie on both the line $x = 3$ and on the parabola C with equation $y^2 = 27x$. The line l_1 is tangent to C at A and the line l_2 is tangent to C at B. Given that at A, $y > 0$,

a find coordinates of A and B.

b Draw a sketch showing the parabola C. Indicate on your sketch the points A and B and the lines l_1 and l_2.

c Find:

i an equation for l_1

ii an equation for l_2, giving your answers in the form $ax + by + c = 0$, where a, b and c are integers to be found.

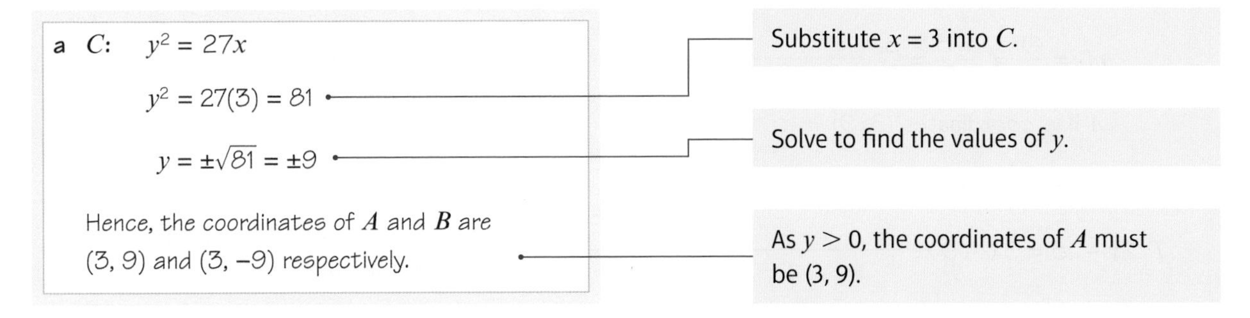

b The tangent line l_1 has a positive gradient.

Hence, at A, the gradient of the curve $\left(\frac{dy}{dx}\right)$ is positive.

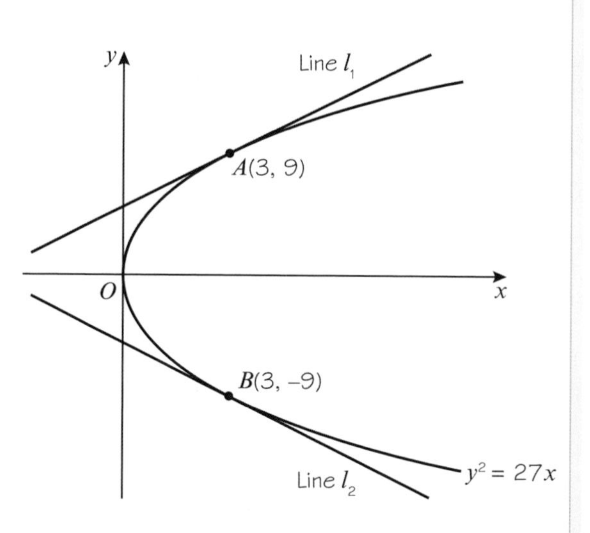

The equation of the curve for $y > 0$ is $y = +\sqrt{27}x^{\frac{1}{2}}$ or $y = +3\sqrt{3}x^{\frac{1}{2}}$.

The equation of the curve for $y < 0$ is $y = -\sqrt{27}x^{\frac{1}{2}}$ or $y = -3\sqrt{3}x^{\frac{1}{2}}$.

The tangent line l_2 has a negative gradient.

Hence, at B, the gradient of the curve $\left(\frac{dy}{dx}\right)$ is negative.

c 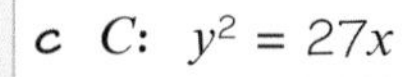 C: $y^2 = 27x$

$$y = \pm\sqrt{27x} = \pm 3\sqrt{3}x^{\frac{1}{2}}$$

Rearrange the equation for C in the form $y = kx^n$.

$$\text{So } \frac{dy}{dx} = \pm 3\sqrt{3}\left(\frac{1}{2}\right)x^{-\frac{1}{2}} = \pm\frac{3\sqrt{3}}{2}x^{-\frac{1}{2}}$$

Differentiate to determine the gradient of C.

$$= \pm\frac{3\sqrt{3}}{2\sqrt{x}}$$

Simplify $\frac{dy}{dx}$.

i At A, $y > 0$ and so

$$m_T = \frac{dy}{dx} = +\frac{3\sqrt{3}}{2\sqrt{x}}$$

At A, $x = 3$ and

$$m_T = \frac{dy}{dx} = \frac{3\sqrt{3}}{2\sqrt{3}} = \frac{3}{2}$$

Subsitute $x = 3$, to calculate the gradient of the tangent to C.

A has coordinates (3, 9)

T: $y - 9 = \frac{3}{2}(x - 3)$

Use $y - y_1 = m(x - x_1)$ to find the equation of the tangent, T. Here $m_T = \frac{3}{2}$ and (x_1, y_1) is (3, 9).

T: $2y - 18 = 3x - 9$

Multiply both sides by 2 and simplify by multiplying out the brackets.

T: $2y - 3x - 9 = 0$ or $3x - 2y + 9 = 0$

Then rearrange into the required form.

Therefore, the equation of the tangent to C at A is $3x - 2y + 9 = 0$.

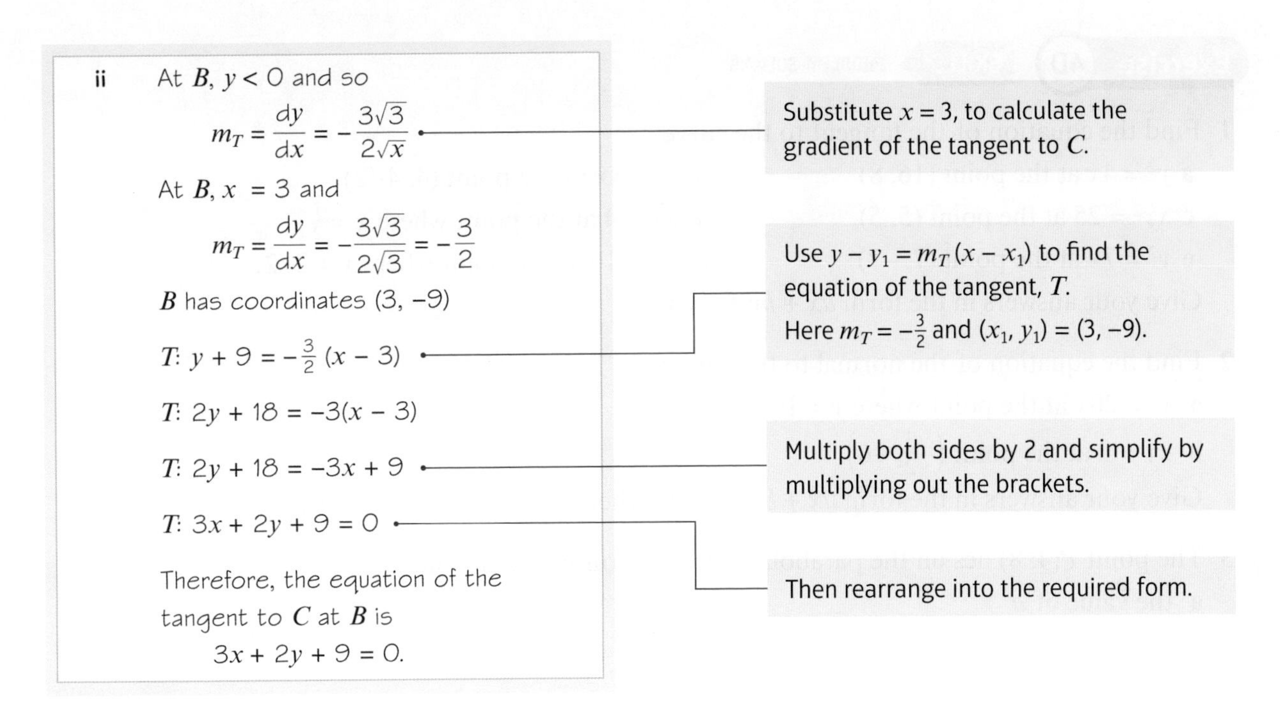

Example 10

The point P with coordinates (3, 6) lies on the parabola C with equation $y^2 = 12x$.
Find the equation of the tangent to C at P, giving your answer in the form $y = mx + c$, where m and c are constants.

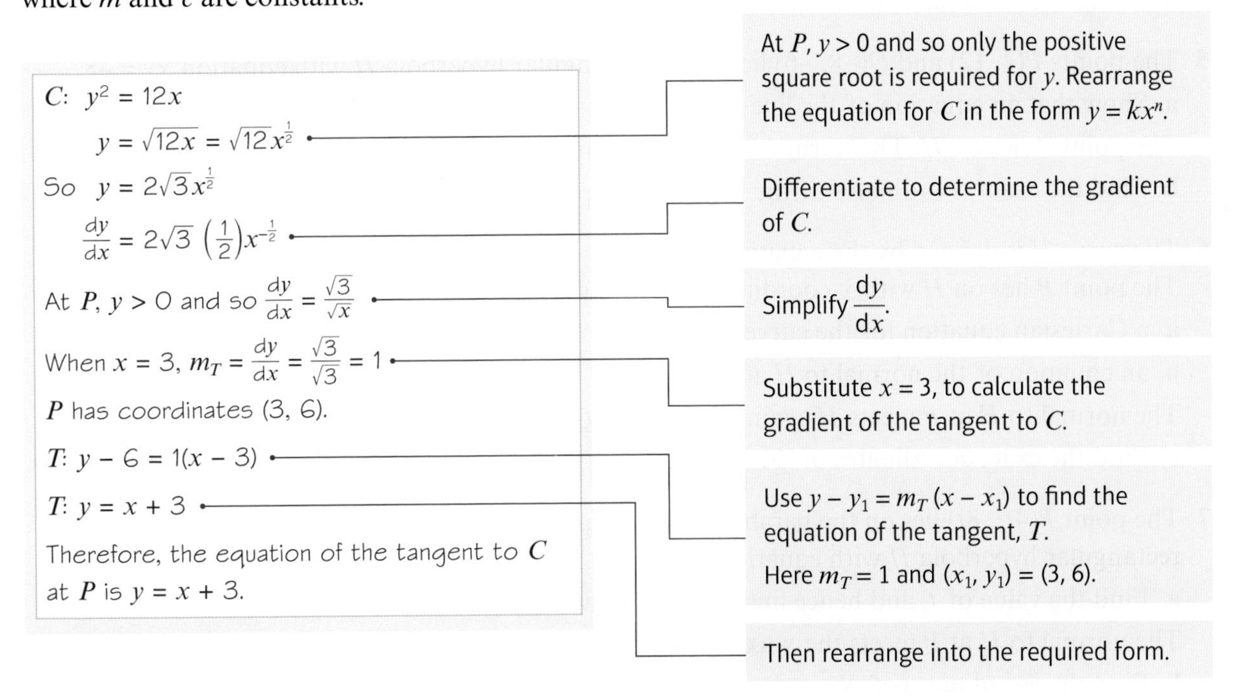

Links In Examples 8, 9 and 10 it is possible for you to find the gradient of a parabola or rectangular hyperbola by implicit differentiation. You will learn these techniques in unit Pure 4.

Exercise 4D SKILLS PROBLEM-SOLVING

1 Find the equation of the tangent to the curve:

a $y^2 = 4x$ at the point $(16, 8)$

b $y^2 = 8x$ at the point $(4, 4\sqrt{2})$

c $xy^2 = 25$ at the point $(5, 5)$

d $xy = 4$ at the point where $x = \frac{1}{2}$

e $y^2 = 7x$ at the point $(7, -7)$

f $xy = 16$ at the point where $x = 2\sqrt{2}$.

Give your answers in the form $ax + by + c = 0$.

2 Find the equation of the normal to the curve:

a $y^2 = 20x$ at the point where $y = 10$

b $xy = 9$ at the point $\left(-\frac{3}{2}, -6\right)$.

Give your answers in the form $ax + by + c = 0$, where a, b and c are integers.

(P) **3** The point $P(4, 8)$ lies on the parabola with equation $y^2 = 4ax$. Find:

a the value of a

b an equation of the normal to C at P

c the coordinates of Q

d the length PQ, giving your answer as a simplified surd.

(P) **4** The point $A(-2, -16)$ lies on the rectangular hyperbola H with equation $xy = 32$.

a Find an equation of the normal to H at A.

The normal to H at A meets H again at the point B.

b Find the coordinates of B.

(P) **5** The points $P(4, 12)$ and $Q(-8, -6)$ lie on the rectangular hyperbola H with equation $xy = 48$.

a Show that an equation of the line PQ is $3x - 2y + 12 = 0$.

The point A lies on H. The normal to H at A is parallel to the chord PQ.

b Find the exact coordinates of the two possible positions of A.

(P) **6** The curve H is defined by the equations $x = \sqrt{3}t$, $y = \frac{\sqrt{3}}{t}$, $t \in \mathbb{R}$, $t \neq 0$.

The point P lies on H with x-coordinate $2\sqrt{3}$. Find:

a a Cartesian equation for the curve H

b an equation of the normal to H at P.

The normal to H at P meets H again at the point Q.

c Find the exact coordinates of Q.

(P) **7** The point $P(4t^2, 8t)$ lies on the parabola C with equation $y^2 = 16x$. The point P also lies on the rectangular hyperbola H with equation $xy = 4$.

a Find the value of t, and hence find the coordinates of P.

The normal to H at P meets the x-axis at the point N.

b Find the coordinates of N.

The tangent to C at P meets the x-axis at the point T.

c Find the coordinates of T.

d Hence, find the area of the triangle NPT.

Example 11 SKILLS INTERPRETATION

The point $P(at^2, 2at)$ lies on the parabola C with equation $y^2 = 4ax$ where a is a positive constant. Show that an equation of the normal P is $y + tx = 2at + at^3$.

C: $y^2 = 4ax$

$y = \sqrt{4ax} = 2\sqrt{a}\sqrt{x}$

So $y = 2\sqrt{a}x^{\frac{1}{2}}$

Rearrange the equation for C in the form $y = kx^n$. $\sqrt{4ax} = 2 \times \sqrt{a} \times x^{\frac{1}{2}}$

$\frac{dy}{dx} = \left(\frac{1}{2}\right)2\sqrt{a}x^{-\frac{1}{2}} = \frac{\sqrt{a}}{\sqrt{x}}$

Differentiate to determine the gradient of C and simplify $\frac{dy}{dx}$.

At P, $x = at^2$ and

$m_T = \frac{dy}{dx} = \frac{\sqrt{a}}{\sqrt{at^2}} = \frac{1}{t}$

Gradient of tangent at P is $m_T = \frac{1}{t}$.

Substitute $x = at^2$ to calculate the gradient of the tangent to C.

So gradient of normal is $m_N = -t$.

Use $m_N = \frac{-1}{\left(\frac{1}{t}\right)} = -t$

P has coordinates $(at^2, 2at)$

N: $y - 2at = -t(x - at^2)$

N: $y - 2at = -tx + at^3$

N: $y + tx = 2at + at^3$

Use the formula $y - y_1 = m(x - x_1)$ to find the equation of the normal, N. Here $m_N = -t$ and $(x_1, y_1) = (at^2, 2at)$.

Therefore, the equation of the normal to C at P is

$y + tx = 2at + at^3$

Multiply out brackets and then rearrange into the required form.

Example 12 SKILLS PROBLEM-SOLVING

The point $P\left(ct, \frac{c}{t}\right)$, $t \neq 0$, lies on the rectangular hyperbola H with equation $xy = c^2$, where c is a positive constant.

a Show that an equation of the tangent to H at P is $x + t^2y = 2ct$.

A rectangular hyperbola G has equation $xy = 9$. The tangent to G at the point A and the tangent to G at the point B meet at the point $(-1, 7)$.

b Find the coordinates of A and B.

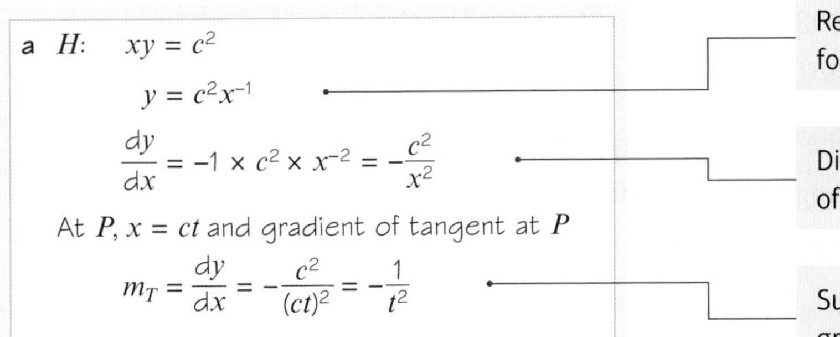

a H: $xy = c^2$

$y = c^2x^{-1}$

Rearrange the equation for H in the form $y = x^n$.

$\frac{dy}{dx} = -1 \times c^2 \times x^{-2} = -\frac{c^2}{x^2}$

Differentiate to determine the gradient of H.

At P, $x = ct$ and gradient of tangent at P

$m_T = \frac{dy}{dx} = -\frac{c^2}{(ct)^2} = -\frac{1}{t^2}$

Substitute $x = ct$, to calculate the gradient of the tangent to H.

T: $y - \frac{c}{t} = -\frac{1}{t^2}(x - ct)$

Use $y - y_1 = m_T(x - x_1)$ to find the equation of the tangent, T.
Here $m_T = -\frac{1}{t^2}$ and $(x_1, y_1) = \left(ct, \frac{c}{t}\right)$.

T: $t^2y - ct = -(x - ct)$

T: $x + t^2y = 2ct$

Therefore, the equation of the tangent to H at P is $x + t^2y = 2ct$.

Rearrange into the required form.

b Compare G: $xy = 9$ with $xy = c^2$

So, as c is positive, $c = 3$.

Tangent to G is $x + t^2y = 6t$ **(1)**

Substitute $c = 3$ into the equation of the tangent derived in **a**.

$-1 + t^2(7) = 6t$

Substitute $x = -1$ and $y = 7$ in **(1)** as tangent goes through point $(-1, 7)$.

$7t^2 - 6t - 1 = 0$

Rearrange into a quadratic equation = 0

$(7t + 1)(t - 1) = 0$

$t = -\frac{1}{7}, 1$

P has coordinates

$\left(ct, \frac{c}{t}\right) = \left(3t, \frac{3}{t}\right)$

Substitute $c = 3$ into the general coordinates of P.

When $t = -\frac{1}{7}$, the coordinates are

$\left(3 \times -\frac{1}{7}, \frac{3}{\frac{-1}{7}}\right) = \left(\frac{-3}{7}, -21\right)$

Substitute $t = -\frac{1}{7}$ into P.

When $t = 1$, the coordinates are

$\left(3 \times 1, \frac{3}{1}\right) = (3, 3)$

Substitute $t = 1$ into P.

Therefore, the coordinates of A and B are $\left(-\frac{3}{7}, -21\right)$ and $(3, 3)$.

Exercise 4E

SKILLS INTERPRETATION, PROBLEM-SOLVING

1 The point $P(3t^2, 6t)$ lies on the parabola C with equation $y^2 = 12x$.

a Show that an equation of the tangent to C at P is $yt = x + 3t^2$.

b Show that an equation of the normal to C at P is $xt + y = 3t^3 + 6t$.

2 The point $P\left(6t, \frac{6}{t}\right)$, $t \neq 0$, lies on the rectangular hyperbola H with equation $xy = 36$.

a Show that an equation of the tangent to H at P is $x + t^2y = 12t$.

b Show that an equation of the normal to H at P is $t^3x - ty = 6(t^4 - 1)$.

(P) **3** The point $P(5t^2, 10t)$ lies on the parabola C with equation $y^2 = 4ax$, where a is a constant and $t \neq 0$.

a Find the value of a.

b Show that an equation of the tangent to C at P is $yt = x + 5t^2$.

The tangent to C at P cuts the x-axis at the point X and the y-axis at the point Y.
The point O is the origin of the coordinate system.

c Find, in terms of t, the area of the triangle OXY.

(P) **4** The point $P(at^2, 2at)$, $t \neq 0$, lies on the parabola C with equation $y^2 = 4ax$, where a is a positive constant.

a Show that an equation of the tangent to C at P is $ty = x + at^2$.

The tangent to C at the point A and the tangent to C at the point B meet at the point with coordinates $(-4a, 3a)$.

b Find, in terms of a, the coordinates of A and the coordinates of B.

(E/P) **5** The point $P\left(4t, \frac{4}{t}\right)$, $t \neq 0$, lies on the rectangular hyperbola H with equation $xy = 16$.

a Show that an equation of the tangent to C at P is $x + t^2y = 8t$. **(3 marks)**

The tangent to H at the point A and the tangent to H at the point B meet at the point X with y-coordinate 5. X lies on the directrix of the parabola C with equation $y^2 = 16x$.

b Write down the coordinates of X. **(1 mark)**

c Find the coordinates of A and B. **(2 marks)**

d **Deduce** the equations of the tangents to H which pass through X.
Give your answers in the form $ax + by + c = 0$, where a, b and c are integers. **(3 marks)**

(E/P) **6** The point $P(at^2, 2at)$ lies on the parabola C with equation $y^2 = 4ax$, where a is a constant and $t \neq 0$. The tangent to C at P cuts the x-axis at the point A.

a Find, in terms of a and t, the coordinates of A. **(3 marks)**

The normal to C at P cuts the x-axis at the point B.

b Find, in terms of a and t, the coordinates of B. **(3 marks)**

c Hence find, in terms of a and t, the area of the triangle APB. **(2 marks)**

(P) **7** The point $P(2t^2, 4t)$ lies on the parabola C with equation $y^2 = 8x$.

a Show that an equation of the normal to C at P is $xt + y = 2t^3 + 4t$.

The normals to C at the points R, S and T meet at the point $(12, 0)$.

b Find the coordinates of R, S and T.

c Deduce the equations of the normals to C which all pass through the point $(12, 0)$.

(P) **8** The point $P(at^2, 2at)$ lies on the parabola C with equation $y^2 = 4ax$, where a is a positive constant and $t \neq 0$. The tangent to C at P meets the y-axis at Q.

a Find in terms of a and t, the coordinates of Q.

The point S is the focus of the parabola.

b State the coordinates of S.

c Show that PQ is perpendicular to SQ.

(E/P) **9** The point $P(6t^2, 12t)$ lies on the parabola C with equation $y^2 = 24x$.

a Show that an equation of the tangent to the parabola at P is $ty = x + 6t^2$. **(5 marks)**

The point X has y-coordinate 9 and lies on the directrix of C.

b State the x-coordinate of X. **(1 mark)**

The tangent at the point B on C goes through point X.

c Find the possible coordinates of B. **(3 marks)**

Chapter review 4

(P) **1** A parabola C has equation $y^2 = 12x$. The point S is the focus of C.

a Find the coordinates of S.

The line l with equation $y = 3x$ intersects C at the point P where $y > 0$.

b Find the coordinates of P.

c Find the area of the triangle OPS, where O is the origin.

(P) **2** A parabola C has equation $y^2 = 24x$. The point P with coordinates $(k, 6)$, where k is a constant, lies on C.

a Find the value of k.

The point S is the focus of C.

b Find the coordinates of S.

The line l passes through S and P and intersects the directrix of C at the point D.

c Show that an equation for l is $4x + 3y - 24 = 0$.

d Find the area of the triangle OPD, where O is the origin.

(P) **3** The parabola C has parametric equations $x = 12t^2$, $y = 24t$. The focus of C is at the point S.

a Find a Cartesian equation of C.

The point P lies on C where $y > 0$. P is 28 units from S.

b Find an equation of the directrix of C.

c Find the exact coordinates of the point P.

d Find the area of the triangle OSP, where O is the origin, giving your answer in the form $k\sqrt{3}$, where k is an integer.

(P) **4** The point $(4t^2, 8t)$ lies on the parabola C with equation $y^2 = 16x$. The line l with equation $4x - 9y + 32 = 0$ intersects the curve at the points P and Q.

a Find the coordinates of P and Q.

b Show that an equation of the normal to C at $(4t^2, 8t)$ is $xt + y = 4t^3 + 8t$.

c Hence, find an equation of the normal to C at P and an equation of the normal to C at Q.

The normal to C at P and the normal to C at Q meet at the point R.

d Find the coordinates of R and show that R lies on C.

e Find the distance OR, where O is the origin, giving your answer in the form $k\sqrt{97}$, where k is an integer.

(P) **5** The point $P(at^2, 2at)$ lies on the parabola C with equation $y^2 = 4ax$, where a is a positive constant. The point Q lies on the directrix of C. The point Q also lies on the x-axis.

a State the coordinates of the focus of C and the coordinates of Q.

The tangent to C at P passes through the point Q.

b Find, in terms of a, the two sets of possible coordinates of P.

(P) **6** The point $P\left(ct, \frac{c}{t}\right)$, $c > 0$, $t \neq 0$, lies on the rectangular hyperbola H with equation $xy = c^2$.

a Show that the equation of the normal to H at P is $t^3x - ty = c(t^4 - 1)$.

b Hence, find the equation of the normal n to the curve V with the equation $xy = 36$, at the point $(12, 3)$. Give your answer in the form $ax + by = d$, where a, b and d are integers.

The line n meets V again at the point Q.

c Find the coordinates of Q.

(E/P) **7** A rectangular hyperbola H has equation $xy = 9$. The lines l_1 and l_2 are tangents to H. The gradients of l_1 and l_2 are both $-\frac{1}{4}$. Find the equations of l_1 and l_2. **(11 marks)**

(E/P) **8** The point P lies on the rectangular hyperbola $xy = c^2$, where $c > 0$. The tangent to the rectangular hyperbola at the point $P\left(ct, \frac{c}{t}\right)$, $t > 0$, cuts the x-axis at the point X and cuts the y-axis at the point Y.

a Find, in terms of c and t, the coordinates of X and Y. **(5 marks)**

Given that the coordinates of point O are $(0, 0)$ and the area of triangle OXY is 144,

b find the exact value of c. **(4 marks)**

(P) **9** A rectangular hyperbola H has Cartesian equation $xy = c^2$, $c > 0$.

The point $\left(ct, \frac{c}{t}\right)$, where $t > 0$, is a general point on H.

a Show that an equation of the tangent to H at $\left(ct, \frac{c}{t}\right)$ is $x + t^2y = 2ct$.

The point P lies on H. The tangent to H at P cuts the x-axis at the point X with coordinates $(2a, 0)$, where a is a constant.

b Use the answer to part **a** to show that P has coordinates $\left(a, \frac{c^2}{a}\right)$.

The point Q, which lies on H, has x-coordinate $2a$.

c Find the y-coordinate of Q.

d Hence, find the equation of the line OQ, where O is the origin.

The lines OQ and XP meet at point R.

e Find, in terms of a, the x-coordinate of R.

Given that the line OQ is perpendicular to the line XP,

f show that $c^2 = 2a^2$

g find, in terms of a, the y-coordinate of R.

Challenge

Find an equation of the line which is a tangent to both the parabola with equation $y^2 = 4ax$ and the parabola with equation $x^2 = 4ay$.

Summary of key points

1. To find the Cartesian equation of a curve given parametrically, you eliminate the parameter t between the parametric equations.

2. A parabola is a set of points which are of equal distance from the focus S and a line called the directrix.

   So, for the **parabola** opposite,

   - $SP = PX$
   - the **focus**, S, has coordinates $(a, 0)$
   - the **directrix** has equation $x + a = 0$.

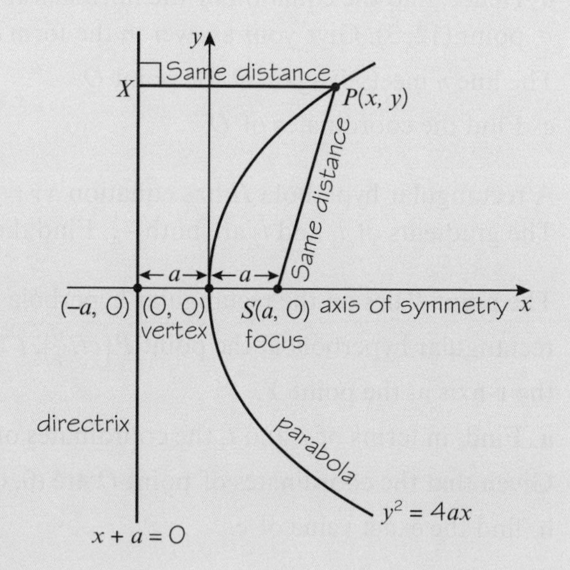

3. The curve opposite is a sketch of a **parabola** with a Cartesian equation of $y^2 = 4ax$, where a is a positive constant.

   This curve has parametric equations:

   $$x = at^2,\ y = 2at,\ t \in \mathbb{R},$$

   where a is a positive constant.

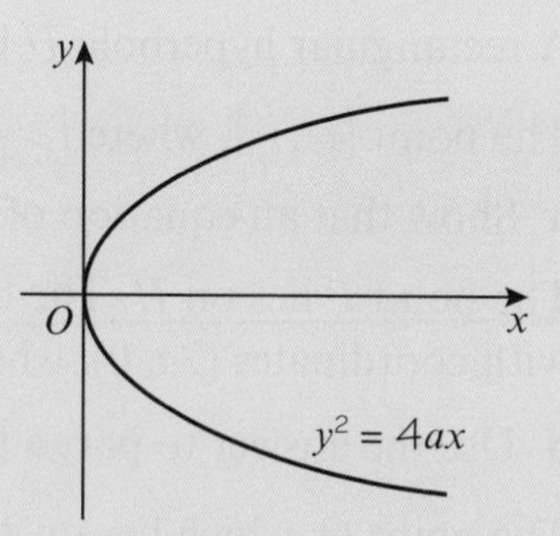

4. The curve opposite is a sketch of a **rectangular hyperbola** with a Cartesian equation of $xy = c^2$, where c is a positive constant.

   This curve has parametric equations:

   $$x = ct,\ y = \frac{c}{t},\ t \in \mathbb{R},\ t \neq 0,$$

   where c is a positive constant.

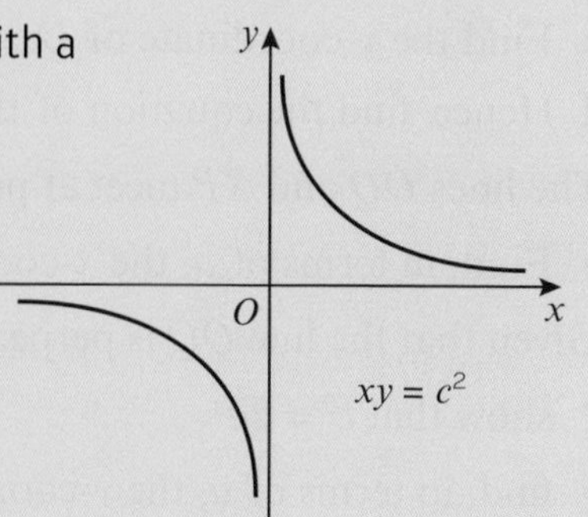

Review exercise 1

E **1** $z_1 = 4 - 5i$ and $z_2 = pi$, where p is a real constant. Find the following, in the form $a + bi$, giving a and b in terms of p:

a $z_1 - z_2$ **(1)**

b $z_1 z_2$ **(1)**

c $\frac{z_1}{z_2}$ **(1)**

← Further Pure 1 Sections 1.1, 1.2, 1.3

E/P **2** $f(z) = z^3 - kz^2 + 3z$ has two imaginary roots.

a Find the range of possible values of k. **(3)**

b Given that $k = 2$, solve the equation $f(z) = 0$. **(3)**

← Further Pure 1 Section 1.1

E **3** The solutions to the quadratic equation $z^2 - 5z + 13 = 0$ are z_1 and z_2. Find z_1 and z_2, giving each answer in the form $a \pm ib$ where $a, b \in \mathbb{R}$. **(3)**

← Further Pure 1 Section 1.1

E **4** The real and imaginary parts of the complex number $z = x + iy$ satisfy the equation $(2 - i)x - (1 + 3i)y - 7 = 0$. Find the values of x and y. **(4)**

← Further Pure 1 Section 1.1

E/P **5** **a** Show that the complex number $\frac{2 + 3i}{5 + i}$ can be expressed in the form $\lambda(1 + i)$, stating the value of λ. **(3)**

b Hence show that $\left(\frac{2 + 3i}{5 + i}\right)^4$ is real and determine its value. **(2)**

← Further Pure 1 Sections 1.2, 1.3

E/P **6** $f(z) = z^3 + 5z^2 + 8z + 6$
Given that $-1 + i$ is a root of the equation $f(z) = 0$, solve $f(z) = 0$ completely. **(4)**

← Further Pure 1 Section 1.8

E/P **7** $f(z) = z^3 - 6z^2 + kz - 26$
Given that $f(2 - 3i) = 0$,

a find the value of k **(2)**

b find the other two roots of the equation $f(z) = 0$. **(3)**

← Further Pure 1 Section 1.8

E **8** $f(z) = z^4 - z^3 - 6z^2 - 20z - 16$

a Write $f(z)$ in the form $(z^2 - 3z - 4)(z^2 + bz + c)$ where b and c are real constants to be found. **(2)**

b Hence find all the solutions to the equation $f(z) = 0$. **(3)**

← Further Pure 1 Section 1.8

E/P **9** $g(z) = z^4 - 8z^3 + 27z^2 - 50z + 50$
Given that $g(1 - 2i) = 0$, find all the roots of the equation $g(z) = 0$. **(5)**

← Further Pure 1 Section 1.8

E/P **10** $f(z) = z^3 + pz^2 + qz - 12$, where p and q are real constants.
Given that α, $\frac{4}{\alpha}$ and $\alpha + \frac{4}{\alpha} + 1$ are the roots of the equation $f(z) = 0$,

a solve completely the equation $f(z) = 0$. **(5)**

b Hence find the values of p and q. **(3)**

← Further Pure 1 Section 1.7

(E) **11** **a** Find, in the form $p + iq$ where p and q are real, the complex number z which satisfies the equation $\frac{3z - 1}{2 - i} = \frac{4}{1 + 2i}$ **(4)**

b Show on a single Argand diagram the points which represent z and z^*. **(2)**

c Express z and z^* in modulus–argument form, giving the arguments to the nearest degree. **(3)**

← Further Pure 1 Sections 1.2, 1.4, 1.6

(E) **12** The complex number z is $-9 + 17i$.

a Show z on an Argand diagram. **(1)**

b Calculate $\arg z$, giving your answer in radians to two decimal places. **(2)**

c Find the complex number w for which $zw = 25 + 35i$, giving your answer in the form $p + iq$, where p and q are real. **(3)**

← Further Pure 1 Sections 1.3, 1.4, 1.5

(E) **13** **a** Given that $z = 2 - i$, show that $z^2 = 3 - 4i$. **(2)**

b Hence, or otherwise, find the roots, z_1 and z_2, of the equation $(z + i)^2 = 3 - 4i$. **(3)**

c Show points representing z_1 and z_2 on a single Argand diagram. **(2)**

d Deduce that $|z_1 - z_2| = 2\sqrt{5}$. **(2)**

e Find the value of $\arg(z_1 + z_2)$. **(2)**

← Further Pure 1 Sections 1.2, 1.4, 1.5

(E) **14** $g(x) = x^3 - x^2 - 1$

a Show that there is a root α of $g(x) = 0$ in the interval $[1.4, 1.5]$. **(2)**

b By considering a change of sign of $g(x)$ in a suitable interval, verify that $\alpha = 1.466$ correct to 3 decimal places. **(3)**

← Further Pure 1 Section 2.1

15 The roots of the quadratic equation $3x^2 + 4x - 1 = 0$ are α and β.

a Without solving the equation, find the value of:

i $\alpha^2 + \beta^2$ **(2)**

ii $\alpha^3 + \beta^3$ **(2)**

b Form an equation with integer coefficients which has roots $\frac{\alpha}{\beta^2}$ and $\frac{\beta}{\alpha^2}$. **(5)**

← Further Pure 1 Sections 2.1, 2.2

16 The roots of the quadratic equation $2x^2 + 5x - 4 = 0$ are α and β.

a Write down the values of $\alpha + \beta$ and $\alpha\beta$. **(1)**

b Form an equation with integer coefficients which has roots $\frac{1}{\beta}$ and $\frac{1}{\alpha}$. **(4)**

← Further Pure 1 Sections 2.1, 2.2

17 The roots of the quadratic equation $x^2 - 3x + 1 = 0$ are α and β.

a Without solving the equation,

i find the value of $\alpha^2 + \beta^2$ **(2)**

ii show that $\alpha^3 + \beta^3 = 18$ **(3)**

iii show that $\alpha^4 + \beta^4 = (\alpha^2 + \beta^2)^2 - 2(\alpha\beta)^2$. **(2)**

b Form an equation with integer coefficients which has roots $(\alpha^3 - \beta)$ and $(\beta^3 - \alpha)$. **(5)**

← Further Pure 1 Sections 2.1, 2.2

(E) **18** A point P with coordinates (x, y) moves so that its distance from the point $(5, 0)$ is equal to its distance from the line with equation $x = -5$.
Prove that the locus of P has an equation of the form $y^2 = 4ax$, stating the value of a. **(7)**

← Further Pure 1 Section 3.2

(E/P) **19** A parabola C has equation $y^2 = 16x$.
The point S is the focus of the parabola.

a Write down the coordinates of S. **(2)**

The point P with coordinates (16, 16) lies on C.

b Find an equation of the line SP, giving your answer in the form $ax + by + c = 0$, where a, b and c are integers to be found. **(3)**

The line SP intersects C at the point Q, where P and Q are distinct points.

c Find the coordinates of Q. **(3)**

← Further Pure 1 Section 3.2

(E) **20** The curve C has equations $x = 3t^2$, $y = 6t$.

a Sketch the graph of the curve C. **(3)**

The curve C intersects the line with equation $y = x - 72$ at the points A and B.

b Find the length AB, giving your answer as a surd in its simplest form. **(6)**

← Further Pure 1 Section 3.2

21 A parabola C has equation $y^2 = 12x$. Points P and Q both lie on the parabola and are both at distance 8 from the directrix of the parabola. Find the length PQ, giving your answer in surd form. **(7)**

← Further Pure 1 Section 3.2

(E/P) **22** The point $P(2, 8)$ lies on the parabola C with equation $y^2 = 4ax$. Find:

a the value of a **(2)**

b an equation of the tangent to C at P. **(5)**

The tangent to C at P cuts the x-axis at the point X and the y-axis at the point Y.

c Find the exact area of the triangle OXY. **(4)**

← Further Pure 1 Section 3.2

(E/P) **23** The point P with coordinates (3, 4) lies on the rectangular hyperbola H with equation $xy = 12$.
The point Q has coordinates (−2, 0).
The points P and Q lie on the line l.

a Find an equation of l, giving your answer in the form $y = mx + c$, where m and c are real constants. **(4)**

The line l cuts H at the point R, where P and R are distinct points.

b Find the coordinates of R. **(4)**

← Further Pure 1 Section 3.3

(E/P) **24** The point $P(12, 3)$ lies on the rectangular hyperbola H with equation $xy = 36$.

a Find an equation of the tangent to H at P. **(5)**

The tangent to H at P cuts the x-axis at the point M and the y-axis at the point N.

b Find the length MN, giving your answer as a simplified surd. **(4)**

← Further Pure 1 Section 3.3

(E/P) **25** The curve with equation $x = 8t$, $y = \frac{16}{t}$ intersects the line with equation $y = \frac{1}{4}x + 4$ at the points A and B.
The midpoint of AB is M. Find the coordinates of M. **(6)**

← Further Pure 1 Section 3.3

(E/P) **26** The point $P(24t^2, 48t)$ lies on the parabola with equation $y^2 = 96x$.
The point P also lies on the rectangular hyperbola with equation $xy = 144$.

a Find the value of t and, hence, the coordinates of P. **(5)**

b Find an equation of the tangent to the parabola at P, giving your answer in the form $y = mx + c$, where m and c are real constants. **(5)**

← Further Pure 1 Section 3.3

(E/P) **27** The points $P(9, 8)$ and $Q(6, 12)$ lie on the rectangular hyperbola H with equation $xy = 72$.

a Show that an equation of the chord PQ of H is $4x + 3y = 60$. **(4)**

The point R lies on H. The tangent to H at R is parallel to the chord PQ.

b Find the exact coordinates of the two possible positions of R. **(6)**

← Further Pure 1 Section 3.2

(E) **28** A rectangular hyperbola H has Cartesian equation $xy = 9$. The point $\left(3t, \frac{3}{t}\right)$ is a general point on H.

a Show that an equation of the tangent to H at $\left(3t, \frac{3}{t}\right)$ is $x + t^2y = 6t$. **(5)**

The tangent to H at $\left(3t, \frac{3}{t}\right)$ cuts the x-axis at A, and the y-axis at B. The point O is the origin of the coordinate system.

b Show that, as t varies, the area of the triangle OAB is constant. **(6)**

← Further Pure 1 Section 3.3

(E) **29** The point $P\left(ct, \frac{c}{t}\right)$ lies on the hyperbola with equation $xy = c^2$, where c is a positive constant.

a Show that an equation of the normal to the hyperbola at P is $t^3x - ty - c(t^4 - 1) = 0$. **(6)**

The normal to the hyperbola at P meets the line $y = x$ at G. Given that $t \neq \pm1$,

b show that $PG^2 = c^2\left(t^2 + \frac{1}{t^2}\right)$. **(4)**

← Further Pure 1 Section 3.3

(E) **30** **a** Show that an equation of the tangent to the rectangular hyperbola with equation $xy = c^2$ at the point $\left(ct, \frac{c}{t}\right)$ is $t^2y + x = 2ct$. **(5)**

Tangents are drawn from the point $(-3, 3)$ to the rectangular hyperbola with equation $xy = 16$.

b Find the coordinates of the points of contact of these tangents with the hyperbola. **(5)**

← Further Pure 1 Section 3.3

(E) **31** The point $P(at^2, 2at)$, where $t > 0$, lies on the parabola with equation $y^2 = 4ax$. The tangent and normal at P cut the x-axis at the points T and N respectively.
Prove that $\frac{PT}{PN} = t$. **(7)**

← Further Pure 1 Section 3.3

(E) **32** The point P lies on the parabola with equation $y^2 = 4ax$, where a is a positive constant.

a Show that an equation of the tangent to the parabola $P(ap^2, 2ap)$, $p > 0$, is $py = x + ap^2$. **(5)**

The tangents at the points $P(ap^2, 2ap)$ and $Q(aq^2, 2aq)$, $p \neq q$, $p > 0$, $q > 0$, meet at the point N.

b Find the coordinates of N. **(4)**

Given further that N lies on the line with equation $y = 4a$,

c find p in terms of q. **(3)**

← Further Pure 1 Section 3.2

(E/P) **33** The point $P(at^2, 2at)$, $t \neq 0$, lies on the parabola with equation $y^2 = 4ax$, where a is a positive constant.

a Show that an equation of the normal to the parabola at P is $y + xt = 2at + at^3$. **(5)**

The normal to the parabola at P meets the parabola again at Q.

b Find, in terms of t, the coordinates of Q. **(4)**

← Further Pure 1 Section 3.2

(E/P) **34** **a** Show that the normal to the rectangular hyperbola, $xy = c^2$, at the point $P\left(ct, \frac{c}{t}\right)$, $t \neq 0$, has equation $y = t^2x + \frac{c}{t} - ct^3$. **(6)**

The normal to the hyperbola at P meets the hyperbola again at the point Q.

b Find, in terms of t, the coordinates of the point Q. **(2)**

Given that the midpoint of PQ is (X, Y) and that $t \neq \pm1$,

c show that $\frac{X}{Y} = -\frac{1}{t^2}$ **(4)**

← Further Pure 1 Section 3.2

Ⓔ **35** The rectangular hyperbola C has equation $xy = c^2$, where c is a positive constant.

a Show that the tangent to C at the point $P\left(cp, \frac{c}{p}\right)$ has equation $p^2y = -x + 2cp$. **(5)**

The point Q has coordinates $\left(cq, \frac{c}{q}\right)$, $q \neq p$. The tangents to C at P and Q meet at N. Given that $p + q \neq 0$,

b show that the y-coordinate of N is $\frac{2c}{p+q}$. **(4)**

The line joining N to the origin O is perpendicular to the chord PQ.

c Find the numerical value of p^2q^2. **(3)**

← Further Pure 1 Section 3.3

Ⓔ **36** The curve C has equation $y^2 = 4ax$, where a is a positive constant.

a Show that an equation of the normal to C at the point $P(ap^2, 2ap)$, $p \neq 0$, is $y + px = 2ap + ap^3$. **(5)**

The normal at P meets C again at the point $Q(aq^2, 2aq)$.

b Find q in terms of p. **(4)**

Given that the midpoint of PQ has coordinates $\left(\frac{125}{18}a, -3a\right)$,

c use your answer to **b**, or otherwise, to find the value of p. **(2)**

← Further Pure 1 Section 3.3

Ⓔ **37** The parabola C has equation $y^2 = 32x$.

a Write down the coordinates of the focus S of C. **(1)**

b Write down the equation of the directrix of C. **(1)**

The points $P(2, 8)$ and $Q(32, -32)$ lie on C.

c Show that the line joining P and Q goes through S. **(5)**

The tangent to C at P and the tangent to C at Q intersect at the point D.

d Show that D lies on the directrix of C. **(4)**

← Further Pure 1 Section 3.2

Challenge

1 Solve the equation $x^2 + 2\text{i}x + 5 = 0$, giving your answer in terms of i.

← Further Pure 1 Section 1.7

2 The roots of a quadratic equation $ax^2 + bx + c = 0$ are α and β where a, b and c are integers.

Given that $\alpha^2 + \beta^2 = -\frac{7}{4}$, $\alpha^4 + \beta^4 = -\frac{79}{16}$, where $\alpha > \beta$ and $\alpha\beta > 0$, find the value of a, the value of b and the value of c.

← Further Pure 1 Section 2.2

5 MATRICES

Learning objectives

After completing this chapter you should be able to:

Prior knowledge check

1 Vectors $\mathbf{a}$ and $\mathbf{b}$ are defined as $\mathbf{a} = \begin{pmatrix} 2 \\ 3 \end{pmatrix}$ and $\mathbf{b} = \begin{pmatrix} 4 \\ -1 \end{pmatrix}$.
Find:

a $\mathbf{a} + \mathbf{b}$ **b** $3\mathbf{a} - 2\mathbf{b}$

c $4(\mathbf{b} - \mathbf{a})$ ← International GCSE Mathematics

Matrices can be used to describe transformations in two and three dimensions. Computer graphics artists use matrices to control the motion of characters in video games and CGI films.

5.1 Introduction to matrices

A matrix is an array (i.e. a collection) of elements (which are usually numbers) set out in a pair of brackets.

Links A **vector** is a simple example of a matrix with just one column.
← International GCSE Mathematics

You describe the **size** of a matrix using the number of rows and columns it contains.

For example, $\begin{pmatrix} 2 & 1 \\ 4 & 0 \end{pmatrix}$ is a 2 × 2 matrix and $\begin{pmatrix} 1 & 4 & -1 & 1 \\ 2 & 3 & 0 & 2 \end{pmatrix}$ is a 2 × 4 matrix. Generally, you refer to a matrix as $n \times m$ where n is the number of rows and m is the number of columns.

- A **square matrix** is one where the number of rows and columns are the same.
- A **zero matrix** is one in which all of the elements are zero. The zero matrix is denoted by $\mathbf{0}$.
- An **identity** matrix is a square matrix in which the elements on the leading diagonal (starting top left) are all 1 and the remaining elements are 0. Identity **matrices** are denoted by $\mathbf{I}_k$ where $\boldsymbol{k}$ describes the size.

  The 2 × 2 identity matrix is $\mathbf{I}_2 = \begin{pmatrix} 1 & 0 \\ 0 & 1 \end{pmatrix}$

Notation Matrices are usually represented with bold capital letters such as $\mathbf{M}$ or $\mathbf{A}$.

Example 1

Write down the size of each matrix in the form $n \times m$.

a $\begin{pmatrix} 2 & -1 \\ 1 & 3 \end{pmatrix}$ **b** $(1 \quad 0 \quad 2)$

c $\begin{pmatrix} 4 \\ -1 \end{pmatrix}$ **d** $\begin{pmatrix} 3 & 2 \\ -1 & 1 \\ 0 & -3 \end{pmatrix}$

a $\begin{pmatrix} 2 & -1 \\ 1 & 3 \end{pmatrix}$ — There are two rows and two columns.

The size is 2 × 2.

b $(1 \quad 0 \quad 2)$ — There is just one row and three columns.

The size is 1 × 3.

c $\begin{pmatrix} 4 \\ -1 \end{pmatrix}$ — There are two rows and one column.

The size is 2 × 1.

d $\begin{pmatrix} 3 & 2 \\ -1 & 1 \\ 0 & -3 \end{pmatrix}$ — There are three rows and two columns.

The size is 3 × 2.

- To add or subtract matrices, you add or subtract the corresponding elements in each matrix. You can add or subtract matrices only when they are the same size.

Notation Matrices which are the same size are said to be **additively conformable**.

Example 2 SKILLS INTERPRETATION

Find: **a** $\begin{pmatrix} 2 & -1 \\ 0 & 3 \end{pmatrix} + \begin{pmatrix} -1 & 4 \\ 5 & 3 \end{pmatrix}$

b $\begin{pmatrix} 1 & -3 & 4 \\ 2 & 1 & 1 \end{pmatrix} - \begin{pmatrix} 0 & 2 & 1 \\ 5 & 2 & 3 \end{pmatrix}$

a $\begin{pmatrix} 2 & -1 \\ 0 & 3 \end{pmatrix} + \begin{pmatrix} -1 & 4 \\ 5 & 3 \end{pmatrix}$

$= \begin{pmatrix} 1 & 3 \\ 5 & 6 \end{pmatrix}$

b $\begin{pmatrix} 1 & -3 & 4 \\ 2 & 1 & 1 \end{pmatrix} - \begin{pmatrix} 0 & 2 & 1 \\ 5 & 2 & 3 \end{pmatrix}$

$= \begin{pmatrix} 1 & -5 & 3 \\ -3 & -1 & -2 \end{pmatrix}$

Top row:
$2 + -1 = 1$
$-1 + 4 = 3$

Bottom row:
$0 + 5 = 5$
$3 + 3 = 6$

Top row:
$1 - 0 = 1$
$-3 - 2 = -5$
$4 - 1 = 3$

Bottom row:
$2 - 5 = -3$
$1 - 2 = -1$
$1 - 3 = -2$

- To multiply a matrix by a **scalar**, you multiply every element in the matrix by that scalar.

Notation A **scalar** is a number, rather than a matrix. In questions on matrices, scalars will be represented by non-bold letters and numbers.

Example 3

$\mathbf{A} = \begin{pmatrix} 1 & 2 \\ -1 & 0 \end{pmatrix}$, $\mathbf{B} = (6 \quad 0 \quad -4)$.

Find: **a** $2\mathbf{A}$ **b** $\frac{1}{2}\mathbf{B}$

c Explain why you cannot work out $\mathbf{A} + \mathbf{B}$.

a $2\mathbf{A} = \begin{pmatrix} 2 & 4 \\ -2 & 0 \end{pmatrix}$

Note that $2\mathbf{A}$ gives the same answer as $\mathbf{A} + \mathbf{A}$.

b $\frac{1}{2}\mathbf{B} = (3 \quad 0 \quad -2)$

c **A** and **B** are not the same size, so you can't add them.

Top row:
$2 \times 1 = 2$
$2 \times 2 = 4$

Bottom row:
$2 \times -1 = -2$
$2 \times 0 = 0$

$\frac{1}{2} \times 6 = 3$
$\frac{1}{2} \times 0 = 0$
$\frac{1}{2} \times (-4) = -2$

You could also say that **A** and **B** are not additively conformable.

Example 4

$\mathbf{A} = \begin{pmatrix} a & 0 \\ 1 & 2 \end{pmatrix}$, $\mathbf{B} = \begin{pmatrix} 1 & b \\ 0 & 3 \end{pmatrix}$, $\mathbf{C} = \begin{pmatrix} 6 & 6 \\ 1 & c \end{pmatrix}$.

Given that $\mathbf{A} + 2\mathbf{B} = \mathbf{C}$, find the values of the constants a, b and c.

$$\begin{pmatrix} a & 0 \\ 1 & 2 \end{pmatrix} + 2\begin{pmatrix} 1 & b \\ 0 & 3 \end{pmatrix} = \begin{pmatrix} 6 & 6 \\ 1 & c \end{pmatrix}$$

$$\begin{pmatrix} a+2 & 2b \\ 1 & 8 \end{pmatrix} = \begin{pmatrix} 6 & 6 \\ 1 & c \end{pmatrix}$$

If two matrices are equal, then all of their corresponding elements are equal.

$a + 2 = 6 \Rightarrow a = 4$

Compare top left elements.

$2b = 6 \Rightarrow b = 3$

Compare top right elements.

$c = 8$

Compare bottom right elements.

Exercise 5A SKILLS INTERPRETATION

1 Write the size of each matrix in the form $n \times m$.

a $\begin{pmatrix} 1 & 0 \\ -1 & 3 \end{pmatrix}$ **b** $\begin{pmatrix} 1 \\ 2 \end{pmatrix}$ **c** $\begin{pmatrix} 1 & 2 & 1 \\ 3 & 0 & -1 \end{pmatrix}$

d $(1 \quad 2 \quad 3)$ **e** $(3 \quad -1)$ **f** $\begin{pmatrix} 1 & 0 & 0 \\ 0 & 1 & 0 \\ 0 & 0 & 1 \end{pmatrix}$

2 Two matrices **A** and **B** are given as:

$\mathbf{A} = \begin{pmatrix} 1 & 3 & a \\ 2 & -1 & 4 \end{pmatrix}$, $\mathbf{B} = \begin{pmatrix} 1 & 3 & 6 \\ b & -1 & 4 \end{pmatrix}$.

If $\mathbf{A} = \mathbf{B}$, write down the values of a and b.

3 For the matrices

$\mathbf{A} = \begin{pmatrix} 2 & -1 \\ 1 & 3 \end{pmatrix}$, $\mathbf{B} = \begin{pmatrix} 4 & 1 \\ -1 & -2 \end{pmatrix}$, $\mathbf{C} = \begin{pmatrix} 6 & 0 \\ 0 & 1 \end{pmatrix}$

find:

a $\mathbf{A} + \mathbf{C}$ **b** $\mathbf{B} - \mathbf{A}$ **c** $\mathbf{A} + \mathbf{B} - \mathbf{C}$

4 For the matrices

$\mathbf{A} = \begin{pmatrix} 1 \\ 2 \end{pmatrix}$, $\mathbf{B} = (1 \quad -1)$, $\mathbf{C} = (-1 \quad 1 \quad 0)$,

$\mathbf{D} = (0 \quad 1 \quad -1)$, $\mathbf{E} = \begin{pmatrix} 3 \\ -1 \end{pmatrix}$, $\mathbf{F} = (2 \quad 1 \quad 3)$

find where possible:

a $\mathbf{A} + \mathbf{B}$ **b** $\mathbf{A} - \mathbf{E}$ **c** $\mathbf{F} - \mathbf{D} + \mathbf{C}$ **d** $\mathbf{B} + \mathbf{C}$

e $\mathbf{F} - (\mathbf{D} + \mathbf{C})$ **f** $\mathbf{A} - \mathbf{F}$ **g** $\mathbf{C} - (\mathbf{F} - \mathbf{D})$

5 Given that $\begin{pmatrix} a & 2 \\ -1 & b \end{pmatrix} - \begin{pmatrix} 1 & c \\ d & -2 \end{pmatrix} = \begin{pmatrix} 5 & 0 \\ 0 & 5 \end{pmatrix}$, find the values of the constants a, b, c and d.

(P) **6** Given that $\begin{pmatrix} 1 & 2 & 0 \\ a & b & c \end{pmatrix} + \begin{pmatrix} a & b & c \\ 1 & 2 & 0 \end{pmatrix} = \begin{pmatrix} c & 5 & c \\ c & c & c \end{pmatrix}$, find the values of a, b and c.

7 Given that $\begin{pmatrix} 5 & 3 \\ 0 & -1 \\ 2 & 1 \end{pmatrix} + \begin{pmatrix} a & b \\ c & d \\ e & f \end{pmatrix} = \begin{pmatrix} 7 & 1 \\ 2 & 0 \\ 1 & 4 \end{pmatrix}$, find the values of a, b, c, d, e and f.

8 For the matrices $\mathbf{A} = \begin{pmatrix} 2 & 0 \\ 4 & -6 \end{pmatrix}$ and $\mathbf{B} = \begin{pmatrix} 1 \\ -1 \end{pmatrix}$, find:

a $3\mathbf{A}$ **b** $\frac{1}{2}\mathbf{A}$ **c** $2\mathbf{B}$

d Explain why it is not possible to find $\mathbf{A} - \mathbf{B}$.

9 The matrices **A** and **B** are defined as:

$\mathbf{A} = \begin{pmatrix} 3 & -2 \\ 1 & 0 \end{pmatrix}$ and $\mathbf{B} = \begin{pmatrix} 2 & 1 \\ -2 & 3 \end{pmatrix}$

Find:

a $3\mathbf{A} + 2\mathbf{B}$ **b** $2\mathbf{A} - 4\mathbf{B}$ **c** $5\mathbf{A} - 2\mathbf{B}$ **d** $\frac{1}{2}\mathbf{A} + \frac{3}{2}\mathbf{B}$

10 Find the value of k and the value of x such that $\begin{pmatrix} 0 & 1 \\ 2 & 0 \end{pmatrix} + k\begin{pmatrix} 0 & 2 \\ -1 & 0 \end{pmatrix} = \begin{pmatrix} 0 & 7 \\ x & 0 \end{pmatrix}$.

11 Find the values of a, b, c and d such that $2\begin{pmatrix} a & 0 \\ 1 & b \end{pmatrix} - 3\begin{pmatrix} 1 & c \\ d & -1 \end{pmatrix} = \begin{pmatrix} 3 & 3 \\ -4 & -4 \end{pmatrix}$.

12 Find the values of a, b, c and d such that $\begin{pmatrix} 5 & a \\ b & 0 \end{pmatrix} - 2\begin{pmatrix} c & 2 \\ 1 & -1 \end{pmatrix} = \begin{pmatrix} 9 & 1 \\ 3 & d \end{pmatrix}$.

(P) **13** Find the value of k such that $\begin{pmatrix} -3 \\ k \end{pmatrix} + k\begin{pmatrix} 2k \\ 2k \end{pmatrix} = \begin{pmatrix} k \\ 6 \end{pmatrix}$.

5.2 Matrix multiplication

Two matrices can be multiplied together. Unlike the operations we have seen so far, this is completely different from normal arithmetic multiplication.

- Matrices can be multiplied together if the number of columns in the first matrix is equal to the number of rows in the second matrix.

Notation If **AB** exists, then matrix **A** is said to be **multiplicatively conformable** with matrix **B**.

The **product matrix** will have the same number of rows as the first matrix, and the same number of columns as the second matrix.

$\mathbf{AB} = \mathbf{C}$ — If $\mathbf{A}$ has size $n \times m$ and $\mathbf{B}$ has size $m \times k$ then the product matrix, $\mathbf{C}$, has size $n \times k$.

The **order** in which you multiply matrices is important. This has two consequences:

- In general, $\mathbf{AB} \neq \mathbf{BA}$ (even if $\mathbf{A}$ and $\mathbf{B}$ are both square matrices).
- If $\mathbf{AB}$ exists, $\mathbf{BA}$ does not necessarily exist (e.g., if $\mathbf{A}$ is a 3×2 matrix and $\mathbf{B}$ is a 2×4 matrix).

■ To find the product of two multiplicatively conformable matrices, you multiply the elements in each row in the left-hand matrix by the corresponding elements in each column in the right-hand matrix, then add the results together.

$$\begin{pmatrix} 5 & -1 & 2 \\ 8 & 3 & -4 \end{pmatrix} \times \begin{pmatrix} 2 & 2 \\ 9 & -3 \\ 7 & 4 \end{pmatrix} = \begin{pmatrix} 15 & 21 \\ 15 & -9 \end{pmatrix}$$

Here, we are multiplying a 2×3 matrix by a 3×2 matrix, so the product matrix will have size 2×2. To find the bottom left element, work out $8 \times 2 + 3 \times 9 + (-4) \times 7 = 16 + 27 - 28 = 15$

Example 5 SKILLS INTERPRETATION

Given that $\mathbf{A} = \begin{pmatrix} 1 & -2 \\ 3 & 4 \end{pmatrix}$ and $\mathbf{B} = \begin{pmatrix} -3 \\ 2 \end{pmatrix}$,

a find $\mathbf{AB}$ **b** explain why it is not possible to find $\mathbf{BA}$.

a First calculate the size of $\mathbf{AB}$.

$(2 \times 2) \times (2 \times 1)$ gives 2×1

$\mathbf{AB} = \begin{pmatrix} 1 & -2 \\ 3 & 4 \end{pmatrix}\begin{pmatrix} -3 \\ 2 \end{pmatrix} = \begin{pmatrix} p \\ q \end{pmatrix}$

$p = 1 \times (-3) + (-2) \times 2 = -7$

$q = 3 \times (-3) + 4 \times 2 = -1$

So $\mathbf{AB} = \begin{pmatrix} -7 \\ -1 \end{pmatrix}$

b $\mathbf{BA}$ cannot be found, since the number of columns in $\mathbf{B}$ is not the same as the number of rows in $\mathbf{A}$.

The number of rows is two from here.

The number of columns is one from here.

The top number is the total of the first row of $\mathbf{A}$ multiplied by the one column of $\mathbf{B}$.

The bottom number is the total of the second row of $\mathbf{A}$ multiplied by the one column of $\mathbf{B}$.

Watch out Remember that order is important. $\mathbf{A}$ is multiplicatively conformable with $\mathbf{B}$, but $\mathbf{B}$ is not multiplicatively conformable with $\mathbf{A}$.

Example 6

Given that $\mathbf{A} = \begin{pmatrix} -1 & 0 \\ 2 & 3 \end{pmatrix}$ and $\mathbf{B} = \begin{pmatrix} 4 & 1 \\ 0 & -2 \end{pmatrix}$, find:

a **AB** **b** **BA**

a **A** is a 2 × 2 matrix and **B** is a 2 × 2 matrix so they can be multiplied and the product will be a 2 × 2 matrix.

This time there are four elements to be found.

$\mathbf{AB} = \begin{pmatrix} -1 & 0 \\ 2 & 3 \end{pmatrix}\begin{pmatrix} 4 & 1 \\ 0 & -2 \end{pmatrix} = \begin{pmatrix} a & b \\ c & d \end{pmatrix}$

$a = (-1) \times 4 + 0 \times 0 = -4$

$b = (-1) \times 1 + 0 \times (-2) = -1$

$c = 2 \times 4 + 3 \times 0 = 8$

$d = 2 \times 1 + 3 \times (-2) = -4$

a is the total of the first row multiplied by the first column.
b is the total of the first row multiplied by the second column.
c is the total of the second row multiplied by the first column.
d is the total of the second row multiplied by the second column.

So $\mathbf{AB} = \begin{pmatrix} -4 & -1 \\ 8 & -4 \end{pmatrix}$

b **BA** will also be a 2 × 2 matrix.

$\begin{pmatrix} 4 & 1 \\ 0 & -2 \end{pmatrix}\begin{pmatrix} -1 & 0 \\ 2 & 3 \end{pmatrix} = \begin{pmatrix} -2 & 3 \\ -4 & -6 \end{pmatrix}$

First row times first column
$4 \times (-1) + 1 \times 2 = -2$

First row times second column
$4 \times 0 + 1 \times 3 = 3$

Second row times second column
$0 \times 0 + (-2) \times 3 = -6$

Second row times first column
$0 \times (-1) + (-2) \times 2 = -4$

Hint You can enter matrices directly into your calculator to multiply them quickly.

Example 7

$\mathbf{A} = \begin{pmatrix} -1 \\ a \end{pmatrix}$ and $\mathbf{B} = (b \quad 2)$

Given that **BA** = (0), find **AB** in terms of a.

(0) is a 1 × 1 zero matrix.
You could also write it as **0**.

$\mathbf{BA} = (b \quad 2)\begin{pmatrix} -1 \\ a \end{pmatrix} = (-b + 2a)$

BA is a 1 × 1 matrix.

So **BA** = (0) implies that $b = 2a$.

$\mathbf{AB} = \begin{pmatrix} -1 \\ a \end{pmatrix}(b \quad 2) = \begin{pmatrix} -b & -2 \\ ab & 2a \end{pmatrix}$

AB is a 2 × 2 matrix.

Substituting $b = 2a$ gives $\mathbf{AB} = \begin{pmatrix} -2a & -2 \\ 2a^2 & 2a \end{pmatrix}$.

Watch out Although you can multiply matrices using a calculator, you need to know how the process works so that you can deal with matrices containing unknowns.

Example 8

$\mathbf{A} = (1 \quad -1 \quad 2)$, $\mathbf{B} = (3 \quad -2)$ and $\mathbf{C} = \begin{pmatrix} 4 \\ 5 \end{pmatrix}$. Find **BCA**.

$\mathbf{BC} = (3 \quad -2)\begin{pmatrix} 4 \\ 5 \end{pmatrix} = (2)$

$(\mathbf{BC})\mathbf{A} = (2)(1 \quad -1 \quad 2) = (2 \quad -2 \quad 4)$

This product could have been calculated by first working out **CA** and then multiplying **B** by this product. In general, matrix multiplication is **associative** (meaning that the bracketing makes no difference provided the order stays the same), so (**BC**)**A** = **B**(**CA**).

Exercise 5B SKILLS INTERPRETATION

1 Given the sizes of the following matrices,

Matrix	A	B	C	D	E
Size	2 × 2	1 × 2	1 × 3	3 × 2	2 × 3

find the sizes of these matrix products:

a **BA** **b** **DE** **c** **CD**

d **ED** **e** **AE** **f** **DA**

2 Use your calculator to find these products:

a $\begin{pmatrix} 1 & 2 \\ 2 & 4 \end{pmatrix}\begin{pmatrix} -1 \\ 2 \end{pmatrix}$ **b** $\begin{pmatrix} 1 & 2 \\ 3 & 4 \end{pmatrix}\begin{pmatrix} 0 & 5 \\ -1 & -2 \end{pmatrix}$

3 The matrix $\mathbf{A} = \begin{pmatrix} -1 & -2 \\ 0 & 3 \end{pmatrix}$ and the matrix $\mathbf{B} = \begin{pmatrix} 1 & 0 & 1 \\ 1 & 1 & 0 \end{pmatrix}$.

Use your calculator to find:

a **AB** **b** $\mathbf{A}^2$

Hint $\mathbf{A}^2$ means $\mathbf{A} \times \mathbf{A}$.

4 The matrices **A**, **B** and **C** are given by:

$\mathbf{A} = \begin{pmatrix} 2 \\ 1 \end{pmatrix}$, $\mathbf{B} = \begin{pmatrix} 3 & 1 \\ -1 & 2 \end{pmatrix}$, $\mathbf{C} = (-3 \quad -2)$.

Without using your calculator, determine whether or not the following products exist and find the products of those that do.

a **AB** **b** **AC** **c** **BC**

d **BA** **e** **CA** **f** **CB**

5 Find $\begin{pmatrix} 2 & a \\ 1 & -1 \end{pmatrix}\begin{pmatrix} 1 & 3 & 0 \\ 0 & -1 & 2 \end{pmatrix}$, giving your answer in terms of a.

6 Find $\begin{pmatrix} 3 & 2 \\ -1 & x \end{pmatrix}\begin{pmatrix} x & -2 \\ 1 & 3 \end{pmatrix}$, giving your answer in terms of x.

7 The matrices **A**, **B** and **C** are defined as:

$\mathbf{A} = \begin{pmatrix} 2 & -1 \\ 3 & 4 \end{pmatrix}$, $\mathbf{B} = \begin{pmatrix} 1 & 0 \\ -3 & 2 \end{pmatrix}$ and $\mathbf{C} = \begin{pmatrix} -3 & 1 \\ 1 & 2 \end{pmatrix}$.

Use your calculator to find:

a **AB** – **C** **b** **BC** + 3**A** **c** 4**B** – 3**CA**

8 The matrices **M** and **N** are defined as:

$\mathbf{M} = \begin{pmatrix} 3 & k \\ k & 1 \end{pmatrix}$ and $\mathbf{N} = \begin{pmatrix} 1 & k \\ k & -1 \end{pmatrix}$. Find, in terms of k:

a **MN** **b** **NM** **c** 3**M** – 2**N** **d** 2**MN** + 3**N**

(P) **9** The matrix $\mathbf{A} = \begin{pmatrix} 1 & 2 \\ 0 & 1 \end{pmatrix}$.

Find:

a $\mathbf{A}^2$

b $\mathbf{A}^3$

c Suggest a form for $\mathbf{A}^k$.

(P) **10** The matrix $\mathbf{A} = \begin{pmatrix} a & 0 \\ b & 0 \end{pmatrix}$.

a Find, in terms of a and b, the matrix $\mathbf{A}^2$.

b Given that $\mathbf{A}^2 = 3\mathbf{A}$, find the value of a.

11 $\mathbf{A} = \begin{pmatrix} -1 & 3 \end{pmatrix}$, $\mathbf{B} = \begin{pmatrix} 2 \\ 1 \\ 0 \end{pmatrix}$, $\mathbf{C} = \begin{pmatrix} 4 & -2 \\ 0 & -3 \end{pmatrix}$.

Find: **a** **BAC** **b** $\mathbf{AC}^2$

12 $\mathbf{A} = \begin{pmatrix} 1 \\ -1 \\ 2 \end{pmatrix}$, $\mathbf{B} = \begin{pmatrix} 3 & -2 & -3 \end{pmatrix}$

Find: **a** **ABA** **b** **BAB**

(P) **13** **a** Write down $\mathbf{I}_2$.

b Given that matrix $\mathbf{A} = \begin{pmatrix} 2 & -2 \\ 1 & 3 \end{pmatrix}$, show that **AI** = **IA** = **A**.

(P) **14** $\mathbf{A} = \begin{pmatrix} 2 & -1 \\ 3 & 2 \end{pmatrix}$, $\mathbf{B} = \begin{pmatrix} 4 & 2 \\ -1 & 0 \end{pmatrix}$ and $\mathbf{C} = \begin{pmatrix} 1 & 2 \\ 0 & -1 \end{pmatrix}$.

Show that **AB** + **AC** = **A**(**B** + **C**).

E/P **15** $\mathbf{A} = \begin{pmatrix} 1 & 2 \\ 3 & 1 \end{pmatrix}$ and $\mathbf{I}$ is the 2 × 2 identity matrix.

Show that $\mathbf{A}^2 = 2\mathbf{A} + 5\mathbf{I}$. **(2 marks)**

E/P **16** $\mathbf{A} = \begin{pmatrix} p & 3 \\ 6 & p \end{pmatrix}$ and $\mathbf{B} = \begin{pmatrix} q & 2 \\ 4 & q \end{pmatrix}$, where p and q are constants. Show that $\mathbf{AB} = \mathbf{BA}$. **(3 marks)**

E/P **17** The matrix $\mathbf{A} = \begin{pmatrix} 3 & p \\ -4 & q \end{pmatrix}$ is such that $\mathbf{A}^2 = \mathbf{I}$. Find the values of p and q. **(3 marks)**

Challenge

SKILLS CREATIVITY

The 2 × 2 matrix $\mathbf{A}$ has the property that $\mathbf{A}^2 = \mathbf{0}$. Find a possible matrix $\mathbf{A}$ such that:

a at least one of the elements in $\mathbf{A}$ is non-zero

b all of the elements in $\mathbf{A}$ are non-zero.

5.3 Determinants

You can calculate the determinant of a square matrix. The determinant is a scalar value associated with that matrix.

- For a 2 × 2 matrix $\mathbf{M} = \begin{pmatrix} a & b \\ c & d \end{pmatrix}$, the determinant of $\mathbf{M}$ is $ad - bc$.
- If $\det \mathbf{M} = 0$ then $\mathbf{M}$ is a **singular** matrix.
- If $\det \mathbf{M} \neq 0$ then $\mathbf{M}$ is a **non-singular** matrix.

Notation You can write the determinant of $\mathbf{M}$ as $\det \mathbf{M}$, $|\mathbf{M}|$ or $\begin{vmatrix} a & b \\ c & d \end{vmatrix}$. It is also sometimes written as Δ.

Links Singular matrices do not have an inverse.
→ Further Pure 1 Section 5.4

Example 9

Given that $\mathbf{A} = \begin{pmatrix} 6 & 5 \\ 1 & 2 \end{pmatrix}$, find $\det \mathbf{A}$.

$\det \mathbf{A} = ad - bc = 6 \times 2 - 5 \times 1 = 12 - 5 = 7$

Example 10

SKILLS PROBLEM-SOLVING

$\mathbf{A} = \begin{pmatrix} 4 & p+2 \\ -1 & 3-p \end{pmatrix}$

Given that $\mathbf{A}$ is singular, find the value of p.

$\det \mathbf{A} = 4(3 - p) - (p + 2)(-1)$

$= 12 - 4p + p + 2 = 14 - 3p$

$\mathbf{A}$ is singular so $\det \mathbf{A} = 0$.

$14 - 3p = 0 \Rightarrow p = \frac{14}{3}$

Watch out Although you can find the determinant using a calculator, you need to know how the process works so that you can deal with matrices containing unknowns.

Exercise 5C

SKILLS PROBLEM-SOLVING

1 Find the determinants of the following matrices.

a $\begin{pmatrix} 3 & 4 \\ -1 & 2 \end{pmatrix}$ **b** $\begin{pmatrix} 4 & 2 \\ 1 & 2 \end{pmatrix}$ **c** $\begin{pmatrix} -2 & 1 \\ 3 & 0 \end{pmatrix}$

d $\begin{pmatrix} -4 & -4 \\ 1 & 1 \end{pmatrix}$ **e** $\begin{pmatrix} 7 & -4 \\ 0 & 3 \end{pmatrix}$ **f** $\begin{pmatrix} -1 & -1 \\ -6 & -10 \end{pmatrix}$

(P) **2** Find the values of a for which these matrices are singular.

a $\begin{pmatrix} a & 1+a \\ 3 & 2 \end{pmatrix}$ **b** $\begin{pmatrix} 1+a & 3-a \\ a+2 & 1-a \end{pmatrix}$ **c** $\begin{pmatrix} 2+a & 1-a \\ 1-a & a \end{pmatrix}$

(E/P) **3** Given that k is a real number and that $\mathbf{M} = \begin{pmatrix} -2 & 1-k \\ k-1 & k \end{pmatrix}$,

find the exact values of k for which $\mathbf{M}$ is a singular matrix. **(3 marks)**

(E/P) **4** $\mathbf{P} = \begin{pmatrix} 3k & 4-k \\ k-2 & -k \end{pmatrix}$, where k is a real constant.

Given that $\mathbf{P}$ is a singular matrix, find the possible values of k. **(3 marks)**

5 The matrix $\mathbf{A} = \begin{pmatrix} a & 2a \\ b & 2b \end{pmatrix}$ and the matrix $\mathbf{B} = \begin{pmatrix} 2b & -2a \\ -b & a \end{pmatrix}$.

a Find det $\mathbf{A}$ and det $\mathbf{B}$.

b Find $\mathbf{AB}$.

(E/P) **6** The matrix $\mathbf{M} = \begin{pmatrix} 1 & -3 \\ 2 & 1 \end{pmatrix}$ and the matrix $\mathbf{N} = \begin{pmatrix} -1 & k \\ 4 & 3 \end{pmatrix}$, where k is a constant.

a Evaluate the determinant of $\mathbf{M}$. **(1 mark)**

b Given that the determinant of $\mathbf{N}$ is 7, find the value of k. **(2 marks)**

c Using the value of k found in part **b**, find $\mathbf{MN}$. **(1 mark)**

d Verify that det $\mathbf{MN}$ = det $\mathbf{M}$ × det $\mathbf{N}$. **(1 mark)**

Challenge

SKILLS INNOVATION

a Find all the possible 2 × 2 singular matrices whose elements are the numbers 1 and –1.

b Find all the possible 2 × 2 singular matrices whose elements are the numbers 1 and 0.

Hint In part **a**, there are eight possible matrices.

5.4 Inverting a 2 × 2 matrix

You can find the inverse of any non-singular matrix.

- The **inverse** of a matrix $\mathbf{M}$ is the matrix $\mathbf{M}^{-1}$ such that $\mathbf{M}\mathbf{M}^{-1} = \mathbf{M}^{-1}\mathbf{M} = \mathbf{I}$.

You can use the following formula to find the inverse of a 2 × 2 matrix.

- If $\mathbf{M} = \begin{pmatrix} a & b \\ c & d \end{pmatrix}$, then $\mathbf{M}^{-1} = \frac{1}{\det \mathbf{M}}\begin{pmatrix} d & -b \\ -c & a \end{pmatrix}$.

Notation If $\det \mathbf{M} = 0$, you will not be able to find the inverse matrix, since $\frac{1}{\det \mathbf{M}}$ is undefined.

Example 11

$\mathbf{A} = \begin{pmatrix} 3 & 2 \\ -1 & 1 \end{pmatrix}$, $\mathbf{B} = \begin{pmatrix} 2 & 1 \\ 2 & 1 \end{pmatrix}$, $\mathbf{C} = \begin{pmatrix} 1 & 3 \\ 2 & 0 \end{pmatrix}$.

For each of the matrices **A**, **B** and **C**, determine whether or not the matrix is singular.
If the matrix is non-singular, find its inverse.

$\mathbf{A} = \begin{pmatrix} 3 & 2 \\ -1 & 1 \end{pmatrix}$ so $\det \mathbf{A} = 3 \times 1 - 2 \times (-1)$

Use the determinant formula with $a = 3$, $b = 2$, $c = -1$ and $d = 1$.

$\det \mathbf{A} = 5$

Since $5 \neq 0$, **A** is non-singular.

Swap a and d and change the signs of b and c.

So $\mathbf{A}^{-1} = \frac{1}{5}\begin{pmatrix} 1 & -2 \\ 1 & 3 \end{pmatrix}$ or $\begin{pmatrix} 0.2 & -0.4 \\ 0.2 & 0.6 \end{pmatrix}$

$\mathbf{A}^{-1}$ can be left in either form.

$\mathbf{B} = \begin{pmatrix} 2 & 1 \\ 2 & 1 \end{pmatrix}$ so $\det \mathbf{B} = 2 \times 1 - 1 \times 2 = 0$

Remember if $\det \mathbf{B} = 0$ then **B** is singular.

So **B** is singular and $\mathbf{B}^{-1}$ cannot be found.

$\mathbf{C} = \begin{pmatrix} 1 & 3 \\ 2 & 0 \end{pmatrix}$ so $\det \mathbf{C} = 1 \times 0 - 3 \times 2 = -6$

Note that a determinant can be a negative number.

This is non-zero and so **C** is a non-singular matrix.

$\mathbf{C}^{-1} = -\frac{1}{6}\begin{pmatrix} 0 & -3 \\ -2 & 1 \end{pmatrix}$ or $\begin{pmatrix} 0 & \frac{1}{2} \\ \frac{1}{3} & -\frac{1}{6} \end{pmatrix}$

Swap a and d and change the signs of b and c. Then multiply by $\frac{1}{\det \mathbf{C}}$.

Hint You can find the inverse of a matrix using your calculator.

- If $\mathbf{A}$ and $\mathbf{B}$ are non-singular matrices, then $(\mathbf{AB})^{-1} = \mathbf{B}^{-1}\mathbf{A}^{-1}$.

Example 12 SKILLS CRITICAL THINKING

$\mathbf{P}$ and $\mathbf{Q}$ are non-singular matrices. Prove that $(\mathbf{PQ})^{-1} = \mathbf{Q}^{-1}\mathbf{P}^{-1}$.

Let $\mathbf{C} = (\mathbf{PQ})^{-1}$. Then $(\mathbf{PQ})\mathbf{C} = \mathbf{I}$. — Use the definition of inverse $\mathbf{A}^{-1}\mathbf{A} = \mathbf{I} = \mathbf{A}\mathbf{A}^{-1}$.

$\mathbf{P}^{-1}\mathbf{PQC} = \mathbf{P}^{-1}\mathbf{I}$ — Multiply on the left by $\mathbf{P}^{-1}$.

$(\mathbf{P}^{-1}\mathbf{P})\mathbf{QC} = \mathbf{P}^{-1}\mathbf{I}$ — Remember $\mathbf{P}^{-1}\mathbf{P} = \mathbf{I}$, $\mathbf{IQ} = \mathbf{Q}$ and $\mathbf{P}^{-1}\mathbf{I} = \mathbf{P}^{-1}$.

So $\mathbf{QC} = \mathbf{P}^{-1}$

$\mathbf{Q}^{-1}\mathbf{QC} = \mathbf{Q}^{-1}\mathbf{P}^{-1}$ — Multiply on the left by $\mathbf{Q}^{-1}$.

$\mathbf{IC} = \mathbf{Q}^{-1}\mathbf{P}^{-1}$ — Use $\mathbf{Q}^{-1}\mathbf{Q} = \mathbf{I}$.

$\mathbf{C} = \mathbf{Q}^{-1}\mathbf{P}^{-1}$

So $(\mathbf{PQ})^{-1} = \mathbf{Q}^{-1}\mathbf{P}^{-1}$ as required.

Example 13 SKILLS CRITICAL THINKING

$\mathbf{A}$ and $\mathbf{B}$ are non-singular 2×2 matrices such that $\mathbf{BAB} = \mathbf{I}$.

a Prove that $\mathbf{A} = \mathbf{B}^{-1}\mathbf{B}^{-1}$.

b Given that $\mathbf{B} = \begin{pmatrix} 2 & 5 \\ 1 & 3 \end{pmatrix}$, find the matrix $\mathbf{A}$ such that $\mathbf{BAB} = \mathbf{I}$.

a $\mathbf{BAB} = \mathbf{I}$

$\mathbf{B}^{-1}\mathbf{BAB} = \mathbf{B}^{-1}\mathbf{I}$ — Multiply on the left by $\mathbf{B}^{-1}$.

$(\mathbf{B}^{-1}\mathbf{B})\mathbf{AB} = \mathbf{B}^{-1}\mathbf{I}$ — Remember $\mathbf{B}^{-1}\mathbf{B} = \mathbf{I}$ and $\mathbf{B}^{-1}\mathbf{I} = \mathbf{B}^{-1}$.

$\mathbf{AB} = \mathbf{B}^{-1}$

$\mathbf{ABB}^{-1} = \mathbf{B}^{-1}\mathbf{B}^{-1}$ — Multiply on the right by $\mathbf{B}^{-1}$ and remember $\mathbf{B}\mathbf{B}^{-1} = \mathbf{I}$.

$\mathbf{AI} = \mathbf{B}^{-1}\mathbf{B}^{-1}$

And hence $\mathbf{A} = \mathbf{B}^{-1}\mathbf{B}^{-1}$ as required.

b $\mathbf{B} = \begin{pmatrix} 2 & 5 \\ 1 & 3 \end{pmatrix}$ so $\det \mathbf{B} = 2 \times 3 - 5 \times 1 = 1$

So $\mathbf{B}^{-1} = \frac{1}{1}\begin{pmatrix} 3 & -5 \\ -1 & 2 \end{pmatrix} = \begin{pmatrix} 3 & -5 \\ -1 & 2 \end{pmatrix}$ — First find $\mathbf{B}^{-1}$.

From part **a**,

$\mathbf{A} = \mathbf{B}^{-1}\mathbf{B}^{-1} = \begin{pmatrix} 3 & -5 \\ -1 & 2 \end{pmatrix}\begin{pmatrix} 3 & -5 \\ -1 & 2 \end{pmatrix}$ — Use the result from part **a** and matrix multiplication to find $\mathbf{A}$.

$\mathbf{A} = \begin{pmatrix} 14 & -25 \\ -5 & 9 \end{pmatrix}$

Exercise 5D

SKILLS CRITICAL THINKING

1 Determine which of these matrices are singular and which are non-singular. For those that are non-singular, find the inverse matrix.

a $\begin{pmatrix} 3 & -1 \\ -4 & 2 \end{pmatrix}$ **b** $\begin{pmatrix} 3 & 3 \\ -1 & -1 \end{pmatrix}$ **c** $\begin{pmatrix} 2 & 5 \\ 0 & 0 \end{pmatrix}$

d $\begin{pmatrix} 1 & 2 \\ 3 & 5 \end{pmatrix}$ **e** $\begin{pmatrix} 6 & 3 \\ 4 & 2 \end{pmatrix}$ **f** $\begin{pmatrix} 4 & 3 \\ 6 & 2 \end{pmatrix}$

2 Find inverses of these matrices, giving your answers in terms of a and b.

a $\begin{pmatrix} a & 1+a \\ 1+a & 2+a \end{pmatrix}$ **b** $\begin{pmatrix} 2a & 3b \\ -a & -b \end{pmatrix}$

(P) 3 **a** Given that $\mathbf{ABC} = \mathbf{I}$, prove that $\mathbf{B}^{-1} = \mathbf{CA}$.

b Given that $\mathbf{A} = \begin{pmatrix} 0 & 1 \\ -1 & -6 \end{pmatrix}$ and $\mathbf{C} = \begin{pmatrix} 2 & 1 \\ -3 & -1 \end{pmatrix}$, find $\mathbf{B}$.

4 **a** Given that $\mathbf{AB} = \mathbf{C}$, find an expression for $\mathbf{B}$, in terms of $\mathbf{A}$ and $\mathbf{C}$.

b Given further that $\mathbf{A} = \begin{pmatrix} 2 & -1 \\ 4 & 3 \end{pmatrix}$ and $\mathbf{C} = \begin{pmatrix} 3 & 6 \\ 1 & 22 \end{pmatrix}$, find $\mathbf{B}$.

5 **a** Given that $\mathbf{BAC} = \mathbf{B}$, where $\mathbf{B}$ is a non-singular matrix, find an expression for $\mathbf{A}$ in terms of $\mathbf{C}$.

b When $\mathbf{C} = \begin{pmatrix} 5 & 3 \\ 3 & 2 \end{pmatrix}$, find $\mathbf{A}$.

(P) 6 The matrix $\mathbf{A} = \begin{pmatrix} 3a & b \\ 4a & 2b \end{pmatrix}$, where a and b are non-zero constants.

a Find $\mathbf{A}^{-1}$, giving your answer in terms of a and b.

The matrix $\mathbf{B} = \begin{pmatrix} -a & b \\ 3a & 2b \end{pmatrix}$ and the matrix $\mathbf{X}$ is given by $\mathbf{B} = \mathbf{XA}$.

b Find $\mathbf{X}$, giving your answer in terms of a and b.

Chapter review 5

(P) 1 The matrix $\mathbf{A} = \begin{pmatrix} 1 & -3 \\ 2 & 1 \end{pmatrix}$ and $\mathbf{AB} = \begin{pmatrix} 4 & 1 & 9 \\ 1 & 9 & 4 \end{pmatrix}$. Find the matrix $\mathbf{B}$.

(E/P) 2 The matrix $\mathbf{A} = \begin{pmatrix} a & b \\ 2a & 3b \end{pmatrix}$, where a and b are non-zero constants.

a Find $\mathbf{A}^{-1}$, giving your answer in terms of a and b. **(2 marks)**

The matrix $\mathbf{Y} = \begin{pmatrix} a & 2b \\ 2a & b \end{pmatrix}$ and the matrix $\mathbf{X}$ is given by $\mathbf{XA} = \mathbf{Y}$.

b Find $\mathbf{X}$, giving your answer in terms of a and b. **(3 marks)**

E/P **3** $\mathbf{A} = \begin{pmatrix} k & -2 \\ -4 & k \end{pmatrix}$ where k is a real constant.

a For which values of k does $\mathbf{A}$ have an inverse? **(2 marks)**

b Given that $\mathbf{A}$ is non-singular, find $\mathbf{A}^{-1}$ in terms of k. **(3 marks)**

E/P **4** Given that $\mathbf{M} = \begin{pmatrix} 2 & -m \\ m & -1 \end{pmatrix}$ where m is a real constant,

a write down two values of m such that $\mathbf{M}$ is singular **(2 marks)**

b find $\mathbf{M}^{-1}$ in terms of m, given that $\mathbf{M}$ is non-singular. **(3 marks)**

E/P **5** The non-singular matrices $\mathbf{A}$ and $\mathbf{B}$ are such that $\mathbf{AB} = \mathbf{BA}$, and $\mathbf{ABA} = \mathbf{B}$.

a Prove that $\mathbf{A}^2 = \mathbf{I}$. **(3 marks)**

Given that $\mathbf{A} = \begin{pmatrix} 0 & 1 \\ 1 & 0 \end{pmatrix}$, by considering a matrix $\mathbf{B}$ of the form $\begin{pmatrix} a & b \\ c & d \end{pmatrix}$,

b show that $a = d$ and $b = c$. **(3 marks)**

E/P **6** Given that $\mathbf{A} = \begin{pmatrix} 4 & p \\ -2 & -2 \end{pmatrix}$ where p is a constant and $p \neq 4$,

a find $\mathbf{A}^{-1}$ in terms of p. **(2 marks)**

b Given that $\mathbf{A} + \mathbf{A}^{-1} = \begin{pmatrix} 5 & \frac{9}{2} \\ -3 & -4 \end{pmatrix}$, find the value of p. **(3 marks)**

E/P **7** $\mathbf{M} = \begin{pmatrix} k & -3 \\ 4 & k+3 \end{pmatrix}$ where k is a real constant.

a Find $\det \mathbf{M}$ in terms of k. **(2 marks)**

b Show that $\mathbf{M}$ is non-singular for all values of k. **(3 marks)**

c Given that $10\mathbf{M}^{-1} + \mathbf{M} = \mathbf{I}$ where $\mathbf{I}$ is the 2×2 identity matrix, find the value of k. **(3 marks)**

E/P **8** $\mathbf{M} = \begin{pmatrix} 2 & 3 \\ k & -1 \end{pmatrix}$ where k is a constant.

a For which values of k does $\mathbf{M}$ have an inverse? **(2 marks)**

b Given that $\mathbf{M}$ is non-singular, find $\mathbf{M}^{-1}$ in terms of k. **(3 marks)**

E/P **9** Given that $\mathbf{A} = \begin{pmatrix} a & 2 \\ 3 & 2a \end{pmatrix}$ where a is a real constant,

a find $\mathbf{A}^{-1}$ in terms of a **(3 marks)**

b write down two values of a for which $\mathbf{A}^{-1}$ does not exist. **(1 mark)**

Challenge

SKILLS CREATIVITY

Given that $\mathbf{A}$ and $\mathbf{B}$ are 2×2 matrices, prove that $\det(\mathbf{AB}) = \det \mathbf{A} \times \det \mathbf{B}$.

Summary of key points

1 A **square matrix** is one where the number of rows and columns are the same.

2 A **zero matrix** is one in which all of the elements are zero. The zero matrix is denoted by $\mathbf{0}$.

3 An **identity matrix** is a square matrix in which the elements in the leading diagonal (starting top left) are 1 and all the remaining elements are 0. Identity matrices are denoted by $\mathbf{I}_k$ where k describes the size. The 2 × 2 identity matrix is $\mathbf{I}_2 = \begin{pmatrix} 1 & 0 \\ 0 & 1 \end{pmatrix}$

4 To add or subtract matrices, you add or subtract the corresponding elements in each matrix. You can add or subtract matrices only when they are the same size.

5 To multiply a matrix by a scalar, you multiply every element in the matrix by that scalar.

6 • Matrices can be multiplied together if the number of columns in the first matrix is equal to the number of rows in the second matrix. In this case the first matrix is said to be **multiplicatively conformable** with the second matrix.

• To find the product of two multiplicatively conformable matrices, you multiply the elements in each row in the left-hand matrix by the corresponding elements in each column in the right-hand matrix, then add the results together.

7 For a 2 × 2 matrix $\mathbf{M} = \begin{pmatrix} a & b \\ c & d \end{pmatrix}$, the determinant of $\mathbf{M}$ is $ad - bc$.

8 • If det $\mathbf{M} = 0$ then $\mathbf{M}$ is a **singular** matrix.

• If det $\mathbf{M} \neq 0$ then $\mathbf{M}$ is a **non-singular** matrix.

9 The **inverse** of a matrix $\mathbf{M}$ is the matrix $\mathbf{M}^{-1}$ such that $\mathbf{M}\mathbf{M}^{-1} = \mathbf{M}^{-1}\mathbf{M} = \mathbf{I}$.

10 If $\mathbf{M} = \begin{pmatrix} a & b \\ c & d \end{pmatrix}$, then $\mathbf{M}^{-1} = \frac{1}{\det\mathbf{M}}\begin{pmatrix} d & -b \\ -c & a \end{pmatrix}$.

11 If $\mathbf{A}$ and $\mathbf{B}$ are non-singular matrices, then $(\mathbf{AB})^{-1} = \mathbf{B}^{-1}\mathbf{A}^{-1}$.

6 TRANSFORMATIONS USING MATRICES

6.1
6.2
6.3
6.4

Learning objectives

After completing this chapter you should be able to:

Prior knowledge check

1 $\mathbf{A} = \begin{pmatrix} 2 & 1 \\ 3 & 2 \end{pmatrix}$, $\mathbf{B} = \begin{pmatrix} -1 & -1 \\ 2 & 3 \end{pmatrix}$. Find:

a $\mathbf{AB}$ **b** $\mathbf{BA}$ ← Further Pure 1 Section 5.2

2 A matrix $\mathbf{M} = \begin{pmatrix} 3 & 1 \\ 4 & -2 \end{pmatrix}$. Find:

a det $\mathbf{M}$ **b** $\mathbf{M}^{-1}$ ← Further Pure 1 Section 5.3

Einstein's theory of relativity relies on matrices which describe the relationship between different frames of reference.

6.1 Linear transformations in two dimensions

You can define a **transformation** in two dimensions by describing how a general point with position vector $\begin{pmatrix} x \\ y \end{pmatrix}$ is transformed. The new point is called the **image**.

Example 1 SKILLS PROBLEM-SOLVING

The three transformations **S**, **T** and **U** are defined as follows. Find the image of the point (2, 3) under each of these transformations.

$$\mathbf{S}: \begin{pmatrix} x \\ y \end{pmatrix} \mapsto \begin{pmatrix} x+4 \\ y-1 \end{pmatrix} \qquad \mathbf{T}: \begin{pmatrix} x \\ y \end{pmatrix} \mapsto \begin{pmatrix} 2x-y \\ x+y \end{pmatrix} \qquad \mathbf{U}: \begin{pmatrix} x \\ y \end{pmatrix} \mapsto \begin{pmatrix} 2y \\ -x^2 \end{pmatrix}$$

$\mathbf{S}: \begin{pmatrix} 2 \\ 3 \end{pmatrix} \mapsto \begin{pmatrix} 2+4 \\ 3-1 \end{pmatrix} = \begin{pmatrix} 6 \\ 2 \end{pmatrix}$. The image is (6, 2)

Substitute $x = 2$ and $y = 3$ into the expressions for the image of the general point.

$\mathbf{T}: \begin{pmatrix} 2 \\ 3 \end{pmatrix} \mapsto \begin{pmatrix} 2 \times 2 - 3 \\ 2+3 \end{pmatrix} = \begin{pmatrix} 1 \\ 5 \end{pmatrix}$. The image is (1, 5)

Substituting $x = 2$ and $y = 3$ into $2x - y$ gives 1 and into $x + y$ gives 5.

$\mathbf{U}: \begin{pmatrix} 2 \\ 3 \end{pmatrix} \mapsto \begin{pmatrix} 2 \times 3 \\ -2^2 \end{pmatrix} = \begin{pmatrix} 6 \\ -4 \end{pmatrix}$. The image is (6, −4)

When $x = 2$, $-x^2 = -2^2 = -4$

A **linear transformation** has the special properties that the transformation involves only linear terms in x and y. In the example above, **T** is a linear transformation while **S** and **U** are not.

Watch out **S** represents a translation, but this is not a linear transformation since you can't write $x + 4$ in the form $ax + by$.

- Linear transformations always map the origin onto itself.
- Any linear transformation can be represented by a matrix.
- The linear transformation $\mathbf{T}: \begin{pmatrix} x \\ y \end{pmatrix} \mapsto \begin{pmatrix} ax+by \\ cx+dy \end{pmatrix}$ can be represented by the matrix $\mathbf{M} = \begin{pmatrix} a & b \\ c & d \end{pmatrix}$

  since $\begin{pmatrix} a & b \\ c & d \end{pmatrix}\begin{pmatrix} x \\ y \end{pmatrix} = \begin{pmatrix} ax+by \\ cx+dy \end{pmatrix}$.

Notation You can transform any point P by multiplying the transformation matrix by the position vector of P.

Example 2

Find matrices to represent these linear transformations.

a $\mathbf{T}: \begin{pmatrix} x \\ y \end{pmatrix} \mapsto \begin{pmatrix} 2y + x \\ 3x \end{pmatrix}$ **b** $\mathbf{V}: \begin{pmatrix} x \\ y \end{pmatrix} \mapsto \begin{pmatrix} -2y \\ 3x + y \end{pmatrix}$

a Transformation $\mathbf{T}$ is equivalent to

$\mathbf{T}: \begin{pmatrix} x \\ y \end{pmatrix} \mapsto \begin{pmatrix} 1x + 2y \\ 3x + 0y \end{pmatrix}$ — Write the transformation in the form $\begin{pmatrix} ax + by \\ cx + dy \end{pmatrix}$

so the matrix is $\begin{pmatrix} 1 & 2 \\ 3 & 0 \end{pmatrix}$. — Use the coefficients of x and y to form the matrix.

b Transformation $\mathbf{V}$ is equivalent to

$\mathbf{V}: \begin{pmatrix} x \\ y \end{pmatrix} \mapsto \begin{pmatrix} 0x - 2y \\ 3x + y \end{pmatrix}$ — Write the transformation in the form $\begin{pmatrix} ax + by \\ cx + dy \end{pmatrix}$

so the matrix is $\begin{pmatrix} 0 & -2 \\ 3 & 1 \end{pmatrix}$. — Use the coefficients of x and y to form the matrix.

Example 3

a The square S has coordinates (1, 1), (3, 1), (3, 3) and (1, 3). Find the **vertices** of the image of S under the transformation given by the matrix $\mathbf{M} = \begin{pmatrix} -1 & 2 \\ 2 & 1 \end{pmatrix}$.

b Sketch S and the image of S on a coordinate grid.

a $\begin{pmatrix} -1 & 2 \\ 2 & 1 \end{pmatrix}\begin{pmatrix} 1 \\ 1 \end{pmatrix} = \begin{pmatrix} 1 \\ 3 \end{pmatrix}$ — Write each point as a column vector and then use the usual rule for multiplying matrices.

$\begin{pmatrix} -1 & 2 \\ 2 & 1 \end{pmatrix}\begin{pmatrix} 3 \\ 1 \end{pmatrix} = \begin{pmatrix} -1 \\ 7 \end{pmatrix}$

$\begin{pmatrix} -1 & 2 \\ 2 & 1 \end{pmatrix}\begin{pmatrix} 3 \\ 3 \end{pmatrix} = \begin{pmatrix} 3 \\ 9 \end{pmatrix}$

$\begin{pmatrix} -1 & 2 \\ 2 & 1 \end{pmatrix}\begin{pmatrix} 1 \\ 3 \end{pmatrix} = \begin{pmatrix} 5 \\ 5 \end{pmatrix}$

The vertices of the image of S, S', lie at (1, 3), (−1, 7), (3, 9) and (5, 5).

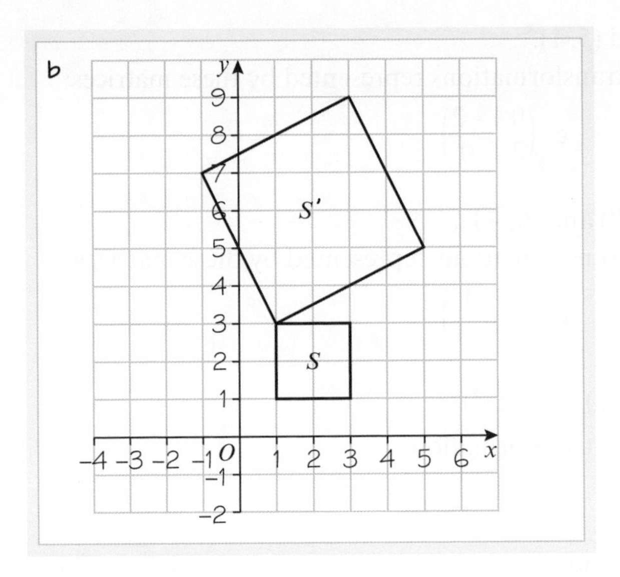

Notation S' is used to denote the image of S after a transformation.

Exercise 6A SKILLS PROBLEM-SOLVING

1 Which of the following transformations are linear transformations?

a $\mathbf{P}: \begin{pmatrix} x \\ y \end{pmatrix} \mapsto \begin{pmatrix} 2x \\ y+1 \end{pmatrix}$ **b** $\mathbf{Q}: \begin{pmatrix} x \\ y \end{pmatrix} \mapsto \begin{pmatrix} x^2 \\ y \end{pmatrix}$ **c** $\mathbf{R}: \begin{pmatrix} x \\ y \end{pmatrix} \mapsto \begin{pmatrix} 2x+y \\ x+xy \end{pmatrix}$

d $\mathbf{S}: \begin{pmatrix} x \\ y \end{pmatrix} \mapsto \begin{pmatrix} y \\ -x \end{pmatrix}$ **e** $\mathbf{T}: \begin{pmatrix} x \\ y \end{pmatrix} \mapsto \begin{pmatrix} y+3 \\ x+3 \end{pmatrix}$ **f** $\mathbf{U}: \begin{pmatrix} x \\ y \end{pmatrix} \mapsto \begin{pmatrix} 2x \\ 3y-2x \end{pmatrix}$

2 Identify which of these are linear transformations and give their matrix representations. Give reasons to explain why the other transformations are not linear.

a $\mathbf{S}: \begin{pmatrix} x \\ y \end{pmatrix} \mapsto \begin{pmatrix} 2x-y \\ 3x \end{pmatrix}$ **b** $\mathbf{T}: \begin{pmatrix} x \\ y \end{pmatrix} \mapsto \begin{pmatrix} 2y+1 \\ x-1 \end{pmatrix}$ **c** $\mathbf{U}: \begin{pmatrix} x \\ y \end{pmatrix} \mapsto \begin{pmatrix} xy \\ 0 \end{pmatrix}$

d $\mathbf{V}: \begin{pmatrix} x \\ y \end{pmatrix} \mapsto \begin{pmatrix} 2y \\ -x \end{pmatrix}$ **e** $\mathbf{W}: \begin{pmatrix} x \\ y \end{pmatrix} \mapsto \begin{pmatrix} y \\ x \end{pmatrix}$

3 Identify which of these are linear transformations and give their matrix representations. Give reasons to explain why the other transformations are not linear.

a $\mathbf{S}: \begin{pmatrix} x \\ y \end{pmatrix} \mapsto \begin{pmatrix} x^2 \\ y^2 \end{pmatrix}$ **b** $\mathbf{T}: \begin{pmatrix} x \\ y \end{pmatrix} \mapsto \begin{pmatrix} -y \\ x \end{pmatrix}$ **c** $\mathbf{U}: \begin{pmatrix} x \\ y \end{pmatrix} \mapsto \begin{pmatrix} x-y \\ x-y \end{pmatrix}$

d $\mathbf{V}: \begin{pmatrix} x \\ y \end{pmatrix} \mapsto \begin{pmatrix} 0 \\ 0 \end{pmatrix}$ **e** $\mathbf{W}: \begin{pmatrix} x \\ y \end{pmatrix} \mapsto \begin{pmatrix} x \\ y \end{pmatrix}$

4 Find matrix representations for these linear transformations:

a $\mathbf{P}: \begin{pmatrix} x \\ y \end{pmatrix} \mapsto \begin{pmatrix} y+2x \\ -y \end{pmatrix}$ **b** $\mathbf{Q}: \begin{pmatrix} x \\ y \end{pmatrix} \mapsto \begin{pmatrix} -y \\ x+2y \end{pmatrix}$

5 The triangle T has vertices at (−1, 1), (2, 3) and (5, 1).
Find the vertices of the image of T under the transformations represented by these matrices:

a $\begin{pmatrix} -1 & 0 \\ 0 & 1 \end{pmatrix}$ **b** $\begin{pmatrix} 1 & 4 \\ 0 & -2 \end{pmatrix}$ **c** $\begin{pmatrix} 0 & -2 \\ 2 & 0 \end{pmatrix}$

6 The square S has vertices at (−1, 0), (0, 1), (1, 0) and (0, −1).
Find the vertices of the image of S under the transformations represented by these matrices:

a $\begin{pmatrix} 2 & 0 \\ 0 & 3 \end{pmatrix}$ **b** $\begin{pmatrix} 1 & -1 \\ 1 & 1 \end{pmatrix}$ **c** $\begin{pmatrix} 1 & 1 \\ 1 & -1 \end{pmatrix}$

7 The rectangle R has vertices at (2, 1), (4, 1), (4, 2) and (2, 2).

a Find the vertices of the image of R under the transformation represented by the matrix $\begin{pmatrix} -1 & 0 \\ 0 & -1 \end{pmatrix}$.

b Sketch R and its image, R', on a coordinate grid.

c Describe fully the transformation that maps R onto R'.

8 A quadrilateral Q has coordinates (1, 0), (4, 2), (3, 4) and (0, 2).

a Find the vertices of the image of Q under the transformation represented by the matrix $\begin{pmatrix} 2 & 0 \\ 0 & 2 \end{pmatrix}$.

b Sketch Q and its image, Q', on a coordinate grid.

c Describe fully the transformation that maps Q onto Q'.

(P) **9** A square S has coordinates (−1, 0), (−3, 0), (−3, 2) and (−1, 2).

a Find the vertices of the image of S under the transformation represented by the matrix $\begin{pmatrix} 0 & 2 \\ 2 & 0 \end{pmatrix}$.

b Sketch S and the image of S on a coordinate grid.

c Describe fully the two transformations that map S onto S'.

10 A triangle T has vertices (4, 1), (4, 3) and (1, 3).

a Find the vertices of the image of T under the transformation represented by the matrix $\begin{pmatrix} 1 & 0 \\ 0 & 1 \end{pmatrix}$.

b Describe the effect of the transformation represented by the matrix $\begin{pmatrix} 1 & 0 \\ 0 & 1 \end{pmatrix}$.

Challenge

SKILLS CREATIVITY

A transformation **T** is defined as **T**: $\begin{pmatrix} x \\ y \end{pmatrix} \mapsto \begin{pmatrix} 2x - 3y \\ x + y \end{pmatrix}$.

a Show that $\mathbf{T}\begin{pmatrix} kx \\ ky \end{pmatrix} = k\mathbf{T}\begin{pmatrix} x \\ y \end{pmatrix}$.

b Show that $\mathbf{T}\left(\begin{pmatrix} x_1 \\ y_1 \end{pmatrix} + \begin{pmatrix} x_2 \\ y_2 \end{pmatrix}\right) = \mathbf{T}\begin{pmatrix} x_1 \\ y_1 \end{pmatrix} + \mathbf{T}\begin{pmatrix} x_2 \\ y_2 \end{pmatrix}$.

Notation $\mathbf{T}\begin{pmatrix} x \\ y \end{pmatrix}$ is used to denote the image of the point $\begin{pmatrix} x \\ y \end{pmatrix}$ after the transformation **T**.

Notation All linear transformations have these properties.

6.2 Reflections and rotations

Any linear transformation can be defined by the effect it has on unit vectors $\begin{pmatrix}1\\0\end{pmatrix}$ and $\begin{pmatrix}0\\1\end{pmatrix}$.

The transformation represented by the matrix $\mathbf{M} = \begin{pmatrix}a & b\\c & d\end{pmatrix}$ will map $\begin{pmatrix}1\\0\end{pmatrix}$ to $\begin{pmatrix}a\\c\end{pmatrix}$ and $\begin{pmatrix}0\\1\end{pmatrix}$ to $\begin{pmatrix}b\\d\end{pmatrix}$.

You can visualise this transformation by considering the unit square:

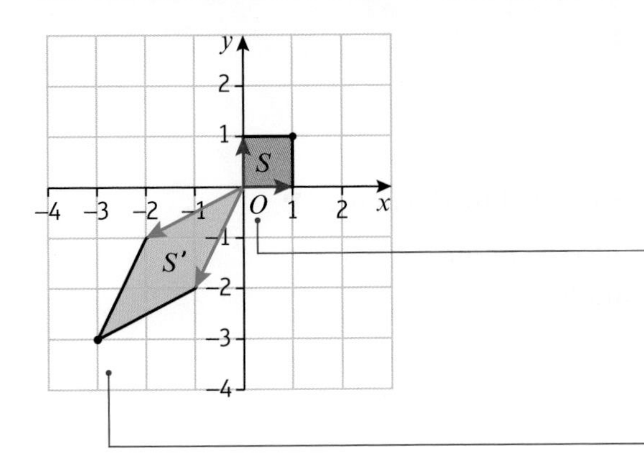

The linear transformation $\begin{pmatrix}-2 & -1\\-1 & -2\end{pmatrix}$ stretches and rotates the unit square S to produce the image S'.

The origin does not change under a linear transformation.

$\mathbf{T}(\mathbf{a} + \mathbf{b}) = \mathbf{Ta} + \mathbf{Tb}$ so you can add the two image vectors to find the fourth vertex of the new shape. This means that the entire transformation can be defined by the images of the unit vectors $\begin{pmatrix}1\\0\end{pmatrix}$ and $\begin{pmatrix}0\\1\end{pmatrix}$.

Points which are mapped onto themselves under a given transformation are called **invariant points.** Lines which map onto themselves are called **invariant lines.**

The only invariant point in the transformation above is the origin, which is always an invariant point of any linear transformation. The transformation above has invariant lines $y = x$ and $y = -x$.

Example 4

The transformation represented by the 2 × 2 matrix $\mathbf{Q}$, is a reflection in the y-axis.

a Write down the matrix $\mathbf{Q}$.

b Write down the equations of three different invariant lines of this transformation.

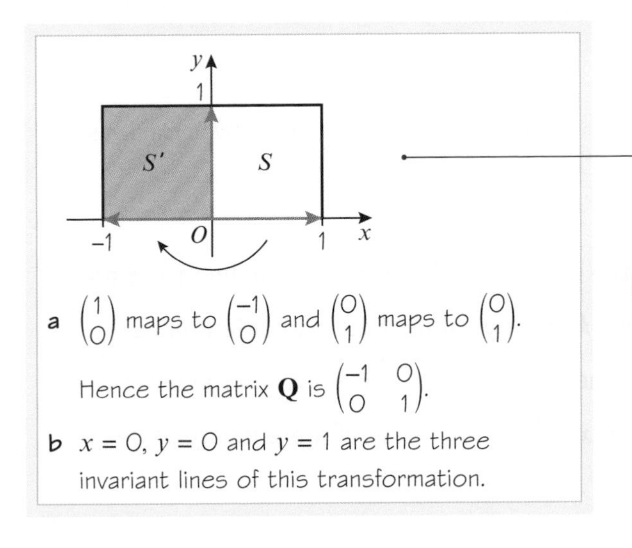

Consider the unit square, and the effect that the transformation has on the unit vectors $\begin{pmatrix}1\\0\end{pmatrix}$ and $\begin{pmatrix}0\\1\end{pmatrix}$. This will completely define the transformation.

a $\begin{pmatrix}1\\0\end{pmatrix}$ maps to $\begin{pmatrix}-1\\0\end{pmatrix}$ and $\begin{pmatrix}0\\1\end{pmatrix}$ maps to $\begin{pmatrix}0\\1\end{pmatrix}$.

Hence the matrix $\mathbf{Q}$ is $\begin{pmatrix}-1 & 0\\0 & 1\end{pmatrix}$.

b $x = 0$, $y = 0$ and $y = 1$ are the three invariant lines of this transformation.

Problem-solving

Each point on the mirror line is invariant, so this line is an invariant line. Any line perpendicular to the mirror line will also be an invariant line, although the points on the line will not, in general, be invariant points.

- A reflection in the y-axis is represented by the matrix $\begin{pmatrix}-1 & 0\\ 0 & 1\end{pmatrix}$. Points on the y-axis are invariant points, and the lines $x = 0$ and $y = k$ for any value of k are invariant lines.
- A reflection in the x-axis is represented by the matrix $\begin{pmatrix}1 & 0\\ 0 & -1\end{pmatrix}$. Points on the x-axis are invariant points, and the lines $y = 0$ and $x = k$ for any value of k are invariant lines.

Example 5 SKILLS INTERPRETATION

$\mathbf{P} = \begin{pmatrix}0 & -1\\ -1 & 0\end{pmatrix}$

a Describe fully the single geometric transformation represented by the matrix **P**.

b Given that **P** maps the point with coordinates (a, b) onto the point with coordinates $(3 + 2a, b + 1)$, find the values of a and b.

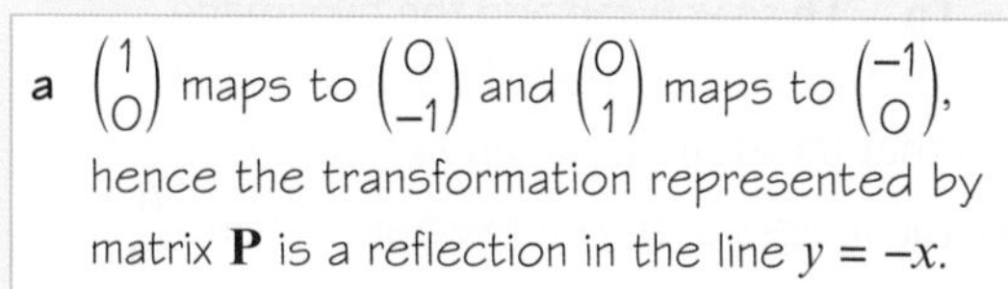

a $\begin{pmatrix}1\\0\end{pmatrix}$ maps to $\begin{pmatrix}0\\-1\end{pmatrix}$ and $\begin{pmatrix}0\\1\end{pmatrix}$ maps to $\begin{pmatrix}-1\\0\end{pmatrix}$, hence the transformation represented by matrix **P** is a reflection in the line $y = -x$.

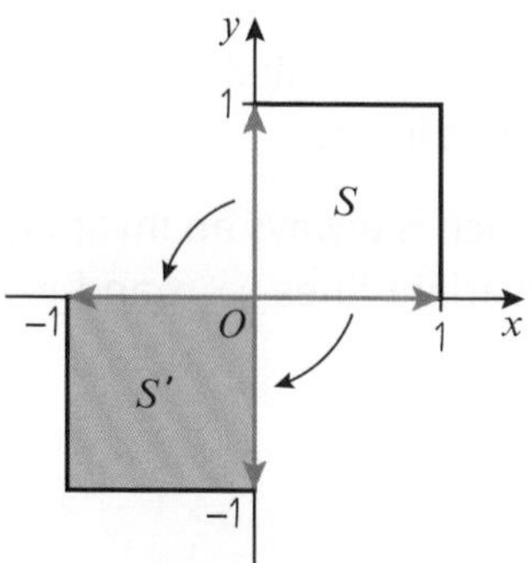

You can visualise the transformation by sketching the effect it has on the unit square.

b $\begin{pmatrix}0 & -1\\ -1 & 0\end{pmatrix}\begin{pmatrix}a\\b\end{pmatrix} = \begin{pmatrix}3 + 2a\\ b + 1\end{pmatrix}$

$\begin{pmatrix}-b\\-a\end{pmatrix} = \begin{pmatrix}3 + 2a\\ b + 1\end{pmatrix}$

So $-b = 3 + 2a$ (1)

and $-a = b + 1$ (2)

Solving simultaneously gives:

$a = -2$ and $b = 1$.

Problem-solving

Write a matrix equation to show the transformation, then solve the resulting equations simultaneously to find the values of a and b.

You can solve two equations in two unknowns quickly using your calculator.

- A reflection in the line $y = x$ is represented by the matrix $\begin{pmatrix}0 & 1\\ 1 & 0\end{pmatrix}$. Points on the line $y = x$ are invariant points, and the lines $y = x$ and $y = -x + k$ for any value of k are invariant lines.
- A reflection in the line $y = -x$ is represented by the matrix $\begin{pmatrix}0 & -1\\ -1 & 0\end{pmatrix}$. Points on the line $y = -x$ are invariant points, and the lines $y = -x$ and $y = x + k$ for any value of k are invariant lines.

Example 6

The transformation represented by the 2×2 matrix **P**, is a rotation of 180° about the point (0, 0).

a Write down the matrix **P**.

b Show that the line $y = 3x$ is invariant under this transformation.

a The given rotation is shown in the diagram:

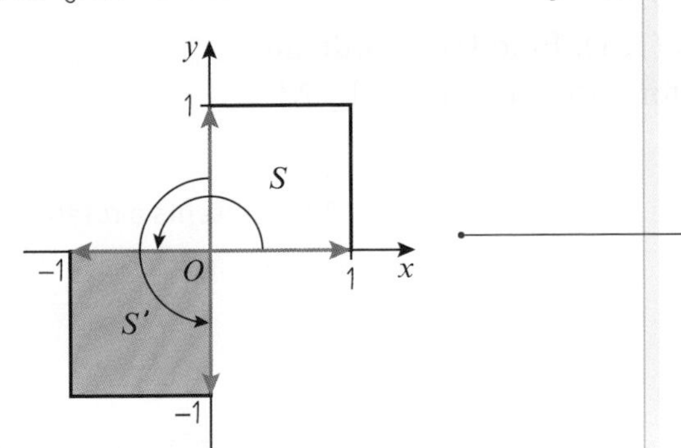

If you need to find the matrix that represents a given transformation, it can help to draw a sketch transforming the unit square. Remember the transformation is defined by its effect on the unit vectors.

$\begin{pmatrix} 1 \\ 0 \end{pmatrix}$ maps to $\begin{pmatrix} -1 \\ 0 \end{pmatrix}$ and $\begin{pmatrix} 0 \\ 1 \end{pmatrix}$ maps to $\begin{pmatrix} 0 \\ -1 \end{pmatrix}$.

Hence the matrix is $\begin{pmatrix} -1 & 0 \\ 0 & -1 \end{pmatrix}$.

b $\begin{pmatrix} -1 & 0 \\ 0 & -1 \end{pmatrix}\begin{pmatrix} x \\ 3x \end{pmatrix} = \begin{pmatrix} -x \\ -3x \end{pmatrix}$

Since $-3x = 3(-x)$, this point lies on the line $y = 3x$.

So, points on the line $y = 3x$ are mapped to points on the line $y = 3x$.

Hence $y = 3x$ is an invariant line.

Problem-solving

Write a general point on the line $y = 3x$ as $\begin{pmatrix} x \\ y \end{pmatrix} = \begin{pmatrix} x \\ 3x \end{pmatrix}$. Apply the transformation to this point, then show that the image also lies on the line. Watch out: although the line is invariant, the only invariant **point** on the line is the origin.

You need to be able to write down the matrix representing a rotation about any angle.

- The matrix representing a rotation through angle θ anticlockwise about the origin is

  $\begin{pmatrix} \cos\theta & -\sin\theta \\ \sin\theta & \cos\theta \end{pmatrix}$

  The only invariant point is the origin (0, 0).

  For $\theta \neq 180°$, there are no invariant lines.

  For $\theta = 180°$, any line passing through the origin is an invariant line.

Hint This general rotation matrix is given in the formulae booklet.

Example 7

$$\mathbf{M} = \begin{pmatrix} -\frac{\sqrt{2}}{2} & -\frac{\sqrt{2}}{2} \\ \frac{\sqrt{2}}{2} & -\frac{\sqrt{2}}{2} \end{pmatrix}$$

a Describe geometrically the rotation represented by **M**.

b A square S has vertices at (1, 0), (2, 0), (2, 1) and (1, 1). Find the coordinates of the vertices of the image of S under the transformation described by **M**.

a $\cos 135° = -\frac{\sqrt{2}}{2}$ and $\sin 135° = \frac{\sqrt{2}}{2}$ so **M** is a rotation, anticlockwise, through 135° about (0, 0).

You are told that **M** represents a rotation, so compare the matrix with $\begin{pmatrix} \cos\theta & -\sin\theta \\ \sin\theta & \cos\theta \end{pmatrix}$

b Apply the matrix to each vertex of S in turn:

$$\begin{pmatrix} -\frac{\sqrt{2}}{2} & -\frac{\sqrt{2}}{2} \\ \frac{\sqrt{2}}{2} & -\frac{\sqrt{2}}{2} \end{pmatrix}\begin{pmatrix} 1 \\ 0 \end{pmatrix} = \begin{pmatrix} -\frac{\sqrt{2}}{2} \\ \frac{\sqrt{2}}{2} \end{pmatrix}$$

$$\begin{pmatrix} -\frac{\sqrt{2}}{2} & -\frac{\sqrt{2}}{2} \\ \frac{\sqrt{2}}{2} & -\frac{\sqrt{2}}{2} \end{pmatrix}\begin{pmatrix} 2 \\ 0 \end{pmatrix} = \begin{pmatrix} -\sqrt{2} \\ \sqrt{2} \end{pmatrix}$$

$$\begin{pmatrix} -\frac{\sqrt{2}}{2} & -\frac{\sqrt{2}}{2} \\ \frac{\sqrt{2}}{2} & -\frac{\sqrt{2}}{2} \end{pmatrix}\begin{pmatrix} 2 \\ 1 \end{pmatrix} = \begin{pmatrix} -\frac{3\sqrt{2}}{2} \\ \frac{\sqrt{2}}{2} \end{pmatrix}$$

$$\begin{pmatrix} -\frac{\sqrt{2}}{2} & -\frac{\sqrt{2}}{2} \\ \frac{\sqrt{2}}{2} & -\frac{\sqrt{2}}{2} \end{pmatrix}\begin{pmatrix} 1 \\ 1 \end{pmatrix} = \begin{pmatrix} -\sqrt{2} \\ 0 \end{pmatrix}$$

Work out the position vector of each vertex in the image of S.

The vertices of S' are $\left(-\frac{\sqrt{2}}{2}, \frac{\sqrt{2}}{2}\right)$, $(-\sqrt{2}, \sqrt{2})$, $\left(-\frac{3\sqrt{2}}{2}, \frac{\sqrt{2}}{2}\right)$ and $(-\sqrt{2}, 0)$.

Online Explore rotations of the unit square using GeoGebra.

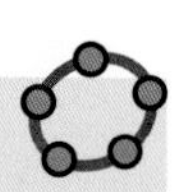

Watch out Read the question carefully. Give your final answers as coordinates, not position vectors.

Exercise 6B

SKILLS INTERPRETATION

1 a Write down the matrix representing a reflection in the x-axis.

A triangle has vertices at $A = (1, 3)$, $B = (3, 3)$ and $C = (3, 2)$.

b Use matrices to show that the images of these vertices after a reflection in the x-axis are $A' = (1, -3)$, $B' = (3, -3)$ and $C' = (3, -2)$.

2 a Write down the matrix representing a reflection in the line $y = -x$.

A rectangle has vertices at $P = (1, 1)$, $Q = (1, 3)$, $R = (2, 3)$ and $S = (2, 1)$.

b Use matrices to show that the images of these vertices after a reflection in the line $y = -x$ are $P' = (-1, -1)$, $Q' = (-3, -1)$, $R' = (-3, -2)$ and $S' = (-1, -2)$.

3 Find the matrices that represent the following rotations.

a 90° anticlockwise about (0, 0)

b 270° anticlockwise about (0, 0)

c 45° anticlockwise about (0, 0)

d 210° anticlockwise about (0, 0)

e 135° clockwise about (0, 0)

Watch out The rotation matrix is for angles measured anticlockwise, so make sure you convert the clockwise angle to its equivalent anticlockwise angle.

4 A triangle has vertices at $A = (1, 1)$, $B = (4, 1)$ and $C = (4, 2)$. Find the exact coordinates of the vertices of the triangle after a rotation through:

a 90° anticlockwise about (0, 0) **b** 150° anticlockwise about (0, 0)

5 A rectangle has vertices at $P = (2, 2)$, $Q = (2, 3)$, $R = (4, 3)$ and $S = (4, 2)$. Find the exact coordinates of the vertices of the rectangle after a rotation through:

a 270° anticlockwise about (0, 0) **b** 135° clockwise about (0, 0)

(E/P) **6** $\mathbf{A} = \begin{pmatrix} 1 & 0 \\ 0 & -1 \end{pmatrix}$ and $\mathbf{B} = \begin{pmatrix} 0 & 1 \\ -1 & 0 \end{pmatrix}$

a Write down fully the transformations represented by the matrices **A** and **B**. **(4 marks)**

b The point (3, 2) is transformed by matrix **A**. Write down the coordinates of the image of this point. **(1 mark)**

c The point (a, b) is transformed onto the point $(a - 3b, 2a - 2b)$ by matrix **B**. Find the values of a and b. **(3 marks)**

(E) **7** $\mathbf{M} = \begin{pmatrix} -\frac{1}{\sqrt{2}} & \frac{1}{\sqrt{2}} \\ -\frac{1}{\sqrt{2}} & -\frac{1}{\sqrt{2}} \end{pmatrix}$

a Write down fully the transformation represented by matrix **M**. **(2 marks)**

b The transformation represented by **M** maps the point (p, q) onto the point C with coordinates $(-\sqrt{2}, -2\sqrt{2})$. Find the values of p and q. **(4 marks)**

c Use your calculator to find $\mathbf{M}^3$. **(1 mark)**

d Point C is mapped onto the point D by the transformation represented by $\mathbf{M}^3$. Find the coordinates of point D and describe fully the transformation represented by $\mathbf{M}^3$. **(2 marks)**

(E/P) **8** **a** Describe fully the transformation represented by the matrix $\mathbf{A} = \begin{pmatrix} 0 & 1 \\ 1 & 0 \end{pmatrix}$. **(2 marks)**

b Write down the equations of three different invariant lines under this transformation. **(2 marks)**

c Write down $\mathbf{A}^{50}$. **(1 mark)**

(E/P) **9** The matrix $\begin{pmatrix} a & b \\ c & -0.5 \end{pmatrix}$ represents an anticlockwise rotation about the origin, through an angle θ.

a Write down the value of a. **(1 mark)**

b Find two possible values of θ, and write down the matrix corresponding to each rotation. **(3 marks)**

(E/P) **10** **a** Write down the matrix representing a rotation through 270° clockwise about (0, 0). **(1 mark)**

b A point (a, b) transformed using this matrix is such that its image is the point $(a - 5b, 4b)$. Find the values of a and b. **(3 marks)**

Challenge

Prove that the general matrix representing a rotation through angle θ anticlockwise about the origin is $\begin{pmatrix} \cos\theta & -\sin\theta \\ \sin\theta & \cos\theta \end{pmatrix}$.

6.3 Enlargements and stretches

You can describe enlargements and stretches using linear transformations.

Example 8

$\mathbf{M} = \begin{pmatrix} 3 & 0 \\ 0 & 2 \end{pmatrix}$

a Find the image T' of a triangle T with vertices (1, 1), (1, 2) and (2, 2) under the transformation represented by $\mathbf{M}$.

b Sketch T and T' on the same set of coordinate axes.

c Describe geometrically the transformation represented by $\mathbf{M}$.

a $\begin{pmatrix} 3 & 0 \\ 0 & 2 \end{pmatrix}\begin{pmatrix} 1 \\ 1 \end{pmatrix} = \begin{pmatrix} 3 \\ 2 \end{pmatrix}$

$\begin{pmatrix} 3 & 0 \\ 0 & 2 \end{pmatrix}\begin{pmatrix} 1 \\ 2 \end{pmatrix} = \begin{pmatrix} 3 \\ 4 \end{pmatrix}$

$\begin{pmatrix} 3 & 0 \\ 0 & 2 \end{pmatrix}\begin{pmatrix} 2 \\ 2 \end{pmatrix} = \begin{pmatrix} 6 \\ 4 \end{pmatrix}$

Use matrix multiplication to find the image of each vertex under the transformation.

The coordinates of the image are (3, 2), (3, 4) and (6, 4).

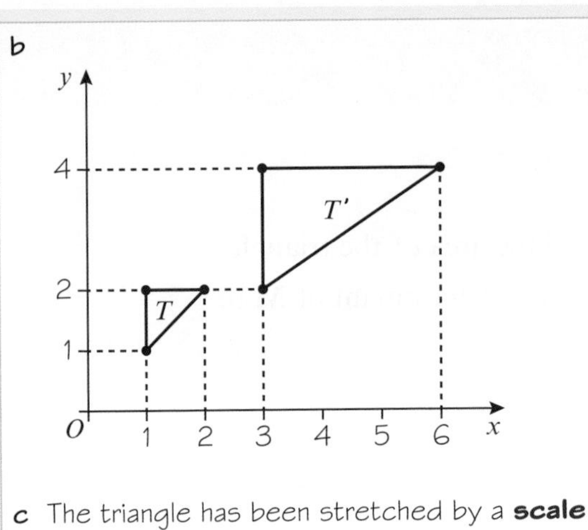

c The triangle has been stretched by a **scale factor** of 3 parallel to the x-axis and by a scale factor of 2 parallel to the y-axis.

Every x-coordinate has been multiplied by 3, and every y-coordinate has been doubled. Note that this is not the same as an enlargement, because the triangle has been stretched by different factors in the x- and y-directions.

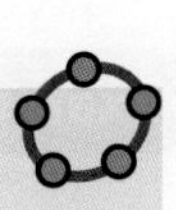

Online Explore enlargements and stretches of triangle T using GeoGebra.

- A transformation represented by the matrix $\begin{pmatrix} a & 0 \\ 0 & b \end{pmatrix}$ is a stretch of scale factor a parallel to the x-axis and a stretch of scale factor b parallel to the y-axis.

  In the case where $a = b$, the transformation is an enlargement with scale factor a.

Notation A stretch parallel to the x-axis only will have matrix $\begin{pmatrix} a & 0 \\ 0 & 1 \end{pmatrix}$.

A stretch parallel to the y-axis only will have matrix $\begin{pmatrix} 1 & 0 \\ 0 & b \end{pmatrix}$.

- For any stretch in this form, the x- and y-axes are invariant lines and the origin is an invariant point.
- For a stretch parallel to the x-axis only, points on the y-axis are invariant points, and any line parallel to the x-axis is an invariant line.
- For a stretch parallel to the y-axis only, points on the x-axis are invariant points, and any line parallel to the y-axis is an invariant line.

Watch out For a stretch in both directions, although the axes are invariant lines, the points on the axes themselves are not invariant. Any point on an axis (apart from the origin) will map to a **different** point on the same axis.

Reflections and rotations of 2D shapes both preserve the area of a shape. When a shape is stretched, its area can increase or decrease. You can use the determinant of the matrix representing this transformation to work out the scale factor for the change in area.

- For a linear transformation represented by matrix $\mathbf{M}$, det $\mathbf{M}$ represents the scale factor for the change in area. This is sometimes called the **area scale factor**.

Watch out If the determinant of the matrix $\mathbf{M}$ is negative, the shape has been reflected.

Example 9 SKILLS INTERPRETATION

$\mathbf{M} = \begin{pmatrix} 2 & 0 \\ 0 & 4 \end{pmatrix}$

a Describe fully the transformation represented by **M**.

b A triangle T has vertices at (1, 0), (4, 0) and (4, 2). Find the area of the triangle.

c The triangle T is transformed using the matrix **M**. Use the determinant of **M** to find the area of the image of T.

a The matrix is of the form $\begin{pmatrix} a & 0 \\ 0 & b \end{pmatrix}$ so it is a stretch parallel to the x-axis, with scale factor 2, and a stretch parallel to the y-axis, with scale factor 4.

b Using $A = \frac{1}{2} \times \text{base} \times \text{height}$:

$$A = \frac{1}{2} \times 3 \times 2 = 3$$

c $\det \mathbf{M} = 8$

$\det \mathbf{M} = 2 \times 4 - 0 \times 0 = 8$

← Further Pure 1 Section 5.3

The area of the image of T is $3 \times 8 = 24$.

The area increases by a factor of $\det \mathbf{M}$.

Exercise 6C SKILLS INTERPRETATION

1 Write down the matrices representing the following linear transformations.

a A stretch with scale factor 4 parallel to the x-axis

b A stretch with scale factor 3 parallel to the y-axis

c An enlargement with scale factor 2

d A stretch with scale factor 5 parallel to the x-axis and scale factor $\frac{1}{2}$ parallel to the y-axis.

2 For each of the transformations in question **1**, write down the area scale factor.

Hint In an enlargement, 'scale factor' refers to the linear scale factor of the enlargement, so the area scale factor will be the square of this value.

3 The unit square is transformed using the matrix $\begin{pmatrix} 3 & 0 \\ 0 & 4 \end{pmatrix}$.

a Write down the coordinates of any invariant points.

b Work out the area of the resulting rectangle.

4 Write down the matrices representing the following transformations.

a A stretch with scale factor −2 parallel to the x-axis

b A stretch with scale factor −3 parallel to the x-axis and scale factor 4 parallel to the y-axis

c An enlargement with scale factor $-\frac{1}{2}$.

E/P **5** $\mathbf{M} = \begin{pmatrix} 2 & 0 \\ 0 & -3 \end{pmatrix}$

a Describe fully the transformation represented by **M**. **(2 marks)**

A 2D shape with area k is transformed using the transformation represented by **M**.

b Given that the image of the shape has area 24, find the value of k. **(2 marks)**

6 A triangle with vertices at (1, 3), (5, 3) and (5, 2) is transformed using the matrix $\begin{pmatrix} 3 & 0 \\ 0 & 3 \end{pmatrix}$.

a Find the coordinates of the vertices of the resulting image.

b Find the area of the new triangle.

7 A rectangle has vertices at (2, 0), (4, 0), (4, 5) and (2, 5). The rectangle is transformed by a stretch with scale factor 2 parallel to the x-axis and scale factor -3 parallel to the y-axis.

a Find the coordinates of vertices of the resulting image.

b Find the area of the new rectangle.

E/P **8** $\mathbf{A} = \begin{pmatrix} 2\sqrt{5} & 0 \\ 0 & 2\sqrt{5} \end{pmatrix}$

a Describe fully the transformation represented by the matrix **A**. **(2 marks)**

b A triangle T has coordinates $(a, 1)$, (4, 1) and (4, 3). Given that T is transformed using matrix **A**, and the area of the resulting triangle is 60, find the value of a. **(3 marks)**

E/P **9** $\mathbf{M} = \begin{pmatrix} p & 1 \\ p & q \end{pmatrix}$ where p and q are constants and $q > 0$.

a Find $\mathbf{M}^2$ in terms of p and q. **(3 marks)**

Given that $\mathbf{M}^2$ represents an enlargement with centre (0, 0) and scale factor 6,

b find the values of p and q. **(3 marks)**

E/P **10** $\mathbf{A} = \begin{pmatrix} 5 & -1 \\ 2 & -2 \end{pmatrix}$ and $\mathbf{B} = \begin{pmatrix} 2 & 1 \\ 2 & 5 \end{pmatrix}$

a Find the matrix **M** where **M** = **AB**. **(1 mark)**

b Describe fully the transformation represented by the matrix **M**. **(3 marks)**

c A triangle T has vertices at (2, 1), (6, 1) and $(6, k)$. Given that T is transformed using matrix **M**, and the resulting triangle has area 320, find the value of k. **(4 marks)**

E **11** $\mathbf{M} = \begin{pmatrix} -1 & \sqrt{2} \\ -\sqrt{2} & -1 \end{pmatrix}$

A pentagon P of area 12 is transformed using matrix **M**. Find the area of the image of the pentagon P'. **(2 marks)**

Hint You will learn how to describe transformations such as this in **Further Pure 1 Section 6.4.**

(E/P) **12** A triangle T has vertices at the points $A = (k, 1)$, $B = (4, 1)$ and $C = (4, k)$ where k is an integer constant.

Triangle T is transformed by the matrix $\begin{pmatrix} 4 & -1 \\ k & 2 \end{pmatrix}$.

Given that triangle T has a right angle at B, and the area of the image triangle T' is 10, find the value of k. **(5 marks)**

(E) **13** A triangle T has vertices at $(0, 0)$, $(7, 7)$ and $(3, -2)$.

a Write down the matrix, **M**, which represents a rotation through 45° anticlockwise about $(0, 0)$. **(1 mark)**

b Find the exact coordinates of the image of T when T is transformed using **M**. **(3 marks)**

c Show that $\det \mathbf{M} = 1$. **(2 marks)**

d Hence find the area of the original triangle T. **(1 mark)**

Challenge

A transformation is represented by the matrix $\mathbf{P} = \begin{pmatrix} 7 & 0 \\ 0 & 0 \end{pmatrix}$.

a Find $\det \mathbf{P}$.

b Show that any point in the xy-**plane** is mapped onto the x-axis by **P**.

Problem-solving

A non-zero singular 2×2 matrix maps any point in the plane onto a straight line through the origin.

6.4 Successive transformations

You can use matrix products to represent combinations of transformations.

- **The matrix PQ represents the transformation Q followed by the transformation P.**

Example 10

a Find the 2×2 matrix **T** that represents a rotation through 90° anticlockwise about the origin followed by reflection in the line $y = x$.

b Describe the single transformation represented by **T**.

a Rotation 90° anticlockwise about the origin: $\begin{pmatrix} 0 & -1 \\ 1 & 0 \end{pmatrix}$

Reflection in the line $y = x$: $\begin{pmatrix} 0 & 1 \\ 1 & 0 \end{pmatrix}$

Write down the two matrices for the single transformations.

$$\mathbf{T} = \begin{pmatrix} 0 & 1 \\ 1 & 0 \end{pmatrix}\begin{pmatrix} 0 & -1 \\ 1 & 0 \end{pmatrix} = \begin{pmatrix} 1 & 0 \\ 0 & -1 \end{pmatrix}$$

Find the matrix product to determine the single matrix that has the same effect as these combined transformations. Make sure you apply the matrix product in the right order.

b **T** represents reflection in the line $y = 0$.

Check where **T** maps the position vectors $\begin{pmatrix} 1 \\ 0 \end{pmatrix}$ and $\begin{pmatrix} 0 \\ 1 \end{pmatrix}$.

Example 11 SKILLS EXECUTIVE FUNCTION

$$\mathbf{M} = \begin{pmatrix} -2\sqrt{2} & -2\sqrt{2} \\ 2\sqrt{2} & -2\sqrt{2} \end{pmatrix}$$

The matrix **M** represents an enlargement with scale factor k followed by rotation through angle θ anticlockwise about the origin.

a Find the value of k.

b Find the value of θ.

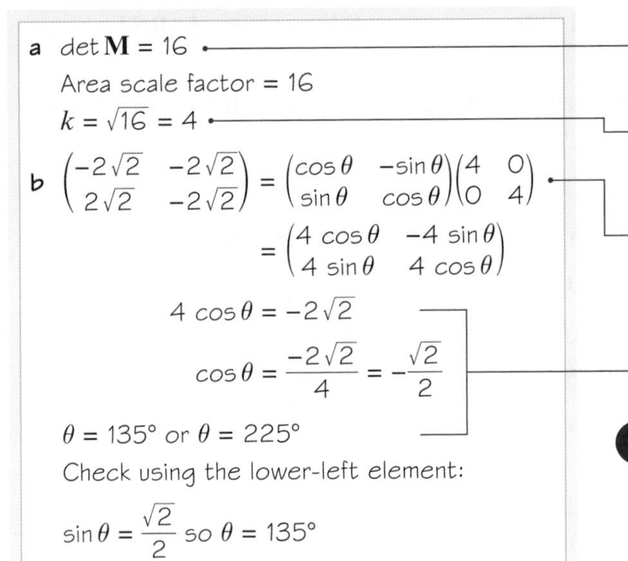

Use your calculator to find det **M**.

k is the linear scale factor. The rotation does not affect area, so $k = \sqrt{\det \mathbf{M}}$.

If the enlargement matrix is **Q** and the rotation matrix is **P** then **M** = **PQ**.

Use one element to find possible values of θ.

Watch out You will need to use one of the sine elements to check which angle is correct. $\sin\theta$ is positive so choose the angle in the second quadrant.

Exercise 6D  EXECUTIVE FUNCTION

1 $\mathbf{A} = \begin{pmatrix} -1 & 0 \\ 0 & -1 \end{pmatrix}$, $\mathbf{B} = \begin{pmatrix} 0 & -1 \\ -1 & 0 \end{pmatrix}$, $\mathbf{C} = \begin{pmatrix} 2 & 0 \\ 0 & 2 \end{pmatrix}$.

Find these matrix products and describe the single transformation represented by each product.

a **AB** **b** **BA** **c** **AC** **d** $\mathbf{A}^2$ **e** $\mathbf{C}^2$

2 **A** = rotation of 90° anticlockwise about (0, 0) **B** = rotation of 180° about (0, 0)
C = reflection in the x-axis **D** = reflection in the y-axis

a Find matrix representations of each of the four transformations **A**, **B**, **C** and **D**.

b Use matrix products to identify the single geometric transformation represented by each of these combinations.

i Reflection in the x-axis followed by rotation of 180° about (0, 0)

ii Rotation of 180° about (0, 0) followed by reflection in the x-axis

iii Reflection in the y-axis followed by reflection in the x-axis

iv Reflection in the y-axis followed by rotation of 90° anticlockwise about (0, 0)

v Rotation of 180° about (0, 0) followed by a second rotation of 180° about (0, 0)

vi Reflection in the x-axis followed by rotation of 90° anticlockwise about (0, 0) followed by reflection in the y-axis

vii Reflection in the y-axis followed by rotation of 180° about (0, 0) followed by reflection in the x-axis.

3 $\mathbf{R} = \begin{pmatrix} 3 & 0 \\ 0 & -2 \end{pmatrix}$, $\mathbf{S} = \begin{pmatrix} 0 & 1 \\ -1 & 0 \end{pmatrix}$ and $\mathbf{T} = \begin{pmatrix} 5 & 0 \\ 0 & 5 \end{pmatrix}$.

Find these matrix products and, where possible, use your knowledge of the standard forms of transformation matrices to find the single transformation represented by the products:

a **RS** **b** **RT** **c** **TS** **d** **TR** **e** **ST** **f** **RST**

4 **A** is a stretch with scale factor 2 parallel to the x-axis, and scale factor 3 parallel to the y-axis.

B is an enlargement with scale factor −2.

C is an enlargement with scale factor 4.

a Write down the matrices representing each of the transformations **A**, **B** and **C**.

b Find the single 2 × 2 matrix representing each of the following combined transformations:

i **B** followed by **A** **ii** **C** followed by **A**

iii **B** followed by **C** **iv** **C** followed by **C**

v **C** followed by **B** followed by **A**

Hint If **C** is represented by matrix **M**, then **C** followed by **C** will be represented by $\mathbf{M}^2$.

(P) **5** Use matrices to show that a reflection in the y-axis followed by reflection in the line $y = -x$ is equivalent to rotation of 90° anticlockwise about (0, 0).

(E/P) **6** A student makes the following claim:

> If **T** is a reflection in the x-axis and **U** is a rotation 90° anticlockwise about the origin then **T** followed by **U** is the same as **U** followed by **T**.

Show, using matrix multiplication, that the student is incorrect. **(4 marks)**

(E/P) **7** $\mathbf{P} = \begin{pmatrix} -4 & 0 \\ 0 & 2 \end{pmatrix}$ and $\mathbf{Q} = \begin{pmatrix} k & 0 \\ 0 & k \end{pmatrix}$, where k is a constant.

a Find the matrix product **PQ**. **(2 marks)**

b Describe, in terms of k, the single transformation represented by **PQ**. **(2 marks)**

c Show that for any value of k, **PQ** = **QP**. **(2 marks)**

(E/P) **8** $\mathbf{A} = \begin{pmatrix} 3 & 0 \\ 0 & 4 \end{pmatrix}$

a Find the matrix $\mathbf{A}^2$. **(1 mark)**

b Describe fully the transformation represented by $\mathbf{A}^2$. **(2 marks)**

$\mathbf{B} = \begin{pmatrix} a & 0 \\ 0 & b \end{pmatrix}$ where a and b are constants.

c Find the general matrix $\mathbf{B}^2$ and state, in terms of a and b, the transformation represented by this matrix. **(3 marks)**

(E/P) **9** The matrix **R** is given by $\begin{pmatrix} \frac{1}{\sqrt{2}} & \frac{-1}{\sqrt{2}} \\ \frac{1}{\sqrt{2}} & \frac{1}{\sqrt{2}} \end{pmatrix}$.

a Find $\mathbf{R}^2$. **(1 mark)**

b Describe the geometric transformation represented by $\mathbf{R}^2$. **(2 marks)**

c Hence describe the geometric transformation represented by **R**. **(1 mark)**

d Write down $\mathbf{R}^8$. **(1 mark)**

(E/P) **10** $\mathbf{M} = \begin{pmatrix} -\frac{3}{\sqrt{2}} & \frac{3}{\sqrt{2}} \\ -\frac{3}{\sqrt{2}} & -\frac{3}{\sqrt{2}} \end{pmatrix}$

The matrix **M** represents an enlargement with scale factor k ($k < 0$) followed by rotation of angle θ anticlockwise about the origin.

a Find the value of k. **(2 marks)**

b Find the value of θ. **(3 marks)**

(E/P) **11** $\mathbf{A} = \begin{pmatrix} 0 & 1 \\ -1 & 0 \end{pmatrix}$ and $\mathbf{B} = \begin{pmatrix} 5 & 0 \\ 0 & 5 \end{pmatrix}$.

A triangle T is transformed using matrix **B**. The image is then transformed using matrix **A**. Given that the area of the image, T' is 75, find the area of T. **(3 marks)**

(E/P) **12** The transformation **T** is a rotation through 225° anticlockwise about the origin.

a Write down the matrix representing this transformation. **(1 mark)**

The transformation **U** is a reflection in the line $y = x$.

b Write down the matrix representing this transformation. **(1 mark)**

c Find the matrix representing the combined transformation of **U** followed by **T**. **(2 marks)**

(E/P) **13** $\mathbf{A} = \begin{pmatrix} k & \sqrt{3} \\ \sqrt{3} & -k \end{pmatrix}$ where k is a constant.

a Find, in terms of k, the matrix $\mathbf{A}^2$. **(2 marks)**

b Describe fully the transformation represented by $\mathbf{A}^2$. **(2 marks)**

(E/P) **14** $\mathbf{P} = \begin{pmatrix} a & b \\ b & -a \end{pmatrix}$ where a and b are constants.

Show that the general matrix $\mathbf{P}^2$ represents an enlargement, and write down, in terms of a and b, the scale factor of the enlargement. **(3 marks)**

Challenge

a Given that $\mathbf{P} = \begin{pmatrix} \cos\theta & -\sin\theta \\ \sin\theta & \cos\theta \end{pmatrix}$, show algebraically that $\mathbf{P}^2 = \begin{pmatrix} \cos 2\theta & -\sin 2\theta \\ \sin 2\theta & \cos 2\theta \end{pmatrix}$.

b Interpret this result geometrically.

6.5 The inverse of a linear transformation

Since $\mathbf{A}^{-1}\mathbf{A} = \mathbf{I}$, you can use inverse matrices to reverse the effect of a linear transformation.

- **The transformation described by the matrix $\mathbf{A}^{-1}$ has the effect of reversing the transformation described by the matrix $\mathbf{A}$.**

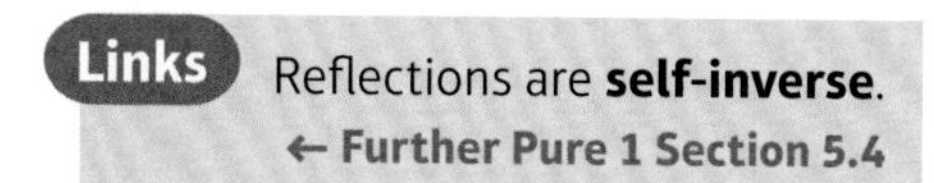

Links Reflections are **self-inverse.**
← **Further Pure 1 Section 5.4**

Example 12 SKILLS PROBLEM-SOLVING

The matrix $\mathbf{A} = \begin{pmatrix} 2 & 4 \\ -2 & -5 \end{pmatrix}$ represents a transformation **T**.

a Given that **T** maps point P with coordinates (x, y) onto the point P' with coordinates $(6, 10)$, find the coordinates of P.

b The matrix **B** represents a transformation **U**. Given that the transformation **T** followed by the transformation **U** is equivalent to a reflection in the line $y = x$, find **B**.

a $\mathbf{A}^{-1} = \begin{pmatrix} 2.5 & 2 \\ -1 & -1 \end{pmatrix}$

This represents the inverse transformation to **T**. You can find it quickly using your calculator.

$\begin{pmatrix} 2 & 4 \\ -2 & -5 \end{pmatrix}\begin{pmatrix} x \\ y \end{pmatrix} = \begin{pmatrix} 6 \\ 10 \end{pmatrix}$

$\mathbf{A}\begin{pmatrix} x \\ y \end{pmatrix} = \begin{pmatrix} 6 \\ 10 \end{pmatrix}$

$\begin{pmatrix} x \\ y \end{pmatrix} = \begin{pmatrix} 2.5 & 2 \\ -1 & -1 \end{pmatrix}\begin{pmatrix} 6 \\ 10 \end{pmatrix}$

$= \begin{pmatrix} 35 \\ -16 \end{pmatrix}$

Left multiply both sides by $\mathbf{A}^{-1}$:
$\mathbf{A}^{-1}\mathbf{A}\begin{pmatrix} x \\ y \end{pmatrix} = \mathbf{A}^{-1}\begin{pmatrix} 6 \\ 10 \end{pmatrix}$ so $\begin{pmatrix} x \\ y \end{pmatrix} = \mathbf{A}^{-1}\begin{pmatrix} 6 \\ 10 \end{pmatrix}$

P has coordinates $(35, -16)$.

b $\mathbf{BA} = \begin{pmatrix} 0 & 1 \\ 1 & 0 \end{pmatrix}$

The matrix representing **T** followed by **U** is **BA**. This is equal to $\begin{pmatrix} 0 & 1 \\ 1 & 0 \end{pmatrix}$, which is the matrix for a reflection in the line $y = x$.

$\mathbf{B} = \begin{pmatrix} 0 & 1 \\ 1 & 0 \end{pmatrix}\begin{pmatrix} 2.5 & 2 \\ -1 & -1 \end{pmatrix} = \begin{pmatrix} -1 & -1 \\ 2.5 & 2 \end{pmatrix}$

Right multiply both sides by $\mathbf{A}^{-1}$. Remember that the order is important.
$(\mathbf{BA})\mathbf{A}^{-1} = \begin{pmatrix} 0 & 1 \\ 1 & 0 \end{pmatrix}\mathbf{A}^{-1}$, so $\mathbf{B} = \begin{pmatrix} 0 & 1 \\ 1 & 0 \end{pmatrix}\mathbf{A}^{-1}$

Exercise 6E SKILLS PROBLEM-SOLVING

1 The matrix $\mathbf{R} = \begin{pmatrix} 0 & -1 \\ 1 & 0 \end{pmatrix}$.

a Give a geometric **interpretation** of the transformation represented by **R**.

b Find $\mathbf{R}^{-1}$.

c Give a geometric interpretation of the transformation represented by $\mathbf{R}^{-1}$.

2 **a** The matrix $\mathbf{S} = \begin{pmatrix} -1 & 0 \\ 0 & -1 \end{pmatrix}$.

i Give a geometric interpretation of the transformation represented by **S**.

ii Show that $\mathbf{S}^2 = \mathbf{I}$.

iii Give a geometric interpretation of the transformation represented by $\mathbf{S}^{-1}$.

b The matrix $\mathbf{T} = \begin{pmatrix} 0 & -1 \\ -1 & 0 \end{pmatrix}$.

i Give a geometric interpretation of the transformation represented by **T**.

ii Show that $\mathbf{T}^2 = \mathbf{I}$.

iii Give a geometric interpretation of the transformation represented by $\mathbf{T}^{-1}$.

c Calculate det **S** and det **T** and comment on their values in the light of the transformations they represent.

3 The matrix **A** represents a reflection in the line $y = x$ and the matrix **B** represents an anticlockwise rotation of 270° about (0, 0).

a Find the matrix $\mathbf{C} = \mathbf{BA}$ and interpret it geometrically.

b Find $\mathbf{C}^{-1}$ and give a geometric interpretation of the transformation represented by $\mathbf{C}^{-1}$.

c Find the matrix $\mathbf{D} = \mathbf{AB}$ and interpret it geometrically.

d Find $\mathbf{D}^{-1}$ and give a geometric interpretation of the transformation represented by $\mathbf{D}^{-1}$.

(E) **4** The matrix $\mathbf{A} = \begin{pmatrix} 1 & 2 \\ -3 & -2 \end{pmatrix}$ represents a transformation **T**.

Given that **T** maps point P with coordinates (x, y) onto the point P' with coordinates (5, 8),

a find the coordinates of P. **(2 marks)**

The matrix **B** represents a transformation **U**. Given that the transformation **T** followed by the transformation **U** is equivalent to a reflection in the line $y = -x$,

b find **B**. **(2 marks)**

(E) **5** $\mathbf{E} = \begin{pmatrix} 4 & 0 \\ 0 & 4 \end{pmatrix}$

a State the transformation represented by matrix **E**. **(1 mark)**

b Use your calculator to find $\mathbf{E}^{-1}$. **(1 mark)**

A triangle T transformed using matrix **E** is such that the coordinates of the image are (4, 6), (9, 7) and (3, 1).

c Using your answer to part **b**, find the coordinates of the vertices of T. **(2 marks)**

(E/P) **6** $\mathbf{M} = \begin{pmatrix} a & 0 \\ 0 & b \end{pmatrix}$ where a and b are non-zero constants.

a Find, in terms of a and b, the matrix $\mathbf{M}^{-1}$. **(2 marks)**

b The point $D = (p, q)$ maps to the point (−6, 8) under the transformation represented by **M**. Find, in terms of a and b, the coordinates of D. **(3 marks)**

(E) **7** $\mathbf{R} = \begin{pmatrix} \frac{\sqrt{3}}{2} & \frac{1}{2} \\ -\frac{1}{2} & \frac{\sqrt{3}}{2} \end{pmatrix}$

a Describe fully the transformation represented by **R**. **(2 marks)**

b The point (p, q) is mapped onto $\left(\frac{1 - 2\sqrt{3}}{2}, \frac{2 + \sqrt{3}}{2}\right)$ under the transformation represented by **R**. Find the values of p and q. **(3 marks)**

(E/P) **8** $\mathbf{A} = \begin{pmatrix} 2 & 4 \\ 1 & 3 \end{pmatrix}$ and $\mathbf{B} = \begin{pmatrix} 0 & -1 \\ -1 & 0 \end{pmatrix}$.

Triangle T is transformed to triangle T' using the matrix **AB**.
Find the matrix **P** such that triangle T' is mapped to T. **(3 marks)**

(E/P) **9** $\mathbf{A} = \begin{pmatrix} 6 & -2 \\ -4 & 1 \end{pmatrix}$

The transformation represented by **A** maps the point P onto the point Q.
Given that Q has coordinates (a, b), find the coordinates of P in terms of a and b. **(4 marks)**

Chapter review 6

(E/P) **1** The matrix **Y** represents an anticlockwise rotation of 90° about (0, 0).

a Find **Y**. **(1 mark)**

The matrices **A** and **B** are such that **AB** = **Y**. Given that $\mathbf{B} = \begin{pmatrix} 3 & 2 \\ 2 & 1 \end{pmatrix}$,

b find **A**. **(3 marks)**

c Describe the transformation represented by the matrix **ABABABAB** and write down its 2 × 2 matrix. **(2 marks)**

(E) **2** The matrix **R** represents a reflection in the x-axis and the matrix **E** represents an enlargement with scale factor 2 and centre (0, 0).

a Find the matrix **C** = **ER** and give a geometric interpretation of the transformation **C** represents. **(4 marks)**

b Find $\mathbf{C}^{-1}$ and give a geometric interpretation of the transformation represented by $\mathbf{C}^{-1}$. **(2 marks)**

(E/P) **3** $\mathbf{P} = \begin{pmatrix} 0 & k \\ k & 0 \end{pmatrix}$

Given that **P** represents a transformation **T** followed by reflection in the line $y = x$,

a find, in terms of k, a matrix representing **T**. **(4 marks)**

b Given that the point (−3, −2) maps to point (9, 6) under **T**, find the value of k. **(2 marks)**

c Given that the line $y = mx$, where m is a real constant, is invariant under **T**, find the two possible values of m. **(3 marks)**

(E/P) **4** The matrix $\mathbf{M} = \begin{pmatrix} 2\sqrt{3} & -2 \\ 2 & 2\sqrt{3} \end{pmatrix}$ represents a rotation followed by an enlargement.

a Find the scale factor of the enlargement. **(2 marks)**

b Find the angle of rotation. **(3 marks)**

A point P is mapped onto a point P' under **M**. Given that the coordinates of P' are (a, b),

c find, in terms of a and b, the coordinates of P. **(4 marks)**

(E) **5** $\mathbf{A} = \begin{pmatrix} 0 & 1 \\ 1 & 0 \end{pmatrix}$ and $\mathbf{B} = \begin{pmatrix} 0 & 1 \\ -1 & 0 \end{pmatrix}$

a Describe fully the transformations represented by the matrices **A** and **B**. **(4 marks)**

b The point (p, q) is transformed by the matrix product **AB**. Give the coordinates of the image of this point in terms of p and q. **(2 marks)**

(E/P) **6** $\mathbf{M} = \begin{pmatrix} -4 & 3 \\ 1 & -2 \end{pmatrix}$

A triangle T has vertices at $(k, 2)$, $(6, 2)$ and $(6, 7)$.

a Given that T is transformed using matrix **M**, and the area of the resulting triangle is 110, find the two possible values of k. **(3 marks)**

b Show that the line $x + 3y = 0$ is invariant under this transformation. **(4 marks)**

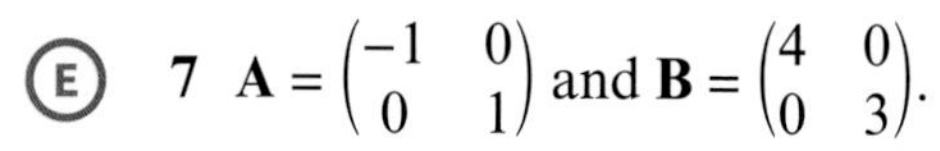

(E) **7** $\mathbf{A} = \begin{pmatrix} -1 & 0 \\ 0 & 1 \end{pmatrix}$ and $\mathbf{B} = \begin{pmatrix} 4 & 0 \\ 0 & 3 \end{pmatrix}$.

a Find the matrix $\mathbf{P} = \mathbf{AB}$. **(1 mark)**

A triangle T is transformed using matrix $\mathbf{P}$.

b Given that the area of T' is 60, find the area of T. **(2 marks)**

(E/P) **8** $\mathbf{P} = \begin{pmatrix} a & 0 \\ 0 & a \end{pmatrix}$ where a is a non-zero constant.

a Find, in terms of a, the matrix $\mathbf{P}^{-1}$. **(2 marks)**

b The point A maps to the point (4, 7) under the transformation represented by $\mathbf{P}$. Find, in terms of a, the coordinates of A. **(3 marks)**

(E/P) **9** The matrix $\mathbf{P} = \begin{pmatrix} -1 & 2 \\ -5 & 8 \end{pmatrix}$ represents a transformation $\mathbf{U}$.

A triangle T is transformed by transformation $\mathbf{U}$ followed by an anticlockwise rotation through 90° about the origin. The resulting image is labelled T'.

Find a matrix $\mathbf{M}$ representing a linear transformation that maps T' back onto T. **(3 marks)**

(E) **10** $\mathbf{A} = \begin{pmatrix} -1 & 0 \\ 0 & -1 \end{pmatrix}$ and $\mathbf{B} = \begin{pmatrix} 4 & -1 \\ 3 & -2 \end{pmatrix}$.

The transformation represented by $\mathbf{B}$ followed by the transformation represented by $\mathbf{A}$ is equivalent to the transformation represented by matrix $\mathbf{P}$.

a Find $\mathbf{P}$. **(1 mark)**

Triangle T is transformed to the triangle T' by the transformation represented by $\mathbf{P}$. Given that the area of triangle T' is 35,

b find the area of triangle T. **(3 marks)**

Triangle T' is transformed to the original triangle T by the transformation represented by matrix $\mathbf{Q}$.

c Find $\mathbf{Q}$. **(2 marks)**

Challenge

a Show that the transformation described by $\begin{pmatrix} 0 & 1 \\ 0 & 1 \end{pmatrix}$ maps any point in the plane onto the line $y = x$.

b Find the matrix representing the linear transformation that maps any point in the plane onto the straight line $y = mx$.

c Explain why, in general, the transformation that maps any point in the plane onto the straight line $ax + by = c$ is **not** a linear transformation.

Summary of key points

1 • Linear transformations always map the origin onto itself.
• Any linear transformation can be represented by a matrix.

2 The linear transformation $\mathbf{T}: \begin{pmatrix} x \\ y \end{pmatrix} \mapsto \begin{pmatrix} ax + by \\ cx + dy \end{pmatrix}$ can be represented by the matrix $\mathbf{M} = \begin{pmatrix} a & b \\ c & d \end{pmatrix}$ since $\begin{pmatrix} a & b \\ c & d \end{pmatrix}\begin{pmatrix} x \\ y \end{pmatrix} = \begin{pmatrix} ax + by \\ cx + dy \end{pmatrix}$.

3 A reflection in the y-axis is represented by the matrix $\begin{pmatrix} -1 & 0 \\ 0 & 1 \end{pmatrix}$. Points on the y-axis are invariant points, and the lines $x = 0$ and $y = k$ for any value of k are invariant lines.

4 A reflection in the x-axis is represented by the matrix $\begin{pmatrix} 1 & 0 \\ 0 & -1 \end{pmatrix}$. Points on the x-axis are invariant points, and the lines $y = 0$ and $x = k$ for any value of k are invariant lines.

5 A reflection in the line $y = x$ is represented by the matrix $\begin{pmatrix} 0 & 1 \\ 1 & 0 \end{pmatrix}$. Points on the line $y = x$ are invariant points, and the lines $y = x$ and $y = -x + k$ for any value of k are invariant lines.

6 A reflection in the line $y = -x$ is represented by the matrix $\begin{pmatrix} 0 & -1 \\ -1 & 0 \end{pmatrix}$. Points on the line $y = -x$ are invariant points, and the lines $y = -x$ and $y = x + k$ for any value of k are invariant lines.

7 The matrix representing a rotation through angle θ anticlockwise about the origin is $\begin{pmatrix} \cos\theta & -\sin\theta \\ \sin\theta & \cos\theta \end{pmatrix}$.
The only invariant point is the origin (0, 0). For $\theta \neq 180°$, there are no invariant lines.
For $\theta = 180°$, any line passing through the origin is an invariant line.

8 A transformation represented by the matrix $\begin{pmatrix} a & 0 \\ 0 & b \end{pmatrix}$ is a stretch of scale factor a parallel to the x-axis and a stretch of scale factor b parallel to the y-axis.
In the case where $a = b$, the transformation is an enlargement with scale factor a.

9 For any stretch of the above form, the x- and y-axes are invariant lines and the origin is an invariant point.

10 For a stretch parallel to the x-axis only, points on the y-axis are invariant points, and any line parallel to the x-axis is an invariant line.

11 For a stretch parallel to the y-axis only, points on the x-axis are invariant points, and any line parallel to the y-axis is an invariant line.

12 For a linear transformation represented by matrix $\mathbf{M}$, det $\mathbf{M}$ represents the scale factor for the change in area. This is sometimes called the **area scale factor**.

13 The matrix $\mathbf{PQ}$ represents the transformation $\mathbf{Q}$ followed by the transformation $\mathbf{P}$.

14 The transformation described by the matrix $\mathbf{A}^{-1}$ has the effect of reversing the transformation described by the matrix $\mathbf{A}$.

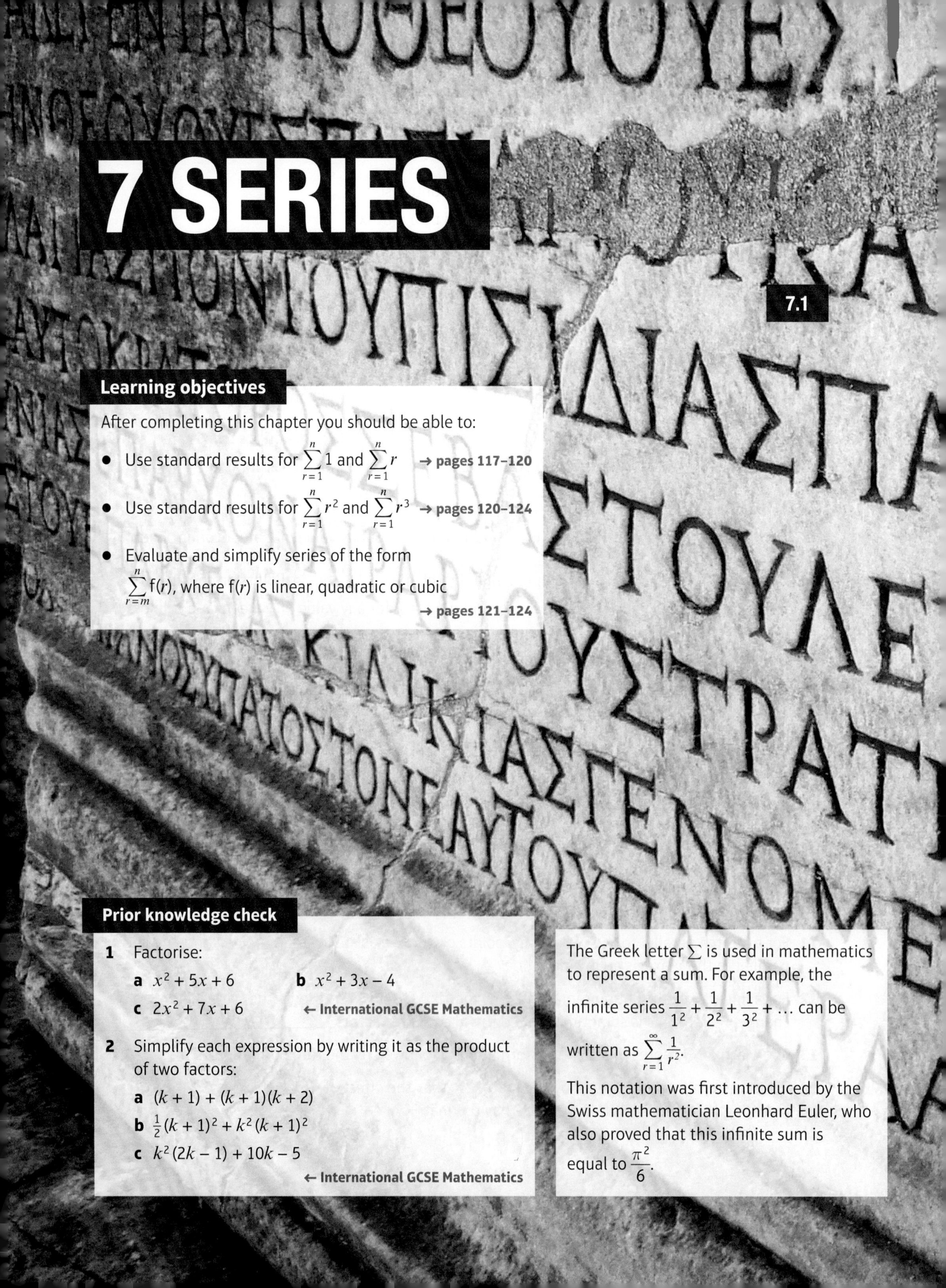

7 SERIES

7.1

Learning objectives

After completing this chapter you should be able to:

- Use standard results for $\sum_{r=1}^{n} 1$ and $\sum_{r=1}^{n} r$ → pages 117–120
- Use standard results for $\sum_{r=1}^{n} r^2$ and $\sum_{r=1}^{n} r^3$ → pages 120–124
- Evaluate and simplify series of the form $\sum_{r=m}^{n} \mathrm{f}(r)$, where $\mathrm{f}(r)$ is linear, quadratic or cubic → pages 121–124

Prior knowledge check

1 Factorise:

a $x^2 + 5x + 6$ **b** $x^2 + 3x - 4$

c $2x^2 + 7x + 6$ ← International GCSE Mathematics

2 Simplify each expression by writing it as the product of two factors:

a $(k + 1) + (k + 1)(k + 2)$

b $\frac{1}{2}(k + 1)^2 + k^2(k + 1)^2$

c $k^2(2k - 1) + 10k - 5$

← International GCSE Mathematics

The Greek letter $\sum$ is used in mathematics to represent a sum. For example, the infinite series $\frac{1}{1^2} + \frac{1}{2^2} + \frac{1}{3^2} + \ldots$ can be written as $\sum_{r=1}^{\infty} \frac{1}{r^2}$.

This notation was first introduced by the Swiss mathematician Leonhard Euler, who also proved that this infinite sum is equal to $\frac{\pi^2}{6}$.

7.1 Sums of natural numbers

You can use **sigma notation** to write **series** clearly and concisely. For example:

$$\sum_{r=1}^{3}(10r - 1) = (10 \times 1 - 1) + (10 \times 2 - 1) + (10 \times 3 - 1)$$
$$= 9 + 19 + 29 = 57$$

$$\sum_{r=1}^{n} r^2 = 1^2 + 2^2 + 3^2 + \ldots + n^2$$

- To find a series of constant terms you can use the formula $\sum_{r=1}^{n} 1 = n$.
- The formula for the sum of the first n natural numbers is $\sum_{r=1}^{n} r = \frac{1}{2}n(n + 1)$.

Links A series is the **sum** of the terms in a **sequence**.
← Pure 2 Section 5.2

Notation The numbers below and above the $\sum$ tell you which value of r to begin at, and which value to end at. You go up in increments of 1 each time.

Notation $\sum_{r=1}^{n} r = 1 + 2 + 3 + \ldots + n$

Example 1

Evaluate: **a** $\sum_{r=1}^{4}(2r - 1)$ **b** $\sum_{r=1}^{50} r$ **c** $\sum_{r=21}^{50} r$

a $\sum_{r=1}^{4}(2r - 1) = (2 \times 1 - 1) + (2 \times 2 - 1) + (2 \times 3 - 1) + (2 \times 4 - 1)$
$= 1 + 3 + 5 + 7$
$= 16$

There are only 4 terms in this series. Write out each one then find the sum.

b $\sum_{r=1}^{50} r = \frac{1}{2} \times 50 \times 51 = 1275$

Substitute $n = 50$ in $\sum_{r=1}^{n} r = \frac{1}{2}n(n + 1)$.

c $\sum_{r=21}^{50} r = \sum_{r=1}^{50} r - \sum_{r=1}^{20} r = 1275 - \frac{1}{2} \times 20 \times 21$
$= 1275 - 210 = 1065$

Problem-solving

$\sum_{r=21}^{50} r = 21 + 22 + 23 + \ldots + 49 + 50$

Find the sum of the natural numbers up to 50, then subtract the sum of the natural numbers up to 20.

- To find a series that does not start at $r = 1$, use $\sum_{r=k}^{n} f(r) = \sum_{r=1}^{n} f(r) - \sum_{r=1}^{k-1} f(r)$.

Watch out You need to subtract the sum up to $k - 1$, not k.

Example 2 SKILLS REASONING/ARGUMENTATION

Show that $\sum_{r=5}^{2N-1} r = 2N^2 - N - 10$, $N \geqslant 3$.

$$\sum_{r=5}^{2N-1} r = \sum_{r=1}^{2N-1} r - \sum_{r=1}^{4} r$$
$$= \frac{1}{2} \times (2N-1)((2N-1)+1) - \frac{1}{2} \times 4 \times 5$$
$$= \frac{1}{2} \times (2N-1)(2N) - 10$$
$$= \frac{1}{2} \times (4N^2 - 2N) - 10$$
$$= 2N^2 - N - 10$$

Substitute $n = 2N - 1$ and $n = 4$ in $\frac{1}{2}n(n+1)$.

You can rearrange expressions involving sigma notation. This allows you to evaluate more complicated series.

- $\sum_{r=1}^{n} k\text{f}(r) = k\sum_{r=1}^{n} \text{f}(r)$
- $\sum_{r=1}^{n} (\text{f}(r) + \text{g}(r)) = \sum_{r=1}^{n} \text{f}(r) + \sum_{r=1}^{n} \text{g}(r)$

Example 3

Evaluate $\sum_{r=1}^{25} (3r+1)$.

$$\sum_{r=1}^{25} (3r+1) = 3\sum_{r=1}^{25} r + \sum_{r=1}^{25} 1$$
$$= 3 \times \frac{1}{2} \times 25 \times 26 + 25$$
$$= 975 + 25 = 1000$$

Use the rules given above to write the expression in terms of $\sum_{r=1}^{25} r$ and $\sum_{r=1}^{25} 1$.

Example 4 SKILLS REASONING/ARGUMENTATION

a Show that $\sum_{r=1}^{n} (7r-4) = \frac{1}{2}n(7n-1)$.

b Hence evaluate $\sum_{r=20}^{50} (7r-4)$.

a $$\sum_{r=1}^{n} (7r-4) = 7\sum_{r=1}^{n} r - 4\sum_{r=1}^{n} 1$$
$$= \frac{7}{2}n(n+1) - 4n$$
$$= \frac{1}{2}n(7n + 7 - 8)$$
$$= \frac{1}{2}n(7n-1)$$

$7\sum_{r=1}^{n} r = 7 \times \frac{1}{2}n(n+1)$ and $4\sum_{r=1}^{n} 1 = 4 \times n$

b $$\sum_{r=20}^{50} (7r-4) = \sum_{r=1}^{50} (7r-4) - \sum_{r=1}^{19} (7r-4)$$
$$= \frac{1}{2} \times 50 \times 349 - \frac{1}{2} \times 19 \times 132$$
$$= 8725 - 1254$$
$$= 7471$$

Use the result from part **a** to find each sum quickly.
$7 \times 50 - 1 = 349$ and $7 \times 19 - 1 = 132$

Exercise 7A SKILLS REASONING/ARGUMENTATION

1 Evaluate:

a $\sum_{r=0}^{3}(2r+1)$ **b** $\sum_{r=1}^{40} r$ **c** $\sum_{r=1}^{20} r$ **d** $\sum_{r=1}^{99} r$

e $\sum_{r=10}^{40} r$ **f** $\sum_{r=100}^{200} r$ **g** $\sum_{r=21}^{40} r$ **h** $\sum_{r=1}^{k} r + \sum_{r=k+1}^{80} r, 0 < k < 80$

E/P **2** Given that $\sum_{r=1}^{n} r = 528$, find the value of n. **(4 marks)**

E/P **3** Given that $\sum_{r=1}^{k} r = \frac{1}{2}\sum_{r=1}^{20} r$, find the value of k. **(4 marks)**

E **4 a** Find an expression for $\sum_{r=1}^{2n-1} r$. **(3 marks)**

b Hence show that $\sum_{r=n+1}^{2n-1} r = \frac{3}{2}n(n-1), n \geqslant 2$. **(3 marks)**

E **5** Show that $\sum_{r=n-1}^{2n} r = \frac{1}{2}(n+2)(3n-1), n \geqslant 1$. **(5 marks)**

6 a Show that $\sum_{r=1}^{n^2} r - \sum_{r=1}^{n} r = \frac{1}{2}n(n^3-1)$.

b Hence evaluate $\sum_{r=10}^{81} r$.

Hint Use your result from part **a**.

7 Calculate the sum of each series:

a $\sum_{r=1}^{55}(3r-1)$ **b** $\sum_{r=1}^{90}(2-7r)$ **c** $\sum_{r=1}^{46}(9+2r)$

8 Show that:

a $\sum_{r=1}^{n}(3r+2) = \frac{1}{2}n(3n+7)$ **b** $\sum_{r=1}^{2n}(5r-4) = n(10n-3)$

c $\sum_{r=1}^{n+2}(2r+3) = (n+2)(n+6)$ **d** $\sum_{r=3}^{n}(4r+5) = (2n+11)(n-2)$

E/P **9 a** Show that $\sum_{r=1}^{k}(4r-5) = 2k^2 - 3k$. **(5 marks)**

b Find the smallest value of k for which $\sum_{r=1}^{k}(4r-5) > 4850$. **(4 marks)**

E/P **10** Given that $f(r) = ar + b$ and $\sum_{r=1}^{n} f(r) = \frac{1}{2}n(7n+1)$, find the constants a and b. **(5 marks)**

E **11 a** Show that $\sum_{r=1}^{4n-1}(3r+1) = 24n^2 - 2n - 1, n \geqslant 1$. **(5 marks)**

b Hence calculate $\sum_{r=1}^{99}(3r+1)$. **(2 marks)**

(E) **12** **a** Show that $\sum_{r=1}^{2k+1}(4-5r) = -(2k+1)(5k+1),\ k \geqslant 0.$ **(5 marks)**

b Hence evaluate $\sum_{r=1}^{25}(4-5r)$. **(2 marks)**

c Find the value of $\sum_{r=1}^{15}(5r-4)$. **(1 mark)**

(P) **13** Given that $\sum_{r=1}^{n}\text{f}(r) = n^2 + 4n$, deduce an expression for f(r) in terms of r.

(E/P) **14** $\text{f}(r) = ar + b$, where a and b are rational constants.

Given that $\sum_{r=1}^{4}\text{f}(r) = 36$ and $\sum_{r=1}^{6}\text{f}(r) = 78$,

a find an expression for $\sum_{r=1}^{n}\text{f}(r)$ **(6 marks)**

b hence calculate $\sum_{r=1}^{10}\text{f}(r)$. **(2 marks)**

Challenge

SKILLS INNOVATION

Given that $\sum_{r=n}^{2n}(12-2r) = 0$, find the value of n.

7.2 Sums of squares and cubes

The expression for $\sum_{r=1}^{n}1$ is linear, and the expression for $\sum_{r=1}^{n}r$ is quadratic. Similarly, you can find a cubic expression for the sum of the **squares** of the first n natural numbers, and a quartic expression for the sum of the **cubes** of the first n natural numbers.

- **The formula for the sum of the squares of the first n natural numbers is $\sum_{r=1}^{n}r^2 = \frac{1}{6}n(n+1)(2n+1)$.**
- **The formula for the sum of the cubes of the first n natural numbers is $\sum_{r=1}^{n}r^3 = \frac{1}{4}n^2(n+1)^2$.**

Links You can prove both of these results using mathematical **induction**.
→ **Further Pure 1 Section 8.1**

Example 5

Evaluate: **a** $\sum_{r=20}^{40}r^2$ **b** $\sum_{r=1}^{25}r^3$

a $\sum_{r=20}^{40}r^2 = \sum_{r=1}^{40}r^2 - \sum_{r=1}^{19}r^2$

$= \frac{1}{6} \times 40(40+1)(80+1)$
$- \frac{1}{6} \times 19(19+1)(38+1)$

$= 22\,140 - 2470 = 19\,670$

Substitute $n = 40$ and $n = 19$ in the formula: $\sum_{r=1}^{n}r^2 = \frac{1}{6}n(n+1)(2n+1)$

b $\sum_{r=1}^{25}r^3 = \frac{1}{4} \times 25^2 \times 26^2 = 105\,625$

Substitute $n = 25$ in $\frac{1}{4}n^2(n+1)^2$

Example 6 SKILLS REASONING/ARGUMENTATION

a Show that $\sum_{r=n+1}^{2n} r^2 = \frac{1}{6}n(2n+1)(7n+1)$.

b Verify that the result is true for $n = 1$ and $n = 2$.

a $\sum_{r=n+1}^{2n} r^2 = \sum_{r=1}^{2n} r^2 - \sum_{r=1}^{n} r^2$

$= \frac{1}{6} \times 2n(2n+1)(4n+1) - \frac{1}{6}n(n+1)(2n+1)$

$= \frac{1}{6}n(2n+1)[2(4n+1) - (n+1)]$

$= \frac{1}{6}n(2n+1)(7n+1)$

b When $n = 1$:

$\sum_{r=n+1}^{2n} r^2 = \sum_{r=2}^{2} r^2 = 2^2 = 4$

$\frac{1}{6}n(2n+1)(7n+1) = \frac{1}{6} \times 3 \times 8 = 4$ ✓

When $n = 2$:

$\sum_{r=n+1}^{2n} r^2 = \sum_{r=3}^{4} r^2 = 3^2 + 4^2 = 25$

$\frac{1}{6}n(2n+1)(7n+1) = \frac{2}{6} \times 5 \times 15 = 25$ ✓

Replace n by $2n$ in $\frac{1}{6}n(n+1)(2n+1)$

Problem-solving

Look for common factors in each part of the expression. Here you can take out a factor of $\frac{1}{6}n(2n+1)$.

Watch out When you have been asked to find a general result for a sum, it is good practice to test it for small values of n. It will not prove that you are correct, but if one value of n does not work, you know that your result is incorrect.

Example 7 SKILLS REASONING/ARGUMENTATION

a Show that $\sum_{r=1}^{n}(r^2 + r - 2) = \frac{1}{3}n(n+4)(n-1)$.

b Hence find the sum of the series $4 + 10 + 18 + 28 + 40 + \ldots + 418$.

a $\sum_{r=1}^{n}(r^2 + r - 2)$

$= \sum_{r=1}^{n} r^2 + \sum_{r=1}^{n} r - 2\sum_{r=1}^{n} 1$

$= \frac{1}{6}n(n+1)(2n+1) + \frac{1}{2}n(n+1) - 2n$

$= \frac{1}{6}n[(n+1)(2n+1) + 3(n+1) - 12]$

$= \frac{1}{6}n(2n^2 + 3n + 1 + 3n + 3 - 12)$

$= \frac{1}{6}n(2n^2 + 6n - 8)$

$= \frac{1}{3}n(n^2 + 3n - 4)$

$= \frac{1}{3}n(n+4)(n-1)$

Use the results for $\sum_{r=1}^{n} r^2$, $\sum_{r=1}^{n} r$ and $\sum_{r=1}^{n} 1$

b $0 + 4 + 10 + 18 + 28 + 40 + \ldots + 418$

$= \sum_{r=1}^{20}(r^2 + r - 2)$

$= \frac{1}{3} \times 20(20 + 4)(20 - 1)$

$= 3040$

Problem-solving

The question says 'hence' so use your answer to part **a**. When $r = 1$, $r^2 + r - 2 = 0$, and when $r = 20$, $r^2 + r - 2 = 418$, so you can write the sum as $\sum_{r=1}^{20}(r^2 + r - 2)$.

Example 8

a Show that $\sum_{r=1}^{n} r(r + 3)(2r - 1) = \frac{1}{6}n(n + 1)(3n^2 + an + b)$, where a and b are integers to be found.

b Hence calculate $\sum_{r=11}^{40} r(r + 3)(2r - 1)$.

a $\sum_{r=1}^{n} r(r + 3)(2r - 1)$ — First multiply out the brackets.

$= \sum_{r=1}^{n}(2r^3 + 5r^2 - 3r)$

$= 2\sum_{r=1}^{n} r^3 + 5\sum_{r=1}^{n} r^2 - 3\sum_{r=1}^{n} r$

Use these rules:

$\sum_{r=1}^{n} k\text{f}(r) = k\sum_{r=1}^{n}\text{f}(r)$

$\sum_{r=1}^{n}(\text{f}(r) + \text{g}(r)) = \sum_{r=1}^{n}\text{f}(r) + \sum_{r=1}^{n}\text{g}(r)$

$= \frac{2}{4}n^2(n + 1)^2 + \frac{5}{6}n(n + 1)(2n + 1) - \frac{3}{2}n(n + 1)$ — Use the results for $\sum_{r=1}^{n} r^3$, $\sum_{r=1}^{n} r^2$ and $\sum_{r=1}^{n} r$.

$= \frac{1}{6}n(n + 1)[3n(n + 1) + 5(2n + 1) - 9]$

$= \frac{1}{6}n(n + 1)(3n^2 + 13n - 4)$

b $\sum_{r=11}^{40} r(r + 3)(2r - 1)$

$= \sum_{r=1}^{40} r(r + 3)(2r - 1) - \sum_{r=1}^{10} r(r + 3)(2r - 1)$

$= \frac{1}{6}(40 \times 41 \times 5316) - \frac{1}{6}(10 \times 11 \times 426)$ — Substitute $n = 40$ and $n = 10$ in the result for **a**.

$= 1\,453\,040 - 7810$

$= 1\,445\,230$

Exercise 7B

SKILLS REASONING/ARGUMENTATION

1 Evaluate:

a $\sum_{r=1}^{4} r^2$ **b** $\sum_{r=1}^{40} r^2$ **c** $\sum_{r=21}^{40} r^2$ **d** $\sum_{r=1}^{99} r^3$

e $\sum_{r=100}^{200} r^3$ **f** $\sum_{r=1}^{k} r^2 + \sum_{r=k+1}^{80} r^2$, $0 < k < 80$

2 Show that:

a $\sum_{r=1}^{2n} r^2 = \frac{1}{3}n(2n + 1)(4n + 1)$ **b** $\sum_{r=1}^{2n-1} r^2 = \frac{1}{3}n(2n - 1)(4n - 1)$

c $\sum_{r=n}^{2n} r^2 = \frac{1}{6}n(n + 1)(14n + 1)$

(P) **3** Show that, for any $k \in \mathbb{N}$, $\sum_{r=1}^{n+k} r^3 = \frac{1}{4}(n+k)^2(n+k+1)^2$.

(E) **4** **a** Show that $\sum_{r=n+1}^{3n} r^3 = n^2(4n+1)(5n+2)$. **(3 marks)**

b Hence evaluate $\sum_{r=11}^{30} r^3$. **(2 marks)**

(E) **5** **a** Show that $\sum_{r=n}^{2n} r^3 = \frac{3}{4}n^2(n+1)(5n+1)$. **(3 marks)**

b Hence evaluate $\sum_{r=30}^{60} r^3$. **(2 marks)**

6 Evaluate:

a $\sum_{m=1}^{30}(m^2-1)$ **b** $\sum_{r=1}^{40} r(r+4)$ **c** $\sum_{r=1}^{80} r(r^2+3)$ **d** $\sum_{r=11}^{35}(r^3-2)$

(E) **7** **a** Show that $\sum_{r=1}^{n}(r+2)(r+5) = \frac{1}{3}n(n^2+12n+41)$. **(4 marks)**

b Hence calculate $\sum_{r=10}^{50}(r+2)(r+5)$. **(3 marks)**

(E) **8** **a** Show that $\sum_{r=1}^{n}(r^2+3r+1) = \frac{1}{3}n(n+a)(n+b)$, where a and b are integers to be found. **(4 marks)**

b Hence evaluate $\sum_{r=19}^{40}(r^2+3r+1)$. **(3 marks)**

(E) **9** **a** Show that $\sum_{r=1}^{n} r^2(r-1) = \frac{1}{12}n(n+1)(3n^2-n-2)$. **(4 marks)**

b Hence show that $\sum_{r=1}^{2n-1} r^2(r-1) = \frac{1}{3}n(2n-1)(6n^2-7n+1)$. **(4 marks)**

(E) **10** **a** Show that $\sum_{r=1}^{n}(r+1)(r+3) = \frac{1}{6}n(2n^2+an+b)$, where a and b are integers to be found. **(4 marks)**

b Hence find an expression, only in terms of n, for $\sum_{r=n+1}^{2n}(r+1)(r+3)$. **(3 marks)**

(E) **11** **a** Show that $\sum_{r=1}^{n}(r+3)(r+4) = \frac{1}{3}n(n^2+an+b)$, where a and b are integers to be found. **(4 marks)**

b Hence find an expression, only in terms of n, for $\sum_{r=n+1}^{3n}(r+3)(r+4)$. **(3 marks)**

(E) **12** **a** Show that $\sum_{r=1}^{n} r(r+3)^2 = \frac{1}{4}n(n+1)(n^2+an+b)$, where a and b are integers to be found. **(5 marks)**

b Hence evaluate $\sum_{r=10}^{20} r(r+3)^2$. **(3 marks)**

(P) **13 a** Show that, for any $k \in \mathbb{N}$, $\sum_{r=1}^{kn}(2r-1) = k^2n^2$.

b Hence find a value of n such that $\sum_{r=1}^{5n}(2r-1) = \sum_{r=1}^{n} r^3$.

(E/P) **14 a** Show that $\sum_{r=1}^{n}(r^3 - r^2) = \frac{1}{12}n(n+1)(n-1)(3n+2)$. **(4 marks)**

b Hence find the value of n that satisfies $\sum_{r=1}^{n}(r^3 - r^2) = \sum_{r=1}^{n} 7r$. **(5 marks)**

Challenge

SKILLS CREATIVITY

a Find polynomials $f_2(x)$, $f_3(x)$, $f_4(x)$ such that for every $n \in \mathbb{N}$:

$$\sum_{r=1}^{n} f_2(r) = n^2, \quad \sum_{r=1}^{n} f_3(r) = n^3, \quad \sum_{r=1}^{n} f_4(r) = n^4$$

b Hence show that for any linear, quadratic, or cubic polynomial $h(x)$ there exists a polynomial $g(x)$ such that $\sum_{r=1}^{n} g(r) = n(h(n))$.

Hint The polynomial $f_1(x) = 1$ satisfies $\sum_{r=1}^{n} f_1(r) = n$

Chapter review 7

Throughout this exercise you may assume the standard results for $\sum_{r=1}^{n} r$, $\sum_{r=1}^{n} r^2$ and $\sum_{r=1}^{n} r^3$.

1 Evaluate:

a $\sum_{r=1}^{10} r$ **b** $\sum_{r=10}^{50} r$ **c** $\sum_{r=1}^{10} r^2$ **d** $\sum_{r=1}^{10} r^3$

e $\sum_{r=26}^{50} r^2$ **f** $\sum_{r=50}^{100} r^3$ **g** $\sum_{r=1}^{60} r + \sum_{r=1}^{60} r^2$

2 Write each of the following as an expression in terms of n.

a $\sum_{r=1}^{n}(3r-5)$ **b** $\sum_{r=1}^{n}(r^2+r)$ **c** $\sum_{r=1}^{n}(3r^2+7r)$

d $\sum_{r=1}^{n}(4r^3+6r^2)$ **e** $\sum_{r=1}^{n}(r^2-2r)$ **f** $\sum_{r=1}^{n}(2r^2-3r)$

g $\sum_{r=1}^{n}(r^2-5)$ **h** $\sum_{r=1}^{n}(2r^3+3r^2+r+4)$

(E) **3** Evaluate $\sum_{r=1}^{30} r(3r-1)$. **(5 marks)**

(E) **4 a** Show that $\sum_{r=1}^{n} r^2(r-3) = \frac{1}{4}n(n+1)(n^2+an+b)$, where a and b are integers to be found. **(4 marks)**

b Hence evaluate $\sum_{r=1}^{20} r^2(r-3)$. **(2 marks)**

(E) **5 a** Show that $\sum_{r=1}^{n}(2r-1)^2 = \frac{1}{3}n(an+b)(an-b)$, where a and b are integers to be found. **(5 marks)**

b Hence find $\sum_{r=1}^{2n}(2r-1)^2$. **(2 marks)**

(E) **6 a** Show that $\sum_{r=1}^{n} r(r+2) = \frac{1}{6}n(n+1)(an+b)$, where a and b are integers to be found. **(4 marks)**

b Hence evaluate $\sum_{r=15}^{30} r(r+2)$. **(3 marks)**

(E/P) **7 a** Show that $\sum_{r=n+1}^{2n} r^2 = \frac{1}{6}n(2n+1)(an+b)$, where a and b are integers to be found. **(4 marks)**

b Hence evaluate $\sum_{r=16}^{30} r^2$. **(2 marks)**

(E/P) **8 a** Show that $\sum_{r=1}^{n}(r^2-r-1) = \frac{1}{3}n(n^2-4)$. **(4 marks)**

b Hence evaluate $\sum_{r=10}^{40}(r^2-r-1)$. **(3 marks)**

c Find the value of n such that $\sum_{r=1}^{n}(r^2-r-1) = \sum_{r=1}^{2n} r$. **(5 marks)**

(E/P) **9 a** Show that $\sum_{r=1}^{n} r(2r^2+1) = \frac{1}{2}n(n+1)(n^2+n+1)$. **(4 marks)**

b Hence show that there are no values of n that satisfy $\sum_{r=1}^{n} r(2r^2+1) = \sum_{r=1}^{n}(100r^2-r)$. **(6 marks)**

(E/P) **10 a** Show that $\sum_{r=1}^{n} r(r+1)^2 = \frac{1}{12}n(n+1)(n+2)(an+b)$, where a and b are integers to be found. **(5 marks)**

b Hence find the value of n that satisfies $\sum_{r=1}^{n} r(r+1)^2 = \sum_{r=1}^{n} 70r$. **(6 marks)**

(E/P) **11** Find the value of n that satisfies $\sum_{r=1}^{n} r^2 = \sum_{r=1}^{n+1}(9r+1)$. **(7 marks)**

Challenge

SKILLS CREATIVITY

Show that:

a $\sum_{i=1}^{n}\left(\sum_{r=1}^{i} r^2\right) = \frac{1}{12}n(n+1)^2(n+2)$

b $\sum_{j=1}^{n}\left(\sum_{i=1}^{j}\left(\sum_{r=1}^{i} r\right)\right) = \frac{1}{24}n(n+1)(n+2)(n+3)$

Summary of key points

1 To find the sum of a series of constant terms you can use the formula $\sum_{r=1}^{n} 1 = n$.

2 The formula for the sum of the first n natural numbers is $\sum_{r=1}^{n} r = \frac{1}{2}n(n+1)$.

3 To find the sum of a series that does not start at $r = 1$, use $\sum_{r=k}^{n} f(r) = \sum_{r=1}^{n} f(r) - \sum_{r=1}^{k-1} f(r)$.

4 You can rearrange expressions involving sigma notation.

- $\sum_{r=1}^{n} kf(r) = k\sum_{r=1}^{n} f(r)$
- $\sum_{r=1}^{n} (f(r) + g(r)) = \sum_{r=1}^{n} f(r) + \sum_{r=1}^{n} g(r)$

5 The formula for the sum of the squares of the first n natural numbers is

$$\sum_{r=1}^{n} r^2 = \frac{1}{6}n(n+1)(2n+1)$$

6 The formula for the sum of the cubes of the first n natural numbers is

$$\sum_{r=1}^{n} r^3 = \frac{1}{4}n^2(n+1)^2$$

8 PROOF

8.1(i)
8.1(ii)
8.1(iii)
8.1(iv)

Learning objectives

After completing this chapter you should be able to:

- Understand the principle of proof by mathematical induction and prove results about sums of series → pages 128–131
- Prove results about divisibility using induction → pages 132–134
- Prove results for a general term in a recurrence relation → pages 134–137
- Prove results about matrices using induction → pages 137–139

Prior knowledge check

1 Write down expressions for:

a $\sum_{r=1}^{n} r$ **b** $\sum_{r=1}^{n+1} r^2$

← Further Pure 1 Sections 7.1, 7.2

2 Prove that for all positive integers n, $3^{n+2} - 3^n$ is divisible by 8.

← International GCSE Mathematics

3 Find the first four terms of the recurrence relationship,

$u_{n+1} = u_n + 3, u_1 = 1$ ← Pure 2 Section 5.7

4 $\mathbf{M} = \begin{pmatrix} k & 2 \\ k+1 & -3 \end{pmatrix}$ and $\mathbf{N} = \begin{pmatrix} 5 & -8 \\ 0 & 2 \end{pmatrix}$

Find $\mathbf{MN}$, giving your answer in terms of k.

← Further Pure 1 Section 5.2

These dominoes are set up so that, as each domino falls, it knocks over the next one. As long as the first domino is pushed over, all the dominos will fall. You can prove mathematical statements in a similar way using mathematical induction.

8.1 Proof by mathematical induction

- You can use proof by induction to prove that a general statement is true for all positive integers.
- **Proof by mathematical induction** usually consists of the following four steps:
  1. **Basis step:** Prove the general statement is true for $n = 1$.
  2. **Assumption step:** Assume the general statement is true for $n = k$.
  3. **Inductive step:** Show that the general statement is then true for $n = k + 1$.
  4. **Conclusion step:** The general statement is then true for all positive integers, n.

Watch out You need to carry out **both** the basis step **and** the inductive step in order to complete the proof: carrying out just one of these is not sufficient to prove the general statement.

This method of proof is often useful for proving results about sums of series.

Links You can prove the general results for $\sum_{r=1}^{n} r$, $\sum_{r=1}^{n} r^2$ and $\sum_{r=1}^{n} r^3$ by induction.

← Further Pure 1 Sections 7.1, 7.2

Example 1

Prove by induction that for all positive integers n, $\sum_{r=1}^{n}(2r - 1) = n^2$.

$n = 1$: LHS $= \sum_{r=1}^{1}(2r - 1) = 2(1) - 1 = 1$

RHS $= 1^2 = 1$

As LHS = RHS, the **summation formula** is true for $n = 1$.

1. Basis step
Substitute $n = 1$ into both the LHS (left-hand side) and RHS (right-hand side) of the formula to check if the formula works for $n = 1$.

Assume that the summation formula is true for $n = k$:

$$\sum_{r=1}^{k}(2r - 1) = k^2$$

2. Assumption step
In this step you assume that the general statement given is true for $n = k$.

With $n = k + 1$ terms the summation formula becomes:

$$\sum_{r=1}^{k+1}(2r - 1) = \sum_{r=1}^{k}(2r - 1) + 2(k + 1) - 1$$

3. Inductive step
Sum to k terms plus the $(k + 1)$th term.

This is the $(k + 1)$th term.

Sum of first k terms is k^2 by assumption.

$$= k^2 + (2(k + 1) - 1)$$
$$= k^2 + (2k + 2 - 1)$$
$$= k^2 + 2k + 1$$
$$= (k + 1)^2$$

This is the same expression as n^2 with n replaced by $k + 1$.

Therefore, the summation formula is true when $n = k + 1$.

If the summation formula is true for $n = k$ then it is shown to be true for $n = k + 1$. As the result is true for $n = 1$, it is now also true for all $n \in \mathbb{Z}^+$ by mathematical induction.

4. Conclusion step
The result is true for $n = 1$ and steps **2** and **3** imply that the result is then true for $n = 2$. Continuing to apply steps **2** and **3** implies result is true for $n = 3, 4, 5, \ldots$ etc.

Notation $\mathbb{Z}^+$ is the set of **positive integers**, 1, 2, 3, …. It is equivalent to $\mathbb{N}$, the set of natural numbers.

Example 2

SKILLS REASONING/ARGUMENTATION

Prove by induction that for all positive integers n, $\sum_{r=1}^{n} r^2 = \frac{1}{6}n(n+1)(2n+1)$.

$n = 1$: LHS $= \sum_{r=1}^{1} r^2 = 1^2 = 1$

RHS $= \frac{1}{6}(1)(2)(3) = \frac{6}{6} = 1$

As LHS = RHS, the summation formula is true for $n = 1$.

1. Basis step
Substitute $n = 1$ into both the LHS and RHS of the formula to check if the formula works for $n = 1$.

Assume that the summation formula is true for $n = k$:

$$\sum_{r=1}^{k} r^2 = \frac{1}{6}k(k+1)(2k+1)$$

2. Assumption step
In this step you assume that the result given in the question is true for $n = k$.

With $n = k + 1$ terms the summation formula becomes:

$$\sum_{r=1}^{k+1} r^2 = \sum_{r=1}^{k} r^2 + (k+1)^2$$

3. Inductive step
Sum to k terms plus the $(k + 1)$th term.

$$= \frac{1}{6}k(k+1)(2k+1) + (k+1)^2$$
$$= \frac{1}{6}(k+1)(k(2k+1) + 6(k+1))$$
$$= \frac{1}{6}(k+1)(2k^2 + k + 6k + 6)$$
$$= \frac{1}{6}(k+1)(2k^2 + 7k + 6)$$
$$= \frac{1}{6}(k+1)(k+2)(2k+3)$$
$$= \frac{1}{6}(k+1)((k+1)+1)(2(k+1)+1)$$

Rearrange to get the same expression as $\frac{1}{6}n(n+1)(2n+1)$ with n replaced by $k + 1$.

Therefore, the summation formula is true when $n = k + 1$.

If the summation formula is true for $n = k$, then it is shown to be true for $n = k + 1$. As the result is true for $n = 1$, it is therefore true for all $n \in \mathbb{Z}^+$ by mathematical induction.

4. Conclusion step
The result is true for $n = 1$ and steps **2** and **3** imply that the result is then true for $n = 2$. Continuing to apply steps **2** and **3** implies result is true for $n = 3$, 4, 5, … etc.

Example 3

Prove by induction that for all positive integers n, $\sum_{r=1}^{n} r2^r = 2(1 + (n-1)2^n)$.

$n = 1$: LHS $= \sum_{r=1}^{1} r\,2^r = 1(2)^1 = 2$

RHS $= 2(1 + (1-1)2^1) = 2(1) = 2$

As LHS = RHS, the summation formula is true for $n = 1$.

1. Basis step

Assume that the summation formula is true for $n = k$:

$$\sum_{r=1}^{k} r\,2^r = 2(1 + (k-1)2^k)$$

2. Assumption step

With $n = k + 1$ terms the summation formula becomes:

$$\sum_{r=1}^{k+1} r2^r = \sum_{r=1}^{k} r2^r + (k+1)2^{k+1}$$
$$= 2(1 + (k-1)2^k) + (k+1)2^{k+1}$$
$$= 2 + 2(k-1)2^k + (k+1)2^{k+1}$$
$$= 2 + (k-1)2^{k+1} + (k+1)2^{k+1}$$
$$= 2 + (k - 1 + k + 1)2^{k+1}$$
$$= 2 + 2k2^{k+1}$$
$$= 2(1 + k2^{k+1})$$
$$= 2(1 + ((k+1) - 1)2^{k+1})$$

Therefore, the summation formula is true when $n = k + 1$.

If the summation formula is true for $n = k$, then it is shown to be true for $n = k + 1$. As the result is true for $n = 1$, it is now also true for all $n \in \mathbb{Z}^+$ by mathematical induction.

3. Inductive step

$2^1 \times 2^k = 2^{k+1}$

This is the same expression as $2(1 + (n-1)2^n)$ with n replaced by $k + 1$.

4. Conclusion step

Watch out Induction can prove that a given statement is true for all $n \in \mathbb{Z}^+$, but it does not help you derive statements.

Exercise 8A

SKILLS REASONING/ARGUMENTATION

(E/P) **1** Prove by induction that for any positive integer n, $\sum_{r=1}^{n} r = \frac{1}{2}n(n+1)$. **(5 marks)**

(E/P) **2** Prove by induction that for any positive integer n, $\sum_{r=1}^{n} r^3 = \frac{1}{4}n^2(n+1)^2$. **(5 marks)**

(E/P) **3** **a** Prove by induction that for any positive integer n:

$$\sum_{r=1}^{n} r(r-1) = \frac{1}{3}n(n+1)(n-1)$$ **(6 marks)**

b Hence deduce an expression, in terms of n, for $\sum_{r=1}^{2n+1} r(r-1)$. **(3 marks)**

(E/P) **4** **a** Prove by induction that for any positive integer n:

$$\sum_{r=1}^{n} r(3r-1) = n^2(n+1)$$ **(6 marks)**

b Hence use the standard result for $\sum_{r=1}^{n} r^3$ to find a value of n such that $\sum_{r=1}^{n} r^3 = 4\sum_{r=1}^{n} r(3r-1)$. **(5 marks)**

(P) **5** Prove by induction that for any positive integer n:

a $\sum_{r=1}^{n} \left(\frac{1}{2}\right)^r = 1 - \frac{1}{2^n}$ **b** $\sum_{r=1}^{n} r(r!) = (n+1)! - 1$ **c** $\sum_{r=1}^{n} \frac{4}{r(r+2)} = \frac{n(3n+5)}{(n+1)(n+2)}$

(P) **6** The box below shows a student's attempts to prove $\left(\sum_{r=1}^{n} r\right)^2 = \sum_{r=1}^{n} r^2$ using induction.

> Let $n = 1$. Then LHS $= \left(\sum_{r=1}^{1} r\right)^2 = (1)^2 = 1$, and RHS $= \sum_{r=1}^{1} r^2 = 1^2 = 1$, so that LHS = RHS (Basis step).
>
> Now we assume the statement is true for $n = k$:
>
> $$\left(\sum_{r=1}^{k} r\right)^2 = \sum_{r=1}^{k} r^2$$
>
> and so for $n = k + 1$ the statement is
>
> $$\left(\sum_{r=1}^{k+1} r\right)^2 = \sum_{r=1}^{k+1} r^2$$
>
> Hence, by the principle of mathematical induction, the statement is true for all $n \in \mathbb{Z}^+$.

a Identify the error made in the proof.

b Give a counter-example (i.e. an example used to prove something is wrong) to show that the original statement is not true.

(P) **7** A student claims that $\sum_{r=1}^{n} r = \frac{1}{2}(n^2 + n + 1)$, and produces the following proof.

> Assume that the statement is true for $n = k$:
>
> $$\sum_{r=1}^{k} r = \frac{1}{2}(k^2 + k + 1)$$
>
> When $n = k + 1$:
>
> $$\sum_{r=1}^{k+1} r = \sum_{r=1}^{k} r + (k + 1)$$
> $$= \frac{1}{2}(k^2 + k + 1) + (k + 1)$$
> $$= \frac{1}{2}(k^2 + k + 1 + 2(k + 1))$$
> $$= \frac{1}{2}((k^2 + 2k + 1) + (k + 1) + 1)$$
> $$= \frac{1}{2}((k + 1)^2 + (k + 1) + 1)$$
>
> This is the original formula but with $n = k + 1$. Hence, by the principle of mathematical induction, the statement is true for all $n \in \mathbb{Z}^+$.

a Identify the error made in the proof.

b Give a counter-example to show that the original statement is not true.

SKILLS CREATIVITY

Challenge

Prove by induction that for all positive integers n,
$\sum_{r=1}^{n} (-1)^r r^2 = \frac{1}{2}(-1)^n n(n + 1)$.

Hint $\sum_{r=1}^{n} (-1)^r r^2 = -1^2 + 2^2 - 3^2 + 4^2 - 5^2 + \dots$

8.2 Proving divisibility results

You can use proof by induction to prove that a given expression is divisible by a certain integer.

Example 4 SKILLS REASONING/ARGUMENTATION

Prove by induction that for all positive integers n, $3^{2n} + 11$ is divisible by 4.

Let $f(n) = 3^{2n} + 11$, where $n \in \mathbb{Z}^+$.
$f(1) = 3^{2(1)} + 11 = 9 + 11 = 20 = 4(5)$,
which is divisible by 4.
$f(n)$ is divisible by 4 when $n = 1$.

1. Basis step

Assume true for $n = k$, so that
$f(k) = 3^{2k} + 11$ is divisible by 4.

2. Assumption step

$f(k + 1) = 3^{2(k+1)} + 11$
$= 3^{2k} \times 3^2 + 11$
$= 9(3^{2k}) + 11$
$f(k + 1) - f(k) = (9(3^{2k}) + 11) - (3^{2k} + 11)$
$= 8(3^{2k})$
$= 4(2(3^{2k}))$
$f(k + 1) = f(k) + 4(2(3^{2k}))$
Therefore $f(n)$ is divisible by 4 when $n = k + 1$.

3. Inductive step

As both $f(k)$ and $4(2(3^{2k}))$ are divisible by 4 then the sum of these two must also be divisible by 4.

If $f(n)$ is divisible by 4 when $n = k$, then it has been shown that $f(n)$ is also divisible by 4 when $n = k + 1$. As $f(n)$ is divisible by 4 when $n = 1$, $f(n)$ is also divisible by 4 for all $n \in \mathbb{Z}^+$ by mathematical induction.

4. Conclusion step

Problem-solving

When proving that an expression $f(n)$ is divisible by r, you can complete the induction step by showing that $f(k + 1) - f(k)$ is divisible by r.

Example 5

Prove by induction that for all positive integers n, $n^3 - 7n + 9$ is divisible by 3.

Let $f(n) = n^3 - 7n + 9$, where $n \in \mathbb{Z}^+$.
$f(1) = 1 - 7 + 9 = 3$, which is divisible by 3.
$f(n)$ is divisible by 3 when $n = 1$.

1. Basis step

Assume true for $n = k$, so that
$f(k) = k^3 - 7k + 9$ is divisible by 3.

2. Assumption step

$f(k + 1) = (k + 1)^3 - 7(k + 1) + 9$
$= k^3 + 3k^2 + 3k + 1 - 7(k + 1) + 9$
$= k^3 + 3k^2 + 3k + 1 - 7k - 7 + 9$
$= k^3 + 3k^2 - 4k + 3$

3. Inductive step

Use the binomial theorem or multiply out the three brackets.

$\mathrm{f}(k+1) - \mathrm{f}(k) = (k^3 + 3k^2 - 4k + 3) - (k^3 - 7k + 9)$
$= 3k^2 + 3k - 6$
$= 3(k^2 + k - 2)$
$\mathrm{f}(k+1) = \mathrm{f}(k) + 3(k^2 + k - 2)$

As both $\mathrm{f}(k)$ and $3(k^2 + k - 2)$ are divisible by 3 then their sum must also be divisible by 3.

Therefore $\mathrm{f}(n)$ is divisible by 3 when $n = k + 1$.
If $\mathrm{f}(n)$ is divisible by 3 when $n = k$, then it has been shown that $\mathrm{f}(n)$ is also divisible by 3 when $n = k + 1$. As $\mathrm{f}(n)$ is divisible by 3 when $n = 1$, $\mathrm{f}(n)$ is also divisible by 3 for all $n \in \mathbb{Z}^+$ by mathematical induction.

4. Conclusion step

Example 6 SKILLS REASONING/ARGUMENTATION

Prove by induction that for all positive integers n, $11^{n+1} + 12^{2n-1}$ is divisible by 133.

Let $\mathrm{f}(n) = 11^{n+1} + 12^{2n-1}$, where $n \in \mathbb{Z}^+$.

1. Basis step

$\mathrm{f}(1) = 11^2 + 12 = 133$, which is divisible by 133.
$\mathrm{f}(n)$ is divisible by 133 when $n = 1$.

Assume true for $n = k$, so that
$\mathrm{f}(k) = 11^{k+1} + 12^{2k-1}$ is divisible by 133.

2. Assumption step

3. Inductive step

$\mathrm{f}(k+1) = 11^{k+1+1} + 12^{2(k+1)-1}$
$= 11^{k+1}(11)^1 + 12^{2k-1}(12)^2$
$= 11(11^{k+1}) + 144(12^{2k-1})$

$12^{2(k+1)-1} = 12^{2k+2-1}$
$= 12^{2k-1+2}$
$= 12^{2k-1}(12)^2$

$\mathrm{f}(k+1) - \mathrm{f}(k) = (11(11^{k+1}) + 144(12^{2k-1})) - (11^{k+1} + 12^{2k-1})$
$= 10(11^{k+1}) + 143(12^{2k-1})$
$= 10(11^{k+1}) + 10(12^{2k-1}) + 133(12^{2k-1})$
$= 10(11^{k+1} + 12^{2k-1}) + 133(12^{2k-1})$

Problem-solving

Always keep an eye on what you are trying to prove. You need to show that this expression is divisible by 133, so write $143(12^{2k-1})$ as $10(12^{2k-1}) + 133(12^{2k-1})$.

$\mathrm{f}(k+1) = \mathrm{f}(k) + 10(11^{k+1} + 12^{2k-1}) + 133(12^{2k-1})$
$= \mathrm{f}(k) + 10\mathrm{f}(k) + 133(12^{2k-1})$
$= 11\mathrm{f}(k) + 133(12^{2k-1})$

As both $11\mathrm{f}(k)$ and $133(12^{2k-1})$ are divisible by 133 then their sum must also be divisible by 133.

Therefore $\mathrm{f}(n)$ is divisible by 133 when $n = k + 1$.
If $\mathrm{f}(n)$ is divisible by 133 when $n = k$, then it has been shown that $\mathrm{f}(n)$ is also divisible by 133 when $n = k + 1$. As $\mathrm{f}(n)$ is divisible by 133 when $n = 1$, $\mathrm{f}(n)$ is also divisible by 133 for all $n \in \mathbb{Z}^+$ by mathematical induction.

4. Conclusion step

Exercise 8B SKILLS REASONING/ARGUMENTATION

(P) **1** Prove by induction that for all positive integers n:

a $8^n - 1$ is divisible by 7
b $3^{2n} - 1$ is divisible by 8
c $5^n + 9^n + 2$ is divisible by 4
d $2^{4n} - 1$ is divisible by 15
e $3^{2n-1} + 1$ is divisible by 4
f $n^3 + 6n^2 + 8n$ is divisible by 3
g $n^3 + 5n$ is divisible by 6
h $2^n(3^{2n}) - 1$ is divisible by 17

(E/P) **2** $f(n) = 13^n - 6^n$

a Show that $f(k + 1) = 6f(k) + 7(13^k)$. **(3 marks)**

b Hence, or otherwise, prove by induction that for all positive integers n, $f(n)$ is divisible by 7. **(4 marks)**

(E/P) **3** $g(n) = 5^{2n} - 6n + 8$

a Show that $g(k + 1) = 25g(k) + 9(16k - 22)$. **(3 marks)**

b Hence, or otherwise, prove by induction that for all positive integers n, $g(n)$ is divisible by 9. **(4 marks)**

(E/P) **4** Prove by induction that for all positive integers n, $8^n - 3^n$ is divisible by 5. **(6 marks)**

(E/P) **5** Prove by induction that for all positive integers n, $3^{2n+2} + 8n - 9$ is divisible by 8. **(6 marks)**

(E/P) **6** Prove by induction that for all positive integers n, $2^{6n} + 3^{2n-2}$ is divisible by 5. **(6 marks)**

8.3 Using mathematical induction to produce a proof for a general term of a recurrence relation

In Pure 2, you saw **recurrence formulae** which allowed you to generate successive terms of a sequence.

Example 7 SKILLS REASONING/ARGUMENTATION

A sequence can be described by the recurrence formula

$$u_{n+1} = 3u_n + 4, n \geqslant 1, u_1 = 1.$$

a Find the first five terms of the sequence.

b Show that the general statement $u_n = 3^n - 2, n \geqslant 1$, gives the same first five terms of the sequence.

a $u_{n+1} = 3u_n + 4, n \geqslant 1, u_1 = 1$

Substituting $n = 1$; $u_2 = 3u_1 + 4 = 3(1) + 4 = 7$

Substituting $n = 2$; $u_3 = 3u_2 + 4 = 3(7) + 4 = 25$

Substituting $n = 3$; $u_4 = 3u_3 + 4 = 3(25) + 4 = 79$

Substituting $n = 4$; $u_5 = 3u_4 + 4 = 3(79) + 4 = 241$

The first five terms of the sequence are 1, 7, 25, 79, 241

b $u_n = 3^n - 2, n \geq 1$

Substituting $n = 1$; $u_1 = 3^1 - 2 = 3 - 2 = 1$

Substituting $n = 2$; $u_2 = 3^2 - 2 = 9 - 2 = 7$

Substituting $n = 3$; $u_3 = 3^3 - 2 = 27 - 2 = 25$

Substituting $n = 4$; $u_4 = 3^4 - 2 = 81 - 2 = 79$

Substituting $n = 5$; $u_5 = 3^5 - 2 = 243 - 2 = 241$

The first five terms of the sequence are 1, 7, 25, 79, 241

Example 8 demonstrates how to apply the method of proof by induction to show that the **general statement** $u_n = 3^n - 2$ is true for the **recurrence formula** $u_{n+1} = 3u_n + 4$ with first term $u_1 = 1$ and $n \geq 1$.

Example 8

Given that $u_{n+1} = 3u_n + 4$, $u_1 = 1$, prove by mathematical induction that $u_n = 3^n - 2$.

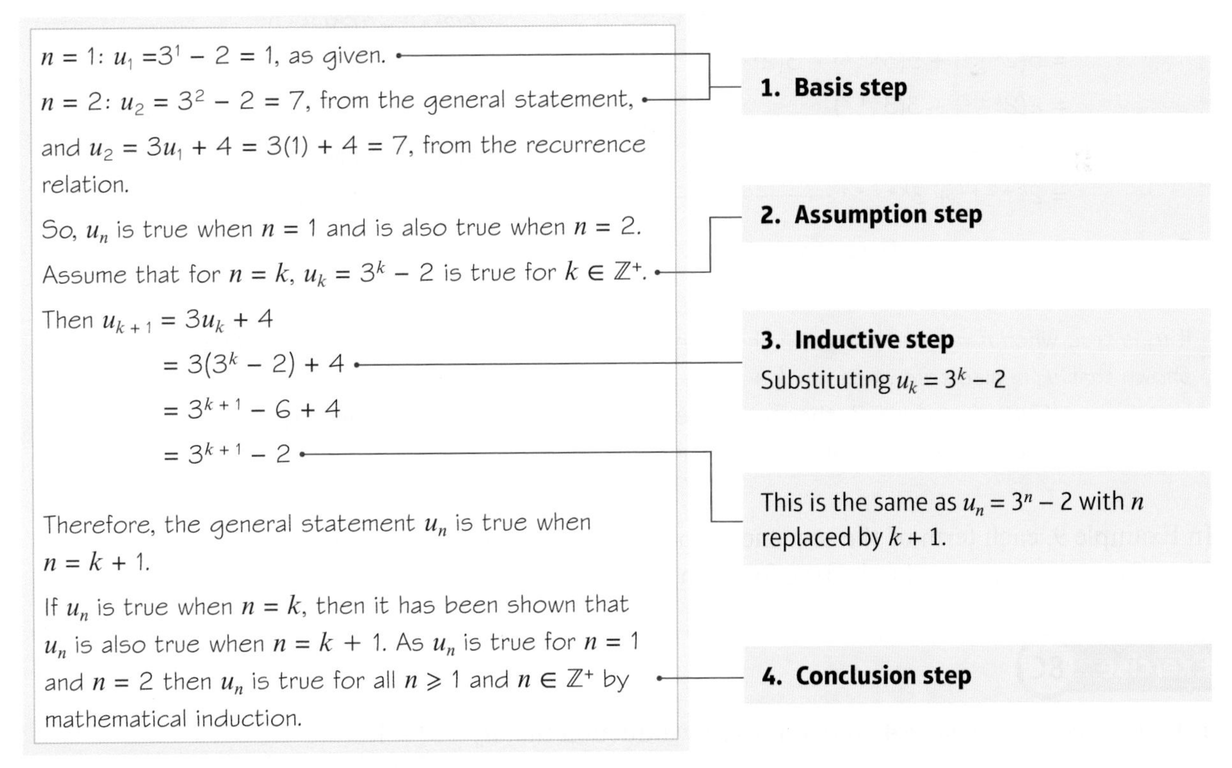

$n = 1$: $u_1 = 3^1 - 2 = 1$, as given.

$n = 2$: $u_2 = 3^2 - 2 = 7$, from the general statement,

and $u_2 = 3u_1 + 4 = 3(1) + 4 = 7$, from the recurrence relation.

So, u_n is true when $n = 1$ and is also true when $n = 2$.

1. Basis step

Assume that for $n = k$, $u_k = 3^k - 2$ is true for $k \in \mathbb{Z}^+$.

2. Assumption step

Then $u_{k+1} = 3u_k + 4$

$= 3(3^k - 2) + 4$

$= 3^{k+1} - 6 + 4$

$= 3^{k+1} - 2$

3. Inductive step

Substituting $u_k = 3^k - 2$

This is the same as $u_n = 3^n - 2$ with n replaced by $k + 1$.

Therefore, the general statement u_n is true when $n = k + 1$.

If u_n is true when $n = k$, then it has been shown that u_n is also true when $n = k + 1$. As u_n is true for $n = 1$ and $n = 2$ then u_n is true for all $n \geqslant 1$ and $n \in \mathbb{Z}^+$ by mathematical induction.

4. Conclusion step

In the **basis step** of the proof, the general statement was checked for both $n = 1$ and $n = 2$. This is because the first application of the recurrence formula $u_{n+1} = 3u_n + 4$ yields u_2 by using the first term $u_1 = 1$.

Example 9

Given that $u_{n+2} = 5u_{n+1} - 6u_n$, $u_1 = 13$, $u_2 = 35$, prove by induction that $u_n = 2^{n+1} + 3^{n+1}$.

$n = 1$; $u_1 = 2^2 + 3^2 = 13$, as given.

$n = 2$; $u_2 = 2^3 + 3^3 = 35$, as given.

$n = 3$; $u_3 = 2^4 + 3^4 = 97$, from the general statement,

and $u_3 = 5u_2 - 6u_1 = 5(35) - 6(13) = 97$,

from the recurrence relation.

So u_n is true when $n = 1$ and $n = 2$ and also when $n = 3$.

1. Basis step

The first application of the recurrence formula yields u_3 by using u_1 and u_2.

So check the general statement for $n = 1, 2, 3$.

Assume that, for $n = k$ and for $n = k + 1$, both

$u_k = 2^{k+1} + 3^{k+1}$ and

$u_{k+1} = 2^{k+1+1} + 3^{k+1+1} = 2^{k+2} + 3^{k+2}$

are true for $k \in \mathbb{Z}^+$.

2. Assumption step

Then $u_{k+2} = 5u_{k+1} - 6u_k$

$= 5(2^{k+2} + 3^{k+2}) - 6(2^{k+1} + 3^{k+1})$

$= 5(2^{k+2}) + 5(3^{k+2}) - 6(2^{k+1}) - 6(3^{k+1})$

$= 5(2^{k+2}) + 5(3^{k+2}) - 3(2^{k+2}) - 2(3^{k+2})$

$= 2(2^{k+2}) + 3(3^{k+2})$

$= 2^{k+3} + 3^{k+3}$

$= 2^{k+2+1} + 3^{k+2+1}$

Therefore, the general statement u_n is true when $n = k + 2$.

3. Inductive step

$6(2^{k+1}) = 3(2)(2^{k+1})$
$= 3(2^{k+2})$

$6(3^{k+1}) = 2(3)(3^{k+1})$
$= 2(3^{k+2})$

This is the same as $u_n = 2^{n+1} + 3^{n+1}$ with n replaced by $k + 1$.

If u_n is true when $n = k$ and $n = k + 1$ then it has been shown that u_n is also true when $n = k + 2$. As u_n is true for $n = 1$, $n = 2$ and $n = 3$, then u_n is true for all $n \geqslant 1$ and $n \in \mathbb{Z}^+$ by mathematical induction.

4. Conclusion step

In Example 9, each term of the sequence depends upon the previous two terms. Therefore u_n was assumed to be true for both $n = k$ and $n = k + 1$.

Exercise 8C

1 Given that $u_{n+1} = 5u_n + 4$, $u_1 = 4$, prove by induction that $u_n = 5^n - 1$.

2 Given that $u_{n+1} = 2u_n + 5$, $u_1 = 3$, prove by induction that $u_n = 2^{n+2} - 5$.

3 Given that $u_{n+1} = 5u_n - 8$, $u_1 = 3$, prove by induction that $u_n = 5^{n-1} + 2$.

4 Given that $u_{n+2} = 5u_{n+1} - 6u_n$, $u_1 = 1$, $u_2 = 5$, prove by induction that $u_n = 3^n - 2^n$.

5 Given that $u_{n+2} = 6u_{n+1} - 9u_n$, $u_1 = -1$, $u_2 = 0$, prove by induction that $u_n = (n - 2)3^{n-1}$.

6 Given that $u_{n+2} = 7u_{n+1} - 10u_n$, $u_1 = 1$, $u_2 = 8$, prove by induction that $u_n = 2(5^{n-1}) - 2^{n-1}$.

7 Given that $u_{n+2} = 6u_{n+1} - 9u_n$, $u_1 = 3$, $u_2 = 36$, prove by induction that $u_n = (3n - 2)3^n$.

8.4 Proving statements involving matrices

You can use matrix multiplication to prove results involving powers of matrices.

Example 10

SKILLS REASONING/ARGUMENTATION

Prove by induction that for all positive integers n, $\begin{pmatrix} 1 & -1 \\ 0 & 2 \end{pmatrix}^n = \begin{pmatrix} 1 & 1 - 2^n \\ 0 & 2^n \end{pmatrix}$.

$n = 1$: LHS $= \begin{pmatrix} 1 & -1 \\ 0 & 2 \end{pmatrix}^1 = \begin{pmatrix} 1 & -1 \\ 0 & 2 \end{pmatrix}$

RHS $= \begin{pmatrix} 1 & 1 - 2^1 \\ 0 & 2^1 \end{pmatrix} = \begin{pmatrix} 1 & -1 \\ 0 & 2 \end{pmatrix}$

As LHS = RHS, the matrix equation is true for $n = 1$.

1. Basis step

Substitute $n = 1$ into both the LHS and RHS of the formula to see if the formula works for $n = 1$.

Assume that the matrix equation is true for $n = k$:

$$\begin{pmatrix} 1 & -1 \\ 0 & 2 \end{pmatrix}^k = \begin{pmatrix} 1 & 1 - 2^k \\ 0 & 2^k \end{pmatrix}$$

2. Assumption step

In this step you assume that the general statement given is true for $n = k$.

With $n = k + 1$ the matrix equation becomes

$$\begin{pmatrix} 1 & -1 \\ 0 & 2 \end{pmatrix}^{k+1} = \begin{pmatrix} 1 & -1 \\ 0 & 2 \end{pmatrix}^k \begin{pmatrix} 1 & -1 \\ 0 & 2 \end{pmatrix}$$

$$= \begin{pmatrix} 1 & 1 - 2^k \\ 0 & 2^k \end{pmatrix} \begin{pmatrix} 1 & -1 \\ 0 & 2 \end{pmatrix}$$

$$= \begin{pmatrix} 1 + 0 & -1 + 2 - 2(2^k) \\ 0 + 0 & 0 + 2(2^k) \end{pmatrix}$$

$$= \begin{pmatrix} 1 & 1 - 2^{k+1} \\ 0 & 2^{k+1} \end{pmatrix}$$

Therefore the matrix equation is true when $n = k + 1$.

3. Inductive step

Use the assumption step.

As this is a proof, you should show your working for each element in the matrix multiplication.

← Further Pure 1 Section 5.2

This is the right-hand side of the original equation with n replaced by $k + 1$.

If the matrix equation is true for $n = k$, then it is shown to be true for $n = k + 1$. As the matrix equation is true for $n = 1$, it is also true for all $n \in \mathbb{Z}^+$ by mathematical induction.

4. Conclusion step

Example 11

Prove by induction that for all positive integers n, $\begin{pmatrix}-2 & 9\\-1 & 4\end{pmatrix}^n = \begin{pmatrix}-3n+1 & 9n\\-n & 3n+1\end{pmatrix}$.

$n = 1$: LHS $= \begin{pmatrix}-2 & 9\\-1 & 4\end{pmatrix}^1 = \begin{pmatrix}-2 & 9\\-1 & 4\end{pmatrix}$

RHS $= \begin{pmatrix}-3(1)+1 & 9(1)\\-(1) & 3(1)+1\end{pmatrix} = \begin{pmatrix}-2 & 9\\-1 & 4\end{pmatrix}$

As LHS = RHS, the matrix equation is true for $n = 1$.

1. Basis step

Substitute $n = 1$ into the LHS and RHS of the formula to see if the formula works for $n = 1$.

Assume that the matrix equation is true for $n = k$:

$\begin{pmatrix}-2 & 9\\-1 & 4\end{pmatrix}^k = \begin{pmatrix}-3k+1 & 9k\\-k & 3k+1\end{pmatrix}$

2. Assumption step

In this step you assume that the general statement given is true for $n = k$.

With $n = k + 1$ the matrix equation becomes

$$\begin{pmatrix}-2 & 9\\-1 & 4\end{pmatrix}^{k+1} = \begin{pmatrix}-2 & 9\\-1 & 4\end{pmatrix}^k\begin{pmatrix}-2 & 9\\-1 & 4\end{pmatrix}$$

$$= \begin{pmatrix}-3k+1 & 9k\\-k & 3k+1\end{pmatrix}\begin{pmatrix}-2 & 9\\-1 & 4\end{pmatrix}$$

3. Inductive step

Use the assumption step.

$$= \begin{pmatrix}6k-2-9k & -27k+9+36k\\2k-3k-1 & -9k+12k+4\end{pmatrix}$$

$$= \begin{pmatrix}-3k-2 & 9k+9\\-k-1 & 3k+4\end{pmatrix}$$

$$= \begin{pmatrix}-3(k+1)+1 & 9(k+1)\\-(k+1) & 3(k+1)+1\end{pmatrix}$$

This is the right-hand side of the original equation with n replaced by $k + 1$.

Therefore the matrix equation is true when $n = k + 1$.

If the matrix equation is true for $n = k$, then it is shown to be true for $n = k + 1$. As the matrix equation is true for $n = 1$, it is also true for all $n \in \mathbb{Z}^+$ by mathematical induction.

4. Conclusion step

Exercise 8D

SKILLS REASONING/ARGUMENTATION

(E/P) **1** Prove by induction that for all positive integers n:

$\begin{pmatrix}1 & 2\\0 & 1\end{pmatrix}^n = \begin{pmatrix}1 & 2n\\0 & 1\end{pmatrix}$ **(6 marks)**

(E/P) **2** Prove by induction that for all positive integers n:

$\begin{pmatrix}3 & -4\\1 & -1\end{pmatrix}^n = \begin{pmatrix}2n+1 & -4n\\n & -2n+1\end{pmatrix}$ **(6 marks)**

(E/P) **3** Prove by induction that for all positive integers n:

$\begin{pmatrix}2 & 0\\1 & 1\end{pmatrix}^n = \begin{pmatrix}2^n & 0\\2^n-1 & 1\end{pmatrix}$ **(6 marks)**

E/P **4 a** Prove by induction that for all positive integers n:

$$\begin{pmatrix} 5 & -8 \\ 2 & -3 \end{pmatrix}^n = \begin{pmatrix} 4n+1 & -8n \\ 2n & 1-4n \end{pmatrix}$$ **(6 marks)**

b Hence find the value of n such that:

$$\begin{pmatrix} -1 & 3 \\ -2 & 5 \end{pmatrix}\begin{pmatrix} 5 & -8 \\ 2 & -3 \end{pmatrix}^n = \begin{pmatrix} 11 & -21 \\ 10 & -19 \end{pmatrix}$$ **(4 marks)**

E/P **5** The matrix $\mathbf{M} = \begin{pmatrix} 2 & 5 \\ 0 & 1 \end{pmatrix}$.

a Prove by induction that for all positive integers n:

$$\mathbf{M}^n = \begin{pmatrix} 2^n & 5(2^n - 1) \\ 0 & 1 \end{pmatrix}$$ **(6 marks)**

b Hence find an expression for $(\mathbf{M}^n)^{-1}$ in terms of n. **(4 marks)**

Chapter review 8

SKILLS REASONING/ARGUMENTATION

E/P **1** Prove by induction that $9^n - 1$ is divisible by 8 for all positive integers n. **(6 marks)**

P **2** The matrix $\mathbf{B}$ is given by $\mathbf{B} = \begin{pmatrix} 1 & 0 \\ 0 & 3 \end{pmatrix}$.

a Find $\mathbf{B}^2$ and $\mathbf{B}^3$.

b Use your answer to part **a** to suggest a general statement for $\mathbf{B}^n$, for all positive integers n.

c Prove by induction that your answer to part **b** is correct.

E/P **3** Prove by induction that for all positive integers n, $\sum_{r=1}^{n}(3r+4) = \frac{1}{2}n(3n+11)$. **(6 marks)**

E/P **4** The matrix $\mathbf{A}$ is given by $\mathbf{A} = \begin{pmatrix} 9 & 16 \\ -4 & -7 \end{pmatrix}$.

a Prove by induction that $\mathbf{A}^n = \begin{pmatrix} 8n+1 & 16n \\ -4n & 1-8n \end{pmatrix}$ for all positive integers n. **(6 marks)**

The matrix $\mathbf{B}$ is given by $\mathbf{B} = (\mathbf{A}^n)^{-1}$.

b Hence find $\mathbf{B}$ in terms of n. **(4 marks)**

E/P **5** The function f is defined by $\mathrm{f}(n) = 5^{2n-1} + 1$, where n is a positive integer.

a Show that $\mathrm{f}(n+1) - \mathrm{f}(n) = \mu(5^{2n-1})$, where μ is an integer to be determined. **(3 marks)**

b Hence prove by induction that $\mathrm{f}(n)$ is divisible by 6. **(4 marks)**

E/P **6** Prove by induction that $7^n + 4^n + 1$ is divisible by 6 for all positive integers n. **(6 marks)**

E/P **7** Prove by induction that for all positive integers n, $\sum_{r=1}^{n} r(r+4) = \frac{1}{6}n(n+1)(2n+13)$. **(6 marks)**

E/P **8 a** Prove by induction that for all positive integers n:

$$\sum_{r=1}^{2n} r^2 = \frac{1}{3}n(2n+1)(4n+1)$$ **(6 marks)**

b Given that $\sum_{r=1}^{2n} r^2 = k\sum_{r=1}^{n} r^2$, show that k must satisfy $n = \dfrac{2-k}{k-8}$. **(5 marks)**

9 Given that $u_{n+1} = 3u_n + 1$, $u_1 = 1$, prove by induction that $u_n = \frac{3^n - 1}{2}$. **(6 marks)**

E/P **10** A sequence $u_1, u_2, u_3, u_4, \ldots\ldots$ is defined by $u_{n+1} = \frac{3u_n - 1}{4}$, $u_1 = 2$.

a Find the first five terms of the sequence. **(3 marks)**

b Prove, by induction for $n \in \mathbb{Z}^+$, that $u_n = 4\left(\frac{3}{4}\right)^n - 1$. **(5 marks)**

E/P **11** The matrix $\mathbf{M} = \begin{pmatrix} 2c & 1 \\ 0 & c \end{pmatrix}$ for some positive constant c.

a Prove by induction that for all positive integers n:

$$\mathbf{M}^n = c^n \begin{pmatrix} 2^n & \frac{2^n - 1}{c} \\ 0 & 1 \end{pmatrix}$$

(7 marks)

b Given that $\det(\mathbf{M}^n) = 50^n$, find the value of c. **(5 marks)**

Challenge

SKILLS CREATIVITY

$$\mathbf{M} = \begin{pmatrix} \cos\theta & -\sin\theta \\ \sin\theta & \cos\theta \end{pmatrix}$$

a Prove by induction that for all positive integers n, $\mathbf{M}^n = \begin{pmatrix} \cos n\theta & -\sin n\theta \\ \sin n\theta & \cos n\theta \end{pmatrix}$.

b Interpret this result geometrically by describing the linear transformations represented by $\mathbf{M}$ and $\mathbf{M}^n$.

Summary of key points

1 You can use **proof by induction** to prove that a general statement is true for all positive integers.

2 Proof by mathematical induction usually consists of the following four steps:

- **Basis:** Show the general statement is true for $n = 1$.
- **Assumption:** Assume that the general statement is true for $n = k$.
- **Inductive:** Show the general statement is true for $n = k + 1$.
- **Conclusion:** State that the general statement is then true for all positive integers, n.

Review exercise 2

(E) **1** $\mathbf{A} = \begin{pmatrix} 3 & 2 & p \\ 0 & 2 & -1 \end{pmatrix}$, $\mathbf{B} = \begin{pmatrix} q & 0 \\ 3 & -1 \end{pmatrix}$, $\mathbf{C} = \begin{pmatrix} 4 \\ -3 \\ 1 \end{pmatrix}$

where p and q are integers.

Determine whether or not the following products exist. Where the product exists, evaluate the product in terms of p and q. Where the product does not exist, give a reason.

a **AB** **(1)**

b **BA** **(1)**

c **BAC** **(1)**

d **CBA** **(1)**

← Further Pure 1 Section 5.2

(E/P) **2** $\mathbf{M} = \begin{pmatrix} 0 & 3 \\ -1 & 2 \end{pmatrix}$, $\mathbf{I} = \begin{pmatrix} 1 & 0 \\ 0 & 1 \end{pmatrix}$, $\mathbf{O} = \begin{pmatrix} 0 & 0 \\ 0 & 0 \end{pmatrix}$

Find the values of the constants, a and b, such that $\mathbf{M}^2 + a\mathbf{M} + b\mathbf{I} = \mathbf{O}$. **(3)**

← Further Pure 1 Sections 5.1, 5.2

(E/P) **3** $\mathbf{A} = \begin{pmatrix} a & b \\ c & d \end{pmatrix}$

Find an expression for λ, in terms of a, b, c and d so that $\mathbf{A}^2 - (a + d)\mathbf{A} = \lambda\mathbf{I}$, where $\mathbf{I}$ is the 2×2 identity matrix. **(3)**

← Further Pure 1 Sections 5.1, 5.2

(E/P) **4** $\mathbf{A} = \begin{pmatrix} 2 & 3 \\ p & -1 \end{pmatrix}$, where p is a real constant.

Given that **A** is singular,

a find the value of p. **(2)**

Given instead that det **A** = 4,

b find the value of p. **(2)**

Using the value of p found in part **b**,

c show that $\mathbf{A}^2 - \mathbf{A} = k\mathbf{I}$, stating the value of the constant k. **(2)**

← Further Pure 1 Section 5.3

(E/P) **5** $\mathbf{A} = \begin{pmatrix} 4 & -1 \\ -6 & 2 \end{pmatrix}$, $\mathbf{B}^{-1} = \begin{pmatrix} 2 & 0 \\ 3 & p \end{pmatrix}$

a Find $\mathbf{A}^{-1}$. **(1)**

b Find $(\mathbf{AB})^{-1}$, in terms of p. **(3)**

Given also that $\mathbf{AB} = \begin{pmatrix} -1 & 2 \\ 3 & -4 \end{pmatrix}$,

c find the value of p. **(2)**

← Further Pure 1 Section 5.4

(E/P) **6** $\mathbf{A} = \begin{pmatrix} 4p & -q \\ -3p & q \end{pmatrix}$, where p and q are non-zero constants.

a Find $\mathbf{A}^{-1}$, in terms of p and q. **(3)**

Given that $\mathbf{AX} = \begin{pmatrix} 2p & 3q \\ -p & q \end{pmatrix}$,

b find **X**, in terms of p and q. **(3)**

← Further Pure 1 Section 5.4

(E) **7** The matrix **A** represents a reflection in the x-axis.

The matrix **B** represents a rotation of 135°, in the anti-clockwise direction, about (0, 0).

Given that **C** = **AB**,

a find the matrix **C** **(2)**

b show that $\mathbf{C}^2 = \mathbf{I}$. **(2)**

← Further Pure 1 Sections 6.2, 6.4

(E/P) **8** The linear transformation **T** is represented by the matrix **M**, where

$$\mathbf{M} = \begin{pmatrix} a & b \\ c & d \end{pmatrix}$$

The transformation **T** maps (1, 0) to (3, 2) and maps (2, 1) to (2, 1).

a Find the values of a, b, c and d. **(4)**

b Show that $\mathbf{M}^2 = \mathbf{I}$. **(2)**

The transformation **T** maps (p, q) to $(8, -3)$.

c Find the value of p and the value of q. **(3)**

← Further Pure 1 Sections 6.1, 6.5

(E) **9** The linear transformation **T** is defined by

$$\mathbf{T}: \begin{pmatrix} x \\ y \end{pmatrix} \mapsto \begin{pmatrix} 2y - x \\ 3y \end{pmatrix}$$

The linear transformation **T** is represented by the matrix **C**.

a Find **C**. **(1)**

The quadrilateral $OABC$ is mapped by **T** to the quadrilateral $OA'B'C'$, where the coordinates of A', B' and C' are (0, 3), (10, 15) and (10, 12) respectively.

b Show that the line $y = 2x$ is invariant under this transformation. **(3)**

c Find the coordinates of A, B and C. **(2)**

d Sketch the quadrilateral $OABC$ and verify that $OABC$ is a rectangle. **(3)**

← Further Pure 1 Sections 6.1, 6.5

(E/P) **10** A triangle T, of area 18 cm², is transformed into a triangle T' by the matrix **A**, where

$$\mathbf{A} = \begin{pmatrix} k & k-1 \\ -3 & 2k \end{pmatrix}, k \in \mathbb{R}.$$

a Find det **A** in terms of k. **(3)**

Given that the area of T' is 198 cm²,

b find the possible values of k. **(3)**

← Further Pure 1 Section 6.3

(E/P) **11** The matrix $\mathbf{M} = \begin{pmatrix} \frac{3}{\sqrt{2}} & -\frac{3}{\sqrt{2}} \\ \frac{3}{\sqrt{2}} & \frac{3}{\sqrt{2}} \end{pmatrix}$ represents a rotation followed by an enlargement.

a Find the scale factor of the enlargement. **(2)**

b Find the angle of rotation. **(3)**

A point P is mapped onto a point P' under **M**. Given that the coordinates of P' are (p, q),

c find, in terms of p and q, the coordinates of P. **(4)**

← Further Pure 1 Sections 6.4, 6.5

(E) **12** Use standard formulae to show that

$$\sum_{r=1}^{n}(2r-1)^2 = \frac{1}{3}n(4n^2-1) \quad \textbf{(4)}$$

← Further Pure 1 Sections 7.1, 7.2

(E) **13** Use standard formulae to show that

$$\sum_{r=1}^{n} r(r^2-3) = \frac{1}{4}n(n+1)(n-2)(n+3) \quad \textbf{(4)}$$

← Further Pure 1 Sections 7.1, 7.2

(E) **14 a** Use standard formulae to show that

$$\sum_{r=1}^{n} r(2r-1) = \frac{n(n+1)(4n-1)}{6} \quad \textbf{(4)}$$

b Hence, evaluate $\sum_{r=11}^{30} r(2r-1)$ **(2)**

← Further Pure 1 Sections 7.1, 7.2

(E) **15 a** Use standard formulae to show that

$$\sum_{r=1}^{n}(6r^2+4r-5) = n(2n^2+5n-2) \quad \textbf{(4)}$$

b Hence calculate the value of

$$\sum_{r=10}^{25}(6r^2+4r-5) \quad \textbf{(2)}$$

← Further Pure 1 Sections 7.1, 7.2

(E) **16 a** Use standard formulae to show that

$$\sum_{r=1}^{n} r(r+1) = \frac{1}{3}n(n+1)(n+2) \quad \textbf{(4)}$$

b Hence, or otherwise, show that

$\sum_{r=n}^{3n} r(r+1) = \frac{1}{3}n(2n+1)(pn+q)$, stating the values of the integers p and q. **(3)**

← Further Pure 1 Sections 7.1, 7.2

(E) **17** Given that

$$\sum_{r=1}^{n} r^2(r-1) = \frac{1}{12}n(n+1)(pn^2+qn+r)$$

a find the values of p, q and r. **(4)**

b Hence evaluate $\sum_{r=50}^{100} r^2(r-1)$ **(2)**

← Further Pure 1 Sections 7.1, 7.2

(E/P) **18** Use the method of mathematical induction to prove, for $n \in \mathbb{Z}^+$, that

$$\sum_{r=1}^{n} r(r+3) = \frac{1}{3}n(n+1)(n+5) \quad \textbf{(6)}$$

← Further Pure 1 Section 8.1

(E/P) **19** Prove by induction that, for all $n \in \mathbb{Z}^+$,

$$\sum_{r=1}^{n} (2r-1)^2 = \frac{1}{3}n(2n-1)(2n+1) \quad \textbf{(6)}$$

← Further Pure 1 Section 8.1

(E/P) **20** The rth term, a_r, in a series is given by

$a_r = r(r+1)(2r+1)$

Prove, by mathematical induction, that the sum of the first n terms of the series is

$\frac{1}{2}n(n+1)^2(n+2)$ **(6)**

← Further Pure 1 Section 8.1

(E/P) **21** Prove, by induction, that for all $n \in \mathbb{Z}^+$,

$$\sum_{r=1}^{n} r^2(r-1) = \frac{1}{12}n(n-1)(n+1)(3n+2)$$ **(6)**

← Further Pure 1 Section 8.1

(E/P) **22** Given that $f(n) = 3^{4n} + 2^{4n+2}$,

a show that, for $k \in \mathbb{Z}^+$, $f(k+1) - f(k)$ is divisible by 15 **(3)**

b prove that, for all $n \in \mathbb{Z}^+$, $f(n)$ is divisible by 5. **(4)**

← Further Pure 1 Section 8.2

(E/P) **23** $f(n) = 24 \times 2^{4n} + 3^{4n}$, where n is a non-negative integer.

a Write down $f(n+1) - f(n)$. **(3)**

b Prove, by induction, that $f(n)$ is divisible by 5 for $n \in \mathbb{Z}^+$. **(4)**

← Further Pure 1 Section 8.2

(E/P) **24** Prove that the expression $7^n + 4^n + 1$ is divisible by 6 for all positive integers n. **(6)**

← Further Pure 1 Section 8.2

(E/P) **25** Prove by induction that $4^n + 6n - 1$ is divisible by 9 for all $n \in \mathbb{Z}^+$. **(6)**

← Further Pure 1 Section 8.2

(E/P) **26** Prove that the expression $3^{4n-1} + 2^{4n-1} + 5$ is divisible by 10 for all positive integers n. **(6)**

← Further Pure 1 Section 8.2

(E/P) **27** $\mathbf{A} = \begin{pmatrix} 1 & c \\ 0 & 2 \end{pmatrix}$, where c is a constant.

Prove by induction that, for all positive integers n,

$$\mathbf{A}^n = \begin{pmatrix} 1 & (2^n - 1)c \\ 0 & 2^n \end{pmatrix} \quad \textbf{(6)}$$

← Further Pure 1 Section 8.3

(E/P) **28** $\mathbf{A} = \begin{pmatrix} 3 & 1 \\ -4 & -1 \end{pmatrix}$

Prove by induction that, for all positive integers n,

$$\mathbf{A}^n = \begin{pmatrix} 2n+1 & n \\ -4n & -2n+1 \end{pmatrix} \quad \textbf{(6)}$$

← Further Pure 1 Section 8.3

(E/P) **29** Manjari is attempting to prove that $2^n + 3$ is divisible by 3 for all positive integers n. She writes the following working:

> Assume true for $n = k$, so $2^k + 3$ is divisible by 3.
> Consider $n = k + 1$:
> $2^{k+1} + 3 = 2 \times 2^k + 3$
> $= 2(2^k + 3) - 3$
> By induction hypothesis $2^k + 3$ is divisible by 3, and 3 is divisible by 3, hence $2^{k+1} + 3$ is divisible by 3.
> Hence by induction $2^n + 3$ is divisible by 3 for all positive integers n.

a Explain the mistake that Manjari has made. **(2)**

b Prove that $2^{2n} - 1$ is divisible by 3 for all positive integers n. **(4)**

← Further Pure 1 Section 8.2

E/P **30** A sequence can be described by the recurrence formula

$u_{n+1} = 2u_n + 1, n \in \mathbb{Z}^+$ and $u_1 = 1$

a Find u_2 and u_3. **(2)**

b Prove by induction that $u_n = 2^n - 1$. **(5)**

← Further Pure 1 Section 8.3

E/P **31** A series of positive integers $u_1, u_2, u_3 \ldots$ is defined by $u_1 = 6$ and $u_{n+1} = 6u_n - 5$ for $n \geqslant 1$.

Prove by induction that $u_n = 5(6^{n-1}) + 1$ for $n \geqslant 1$. **(5)**

← Further Pure 1 Section 8.3

Challenge

1 The diagram below shows two different ways in which four non-parallel lines can divide the plane into regions:

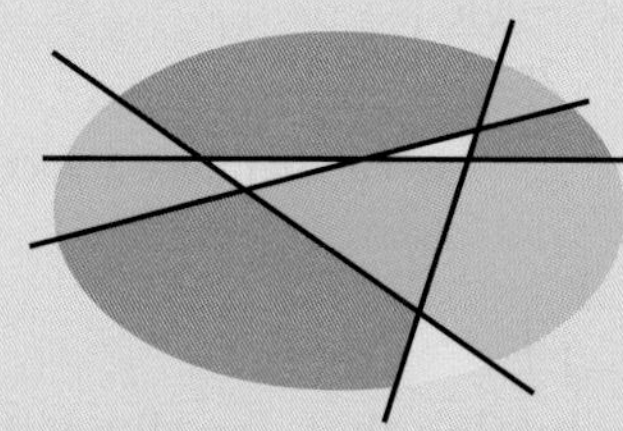

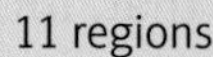

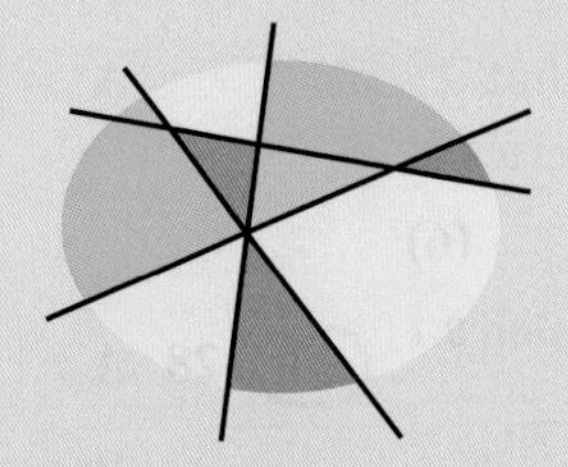

10 regions

Prove that if n non-parallel lines divide the plane into r regions, then $2n \leqslant r \leqslant \frac{1}{2}(n^2 + n + 2)$.

← Further Pure 1 Section 8.1

2 The simultaneous equations $4x + 3y = 8$ and $3x - y = -7$ can be modelled using matrices as follows: $\begin{pmatrix} 4 & 3 \\ 3 & -1 \end{pmatrix}\begin{pmatrix} x \\ y \end{pmatrix} = \begin{pmatrix} 8 \\ -7 \end{pmatrix}$.

Using these matrices, find the value of x and the value of y.

← Further Pure 1 Section 5.2

Hint Use $\mathbf{AA}^{-1} = \begin{pmatrix} 1 & 0 \\ 0 & 1 \end{pmatrix}$

Exam practice

Mathematics International Advanced Subsidiary/ Advanced Level Further Pure Mathematics 1

Time: 1 hour 30 minutes
You must have: Mathematical Formulae and Statistical Tables, Calculator
Answer ALL questions

1 $\mathbf{P} = \begin{pmatrix} 1 & 1 \\ 2 & 4 \end{pmatrix}$

a Show that **P** is non-singular. **(2)**

b Find matrix **Q** such that $\mathbf{PQ} = \begin{pmatrix} 2 \\ 14 \end{pmatrix}$. **(3)**

2 Given that $z_1 = 2 - \text{i}$,

a find, in degrees to 1 decimal place, $\arg(z_1)$. **(2)**

Given also that $z_2 = 3 + 2\text{i}$, find in the form $a + b\text{i}$, where $a, b \in \mathbb{R}$:

b $z_1 z_2$ **(2)**

c $\frac{z_1}{z_2}$ **(2)**

3 The rectangular hyperbola H has parametric equations $x = 5t$, $y = \frac{5}{t}$, $t \neq 0$.

a Write down the Cartesian equation of H in the form $xy = c^2$, where c is an integer. **(2)**

Points A and B lie on H and have parameters $t = 1$ and $t = 5$ respectively.

b Find the coordinates of the midpoint of AB. **(2)**

4 Parabola C has equation $y^2 = 16x$.

a Find the equation of the normal to C at the point P with coordinates $(1, 4)$. **(6)**

The normal at P meets the directrix of the parabola at the point Q.

b Find the coordinates of Q. **(3)**

c Find the coordinates of the point R on C which is the same distance from the point Q and from the focus of C. **(2)**

5 $\mathbf{A} = \begin{pmatrix} \frac{\sqrt{3}}{2} & \frac{-1}{2} \\ \frac{1}{2} & \frac{\sqrt{3}}{2} \end{pmatrix}$ $\mathbf{B} = \begin{pmatrix} 0 & -1 \\ -1 & 0 \end{pmatrix}$

a Describe fully the geometric transformation represented by:

i **A**

ii **B** **(4)**

The matrix $\mathbf{M} = \begin{pmatrix} 1 & 1 & 2 \\ 1 & 3 & 2 \end{pmatrix}$ is transformed by the transformation represented by **A**, followed by the transformation represented by **B** to give the matrix **N**.

b Find the matrix **N**, giving your values correct to 2 decimal places. **(4)**

6 $f(x) = x^3 - 4x - 2$

a Show that the equation $f(x) = 0$ has a root α in the interval $[-1, 0]$. **(1)**

b Use linear interpolation to find α correct to 1 decimal place. **(3)**

The equation $f(x) = 0$ has another root β in the interval $[2, 3]$.

c Taking $x_0 = 2$, use the Newton-Raphson method to find an approximation of β. Give your answer correct to 3 s.f. **(4)**

d Verify that the answer you obtained for β is an accurate estimate to 3 s.f. **(2)**

7 $f(x) = 2x^2 - 5x - 4$

The equation $f(x) = 0$ has roots α and β. Without solving the equation,

a find the value of $\alpha^2 + \beta^2$ **(2)**

b show that $\alpha^3 + \beta^3 = \frac{245}{8}$ **(3)**

c form a quadratic equation with integer coefficients which has roots:

$\left(\alpha + \frac{1}{\alpha^2}\right)$ and $\left(\beta + \frac{1}{\beta^2}\right)$ **(6)**

8 $f(z) = z^4 + az^3 + bz^2 + cz + d$, where a, b, c and d are real numbers.

Given that $2 - \mathrm{i}$ and $2\mathrm{i}$ are roots of the equation $f(z) = 0$,

a write down the other roots of $f(z) = 0$. **(2)**

b Find the value of a, the value of b, the value of c and the value of d. **(7)**

9 Use the standard results for $\sum_{r=1}^{n} r^3$ and $\sum_{r=1}^{n} r$ to show that:

$$\sum_{r=1}^{n} (r^3 + 6r - 3) = \tfrac{1}{4}n^2(n^2 + 2n + 13)$$ **(5)**

10 $f(n) = 2^n + 6^n$

a Show that $f(k + 1) = 6f(k) - 4(2^k)$. **(2)**

b Hence, or otherwise, prove by induction that for $n \in \mathbb{Z}^+$, $f(n)$ is divisible by 8. **(4)**

TOTAL FOR PAPER: 75 MARKS

GLOSSARY

acute (**angle**) an angle less than 90°

algebraic representing mathematical information symbolically, using letters and numbers

Argand diagram a diagram using Cartesian axes on which complex numbers are represented geometrically

argument (of a **complex number**) gives the angle between the positive real axis and the line joining the point to the origin

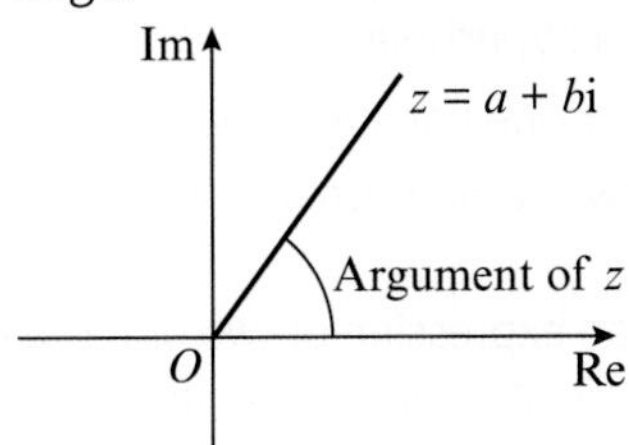

approximation a number that is not exact

asymptote a line that a curve approaches but never quite reaches

axis (plural **axes**) either of the two lines by which the positions of points are measured in a graph

binomial an algebraic expression of the **sum** or difference of two terms. For example, $(a + b)^n$ is the general form of a binomial expression

Cartesian coordinates A unique point in a plane specified by a pair of numerical coordinates

coefficient a number placed before and multiplying the variable in an **algebraic expression**.

For example, 4 is the coefficient of $4x^3$

complex conjugate each of a pair of **complex** numbers having their **real** parts identical and their **imaginary** parts of equal magnitude but opposite sign.

If $z = a + b\mathrm{i}$, then $z^* = a - b\mathrm{i}$, where $a, b \in \mathbb{R}$

complex number a number that can be expressed in the form $a + b\mathrm{i}$, where a and b are **real** numbers, and i is a solution of the equation $x^2 = -1$, which is called an **imaginary number** because there is no **real number** that satisfies this equation

constant a term that does not include a variable. In the expression $x^2 + 3x - 6$, the constant term is -6

converge (of a series) approaching a limiting value as the number of terms increases.

For example, the terms in the sequence $1, \frac{1}{2}, \frac{1}{4}, \frac{1}{8}, \frac{1}{16}, \frac{1}{32}, \ldots, \left(\frac{1}{2}\right)^{n-1}$ converge toward a value of zero as n tends toward infinity

coordinates a set of values that show an exact position. In a two-dimensional grid, the first number represents a point on the x-axis and the second number represents a point on the y-axis

corresponding an equivalent; connected with what you have just mentioned

cubic algebraic expressions in which the highest power is 3

deduce to conclude from a known or assumed fact

derivative a way to represent the rate of change, in other words, $\frac{\mathrm{d}y}{\mathrm{d}x}$ is the first derivative, and $\frac{\mathrm{d}^2y}{\mathrm{d}x^2}$ is the second derivative

determinant the value calculated from the elements of a matrix.

For the matrix $\begin{pmatrix} a & b \\ c & d \end{pmatrix}$ the determinant is $ad - bc$

differentiate the process of finding the instantaneous rate of change of a function with respect to one of its variables

differentiation the instantaneous rate of change of a function with respect to one of its variables

directrix a fixed line used to define a **parabola**

discriminant an expression that allows you to determine if a quadratic equation has one or two real **roots**, or two complex roots

enlargement a transformation of a shape that involves increasing or decreasing the length of each side by a scale factor

equation a statement where values of two mathematical **expressions** are equal. Solving an equation consists of determining the value(s) of the variable

expression any group of **algebraic** terms. For example, $2x + 6y + 3z$ is an algebraic expression

factorise to rewrite an expression using brackets. We factorise $x^2 + 3x + 2$ to get $(x + 1)(x + 2)$

focus lies on the **axis** of **symmetry** of a **parabola**

function the relationship between a set of inputs and a set of outputs, where each input is related to exactly one output

gradient the slope of a line

hyperbola (rectangular) a symmetrical curve which has **parametric** equations $x = ct$, $y = \frac{c}{t}$, $t \in \mathbb{R}$, or in **Cartesian** form $xy = c^2$ where c is a positive constant

imaginary number a number that is expressed in terms of the square root of a negative number a, where $a \in \mathbb{R}$, and $i = \sqrt{-1}$, so $i^2 = -1$

induction see **proof by mathematical induction**

integer a whole number. The symbol for integers is $\mathbb{Z}$

interpretation understanding of a concept

intersect to meet or cross at a certain point

intersection the point at which two or more lines or curves cross (**intersect**)

interval the range of a set of numbers. For example, 3, 4, 5, 6, 7 are the members of the set of numbers satisfied by the interval $2 < x < 8$, where x is an integer

invariant (**point** or **line**) a fixed point or line that does not move under a **transformation**

inverse operations that reverse each other. For example, the inverse of $y = x^2$ is $x = \sqrt{y}$

inverse operation an opposite operation. For example, **differentiation** is the **inverse** of integration and vice versa

linear transformation a transformation moving all points (x, y) in a plane according to some rule

locus (of a set of points) the set of points that satisfies given conditions or a rule

matrix (plural matrices) a rectangular array of numbers or other mathematical expressions for which operations such as addition and multiplication are defined.
For example, $\begin{pmatrix} a & b & c \\ d & e & f \end{pmatrix}$ is a 2×3 matrix with 2 rows and 3 columns

midpoint (of a **line segment**) a point on a line that divides it into two equal parts

modulus (of a complex number) the distance of a point from the origin. For any **complex number** $z = a + bi$, $|z| = \sqrt{a^2 + b^2}$

Natural numbers positive **integers** 1, 2, 3, etc. The symbol for natural numbers is $\mathbb{N}$

normal a line that is perpendicular to another line. For example, the normal to a curve at a given point is perpendicular to the tangent to the curve at the same given point

non-singular (**matrix**) a matrix is non-singular if a **determinant** can be found. That is, the determinant $\neq 0$

numerical relating to numbers

origin the point where the y-axis and x-axis intersect on a flat coordinate plane

parabola the locus of points such that the distance to the **focus** equals the distance to the **directrix**. The parametric equations of a parabola are $x = at^2$, $y = 2at$, $t \in \mathbb{R}$, or in **Cartesian** form: $y^2 = 4ax$ where a is a positive constant

parallel two lines side-by-side, the same distance apart at every point

parametric equation a set of equations that express a set of quantities as functions of a number of independent variables, known as **parameters**.
For example, $x = ct$, $y = \frac{c}{t}$, $t \in \mathbb{R}$ are the parametric equations of a hyperbola

perpendicular one line meeting another at 90°

plane a flat two-dimensional surface extending into **infinity**

polynomial an expression of two or more **algebraic** terms with positive whole number indices. For example, $2x + 6x^2 + 7x^6$ is a polynomial

product $2 \times 3 = 6$, so 6 is the product of 2 and 3

proof by mathematical induction a special form of deductive reasoning (i.e. using the information you have to form an opinion). It is used to prove a fact about all the elements in an **infinite** set by performing a **finite** number of steps

quadrant the area of a two-dimensional **axes** is divided into four quadrants

y
Second quadrant | First quadrant
O x
Third quadrant | Fourth quadrant

quadratic expressions such as $x^2 + 3x$ are quadratic, where the highest power of any variable is 2

quartic expression algebraic expression in which the highest power of any variable is 4

radian describes an angle subtended by a circular arc as the length of the arc divided by the radius of the arc. One **radian** is the angle subtended at the centre of a circle by an arc that is equal in length to the radius of the circle

rational number a number that can be expressed as an integer or fraction

real number a value that can be represented along a number line and includes all **rational** and **irrational** numbers. The symbol for real numbers is $\mathbb{R}$

reciprocal of a number x is $\frac{1}{x}$. Every number has a reciprocal apart from 0, as $\frac{1}{0}$ is not defined

recurrence relationship or **formula** a formula for a sequence that defines the next term in the sequence using the previous term.

For example, for the sequence 5, 7, 9, 11, … the rule can be described as 'add 2 to the previous term' or given that $u_{k+1} = u_k + 2, u_1 = 5$

roots the set of all possible solutions. A quadratic equation has up to 2 roots

scalar a **real** number that is not a **vector** or a **matrix**

scale factor a number which multiplies some quantity. The ratio of any two corresponding lengths in two **similar** geometric figures is called a **scale factor**

series the sum of terms in a sequence

sequence a series of numbers following a set rule. 4, 9, 14, 19, … is an example of an **arithmetic** sequence

sigma notation (Σ) the symbol used to express a sum of the values in a series

singular matrix a matrix in which the **determinant** = 0, and therefore the inverse of the **matrix** does not exist.

For example, $\mathbf{A} = \begin{pmatrix} 4 & 2 \\ 2 & 1 \end{pmatrix}$

$\det\mathbf{A} = 4 \times 1 - 2 \times 2 = 0$ and $\frac{1}{0}$ is not defined so $\mathbf{A}$ is singular

stationary point the point on the graph of a function where the gradient is zero

stretch a transformation that stretches or shrinks a two-dimensional shape in the x or y direction independently

successive coming or following on after the other

sum the addition of two or more numbers.
For example, 2 + 3 = 5, so 5 is the sum of 2 and 3.

summation formula (sigma notation Σ) a useful and concise way to define a series.
For example,

$$\sum_{r=1}^{5} 2r = 2 \times 1 + 2 \times 2 + 2 \times 3 + 2 \times 4 + 2 \times 5 = 30$$

surd a number that cannot be simplified to remove a square root. For example, $\sqrt{2}$ is a surd because it is an irrational number, whereas $\sqrt{4} = \pm 2$ which is a rational number and not a surd

symmetrical when a shape looks the same following a transformation such as reflection or rotation

tangent

- **i** a trigonometric function that is equal to the ratio of the side opposite to an acute angle (in a right-angled triangle) to the adjacent side
- **ii** a line that touches a curve at a point without crossing over and matches the gradient of the curve at that point

transformation a linear mapping that is either a reflection, rotation or stretch

vector an object that has both a magnitude and a direction. Geometrically, a vector is a directed line segment, whose length is the magnitude of the vector and with an arrow indicating the direction

vertex (plural vertices) where two lines meet at an angle, especially in a polygon

ANSWERS

CHAPTER 1

Prior knowledge check

1 a $5\sqrt{2}$ b $6\sqrt{3}$ c $6\sqrt{5}$
2 a 0 b 2 c 1
3 a 13 cm b 67.4°
4 $4 \pm \sqrt{10}$
5 $\frac{28}{13} + \frac{7\sqrt{3}}{13}$

Exercise 1A

1 a 3i b 7i c 11i d 100i
e 15i f $\text{i}\sqrt{5}$ g $2\text{i}\sqrt{3}$ h $3\text{i}\sqrt{5}$
i $10\text{i}\sqrt{2}$ j $7\text{i}\sqrt{3}$
2 a $13 + 11\text{i}$ b $5 + 2\text{i}$ c $4 + \text{i}$ d $3 + 2\text{i}$
e $9 + 9\text{i}$ f $7 - 4\text{i}$ g $4 + 2\text{i}$ h $2\sqrt{2} + 2\text{i}$
i $11 - 5\text{i}$ j 0
3 a $14 + 4\text{i}$ b $24 - 12\text{i}$ c $12 + 5\text{i}$ d $24 + 7\text{i}$
e $3 - 2\text{i}$ f $3 + 5\text{i}$ g $3 + \frac{11}{3}\text{i}$ h $-\frac{11}{2} + \frac{7}{4}\text{i}$
4 a $2\sqrt{2} - \text{i}\sqrt{2}$ b $(-1 + \sqrt{3}) + (3 - 3\sqrt{3})\text{i}$
5 a -12i b 14
6 $a = -10, b = 11$
7 a $-3 + 4\text{i}$ b $28 - 12\text{i}$ c $43 - 13\text{i}$
8 a $z + w = (a + b\text{i}) + (a - b\text{i}) = 2a$
b $z - w = (a + b\text{i}) - (a - b\text{i}) = 2b\text{i}$

Exercise 1B

1 a $z = \pm 11\text{i}$ b $z = \pm 2\text{i}\sqrt{10}$ c $z = \pm 2\text{i}\sqrt{15}$
d $z = \pm 2\text{i}\sqrt{7}$ e $z = \pm 2\text{i}\sqrt{6}$ f $z = \pm\frac{1}{2}\text{i}$
2 a $z = 3 \pm \text{i}\sqrt{7}$ b $z = 7 \pm 2\text{i}\sqrt{3}$ c $z = -1 \pm \frac{3}{4}\text{i}$
3 a $z = -1 \pm 2\text{i}$ b $z = 1 \pm 3\text{i}$ c $z = -2 \pm 5\text{i}$
d $z = -5 \pm \text{i}$ e $z = -\frac{5}{2} \pm \frac{5\text{i}\sqrt{3}}{2}$ f $z = \frac{-3 \pm \text{i}\sqrt{11}}{2}$
4 a $z = -\frac{5}{4} \pm \frac{\sqrt{7}}{4}\text{i}$ b $z = \frac{3}{14} \pm \frac{5\sqrt{3}}{14}\text{i}$ c $z = \frac{1}{10} \pm \frac{\sqrt{59}}{10}\text{i}$
5 $z_1 = 4 + \sqrt{5}\text{i}$ and $z_2 = 4 - \sqrt{5}\text{i}$
6 $-\sqrt{44} < b < \sqrt{44}$ or $-2\sqrt{11} < b < 2\sqrt{11}$

Exercise 1C

1 a $11 + 23\text{i}$ b $36 + 33\text{i}$ c $15 + 23\text{i}$ d $2 - 110\text{i}$
e $-5 - 25\text{i}$ f $39 + 80\text{i}$ g $-77 - 36\text{i}$ h 10i
i $54 - 62\text{i}$ j $-46 + 9\text{i}$
2 a 41 b 53 c They are both real
d $(a + b\text{i})(a - b\text{i}) = a^2 + b^2$, which is real.
3 $a = 7, b = -6$ or $a = 18, b = -\frac{7}{3}$
4 a -1 b 81 c 2i d -60i
5 $-8\text{i}, a = 0, b = -8$
6 $-119 - 120\text{i}$, so real part is -119
7 a -2i b $-49 - 66\text{i}$
8 Substitute $z = 1 - 4\text{i}$ into $\text{f}(z)$ to get $\text{f}(z) = 0$.
9 a $\text{i}^3 = -\text{i}, \text{i}^4 = 1$ b $\text{i}^5 = \text{i}, \text{i}^6 = -1, \text{i}^7 = -\text{i}, \text{i}^8 = 1$
c i 1 ii i iii i

Challenge

a $(a + b\text{i})^2 = a^2 - b^2 + 2ab\text{i}$
b $a^2 - b^2 = 40, 2ab = -42 \Rightarrow a = \frac{-21}{b}$
$b^4 + 40b^2 - 441 = 0 \Rightarrow (b^2 - 9)(b^2 + 49) = 0$
$b = -3\ (b < 0) \Rightarrow a = 7, b^2 \neq -49 \Rightarrow 7 - 3\text{i}$

Exercise 1D

1 a $8 - 2\text{i}$ b $6 + 5\text{i}$ c $\frac{2}{3} + \frac{1}{2}\text{i}$ d $\sqrt{5} - \text{i}\sqrt{10}$
2 a $z + z^* = 12, zz^* = 45$ b $z + z^* = 20, zz^* = 125$
c $z + z^* = \frac{3}{2}, zz^* = \frac{5}{8}$ d $z + z^* = 2\sqrt{5}, zz^* = 50$
3 a $-\frac{6}{5} - \frac{7}{5}\text{i}$ b $-\frac{11}{50} + \frac{27}{50}\text{i}$ c $\frac{31}{2} + \frac{25}{2}\text{i}$ d $\frac{6}{17} - \frac{7}{17}\text{i}$
4 $-\frac{31}{2} - \frac{17}{2}\text{i}$
5 a $\frac{3}{5} + \frac{4}{5}\text{i}$ b $\frac{7}{2} + \frac{1}{2}\text{i}$ c $\frac{41}{5} - \frac{3}{5}\text{i}$
6 $\frac{8}{5} + \frac{9}{5}\text{i}$
7 $6 + 8\text{i}$
8 $\frac{4}{8 - \text{i}\sqrt{2}} \times \frac{8 + \text{i}\sqrt{2}}{8 + \text{i}\sqrt{2}} = \frac{32 + 4\text{i}\sqrt{2}}{66} = \frac{16}{33} + \frac{2\sqrt{2}}{33}\text{i}$
9 $\frac{1}{1 - 9\text{i}} \times \frac{1 + 9\text{i}}{1 + 9\text{i}} = \frac{1}{82} + \frac{9}{82}\text{i}$
10 $\frac{z + 4}{z - 3} = \frac{(8 - \text{i}\sqrt{2})(1 + \text{i}\sqrt{2})}{(1 - \text{i}\sqrt{2})(1 + \text{i}\sqrt{2})} = \frac{10}{3} + \frac{7\sqrt{2}}{3}\text{i}$
11 $z - 2\text{i} = \frac{(6 - 4\text{i})(4 - 2\text{i})}{(4 + 2\text{i})(4 - 2\text{i})} = \frac{16 - 28\text{i}}{20}$
$= \frac{4}{5} - \frac{7}{5}\text{i} \Rightarrow z = \frac{4}{5} + \frac{3}{5}\text{i}$
12 $\frac{(p - 7\text{i})(2 - 5\text{i})}{(2 + 5\text{i})(2 - 5\text{i})} = \frac{2p - 35}{29} + \frac{-5p - 14}{29}\text{i}$
13 $\frac{z}{z^*} = \frac{\sqrt{5} + 4\text{i}}{\sqrt{5} - 4\text{i}} \times \frac{\sqrt{5} + 4\text{i}}{\sqrt{5} + 4\text{i}} = -\frac{11}{21} + \frac{8\sqrt{5}}{21}\text{i}$
14 a $\frac{p + 5\text{i}}{p - 2\text{i}} \times \frac{p + 2\text{i}}{p + 2\text{i}} = \frac{p^2 - 10 + 7p\text{i}}{p^2 + 4}$
$2p^2 - 20 = p^2 + 4$
$p^2 = 24 \Rightarrow p = 2\sqrt{6}$
b $\frac{1}{2} + \frac{\sqrt{6}}{2}\text{i}$

Exercise 1E

1
Im
d (−2, 5)
e (0, 3)
g $\left(\frac{-1}{2}, \frac{5}{2}\right)$
f ($\sqrt{2}$, 2)
a (7, 2)
h (−4, 0)
O
Re
c (−6, −1)
b (5, −4)

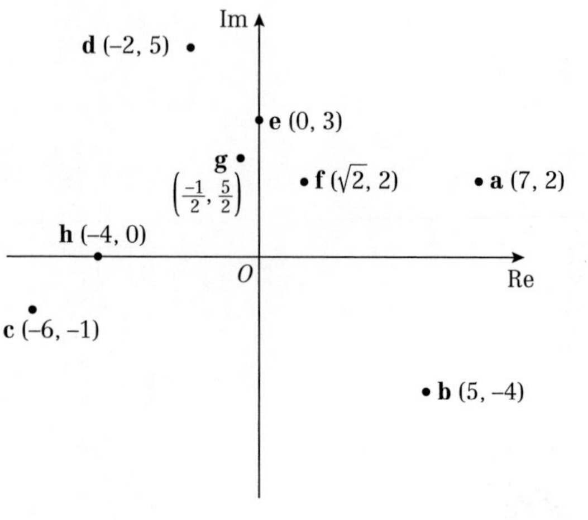

2

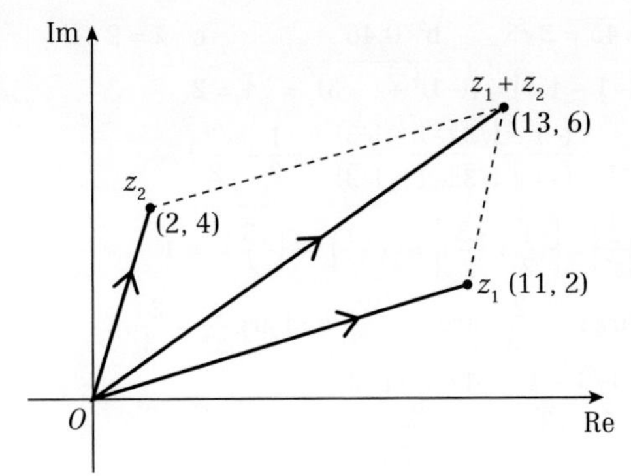

3

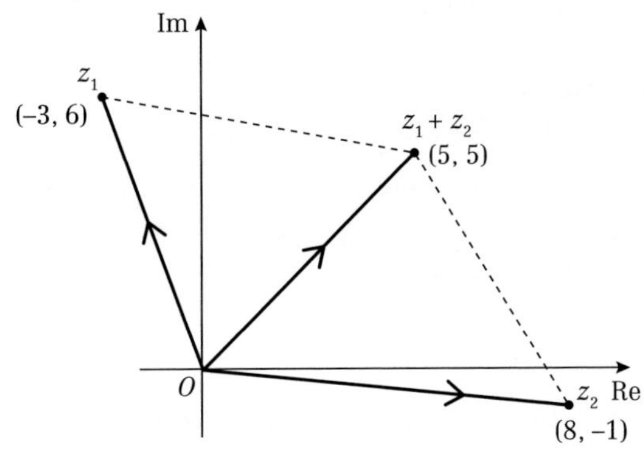

4

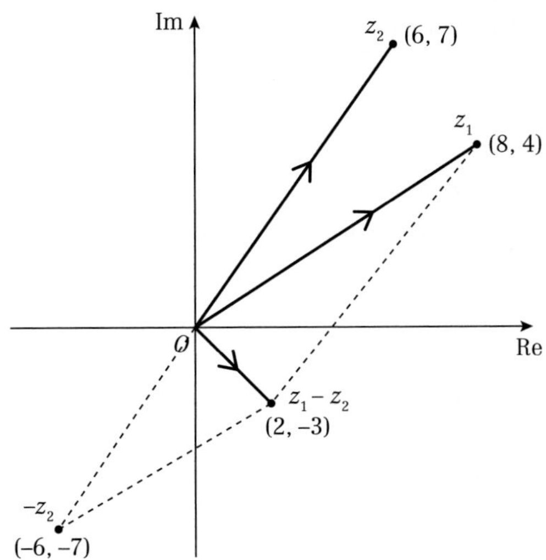

5

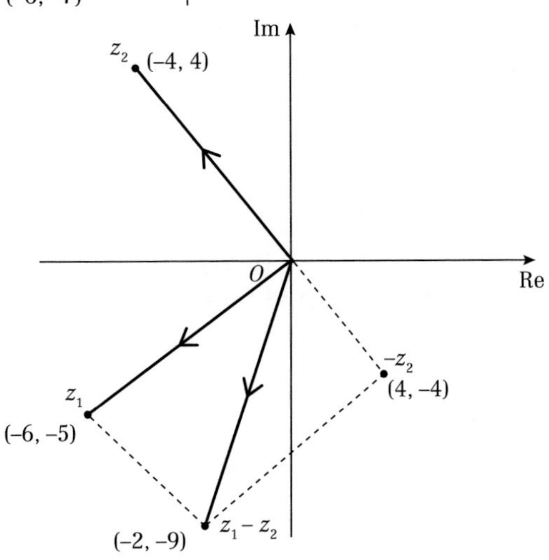

6 **a** $a = -10, b = 7$

b

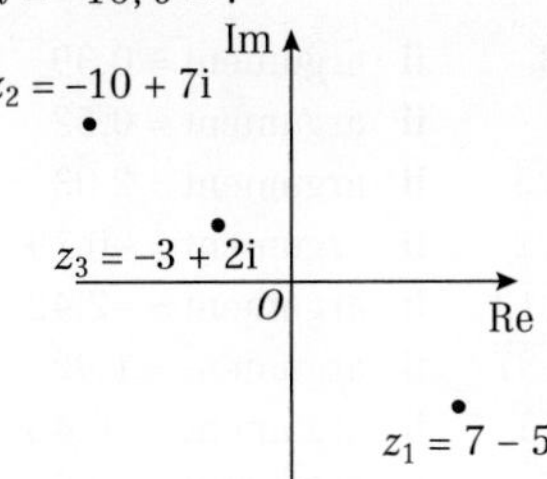

7 **a** $p = -17, q = 10$

b

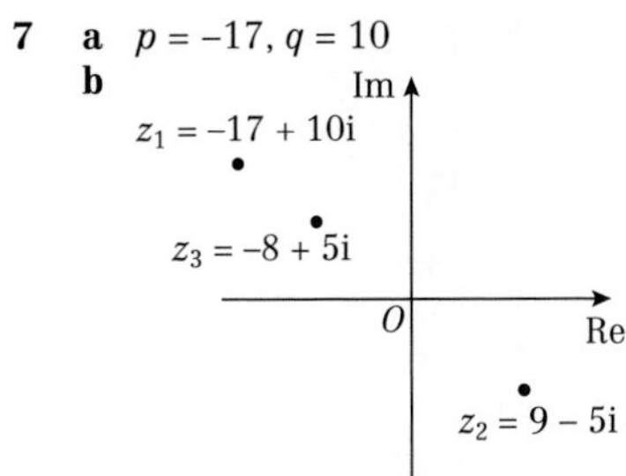

8 **a** $z_1 = 3 + i$ and $z_2 = 3 - i$. Other way round acceptable.

b

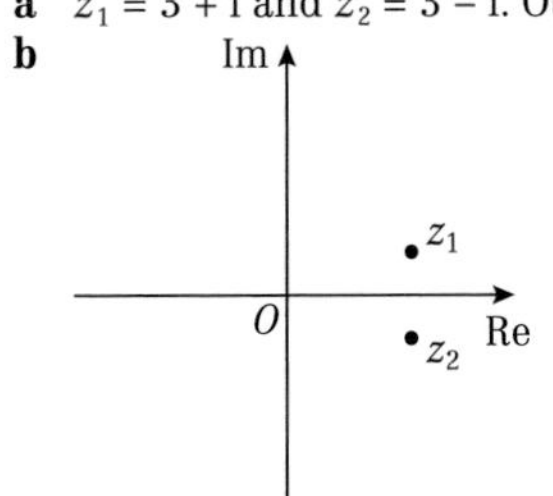

9 **a** $2\left(\frac{3}{2}\right)^3 - 19\left(\frac{3}{2}\right)^2 + 64\left(\frac{3}{2}\right) - 60 = 0$

b $\frac{3}{2}$, $4 + 2i$, $4 - 2i$.

c

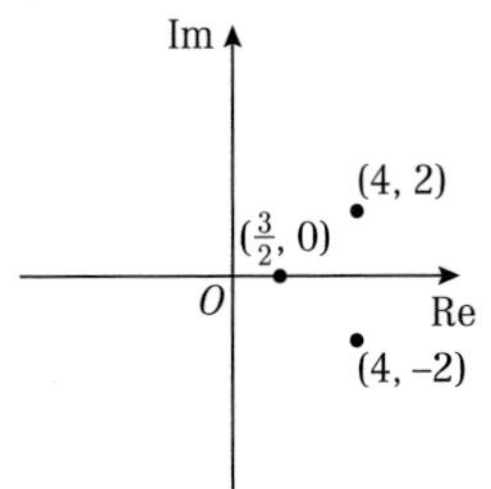

Challenge

a $1, -1, -\frac{1}{2} + \frac{\sqrt{3}}{2}i, -\frac{1}{2} - \frac{\sqrt{3}}{2}i, \frac{1}{2} + \frac{\sqrt{3}}{2}i$ and $\frac{1}{2} - \frac{\sqrt{3}}{2}i$

b

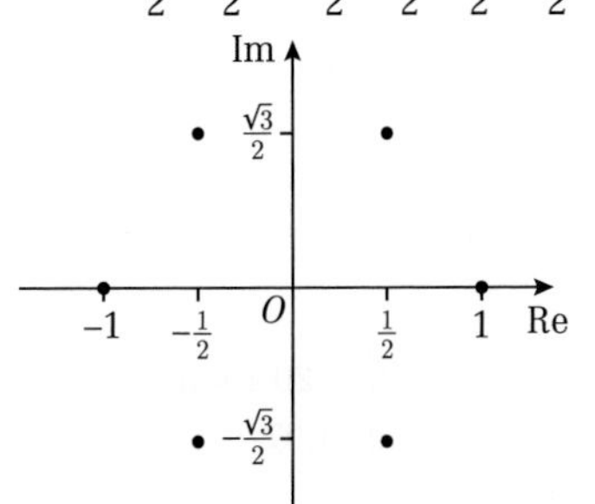

c (0, 1) and (0, −1) are on the unit circle.

$\left(-\frac{1}{2}\right)^2 + \left(\frac{\sqrt{3}}{2}\right)^2 = 1$, so other 4 points also lie on the unit circle.

Exercise 1F

1 a i Modulus = 13 ii argument = 0.39
b i Modulus = 2 ii argument = 0.52
c i Modulus = $3\sqrt{5}$ ii argument = 2.03
d i Modulus = $2\sqrt{2}$ ii argument = −0.79
e i Modulus = $\sqrt{113}$ ii argument = −2.42
f i Modulus = $\sqrt{137}$ ii argument = 1.92
g i Modulus = $\sqrt{15}$ ii argument = −0.46
h i Modulus = 17 ii argument = −2.06

2 a i $\sqrt{8} = 2\sqrt{2}$ ii $\frac{\pi}{4}$
b i $\sqrt{50} = 5\sqrt{2}$ ii $\frac{\pi}{4}$
c i $\sqrt{72} = 6\sqrt{2}$ ii $\frac{3\pi}{4}$
d i $\sqrt{2a^2} = a\sqrt{2}$ ii $-\frac{3\pi}{4}$

3 a $\sqrt{34}$ b $-16 + 30\text{i}$
c $(\sqrt{34})^2 = 34 \quad |(z_1)^2| = 34$

4 a $6 - 4\text{i}$ b $2\sqrt{13}$
c 13 d ±12

5 a

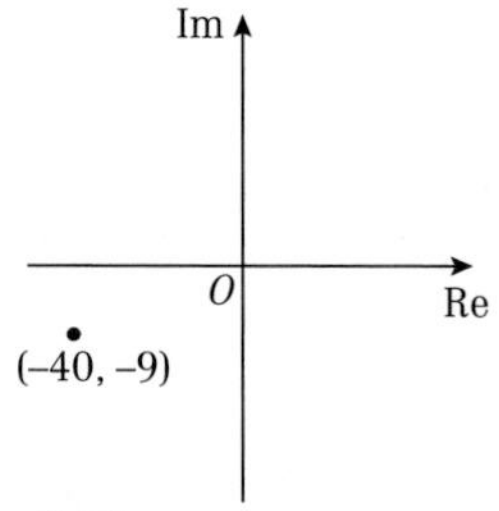

Im
O Re
(−40, −9)

b −2.92

6 a $z^2 = (3 + 4\text{i})(3 + 4\text{i}) = 9 + 16\text{i}^2 + 24\text{i} = -7 + 24\text{i}$
b 25 c 1.85
d

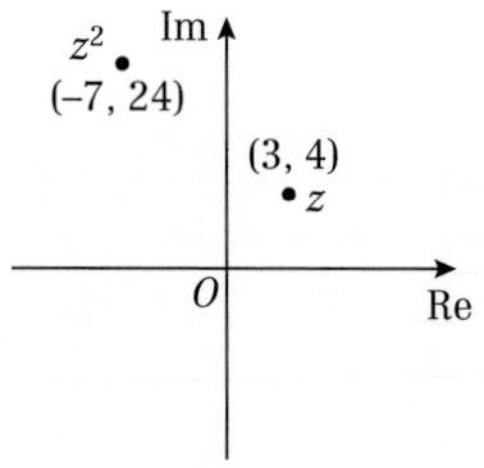

z^2 Im
(−7, 24)
(3, 4) z
O Re

7 a $\frac{z_1}{z_2} = \frac{(4 + 6\text{i})(1 - \text{i})}{(1 + \text{i})(1 - \text{i})} = \frac{10 + 2\text{i}}{2} = 5 + \text{i}$
b $\sqrt{26}$ c 0.20

8 a $\left(\frac{3 - 2p}{2}\right) + \left(\frac{3 + 2p}{2}\right)\text{i}$ b $p = 1$
c $\frac{\sqrt{26}}{2}$
d Argand diagram showing $z_1 = 3 + \text{i}$, $z_2 = \frac{1}{2} + \frac{5}{2}\text{i}$ and $\frac{z_1}{z_2} = 1 - \text{i}$

9 a $4 + 6\text{i}$ b $-20 + 48\text{i}$
c $\sqrt{52} = 2\sqrt{13}$ d 1.97

10 a $|6 + 6\text{i}| = \sqrt{72} = 6\sqrt{2}$
b $\frac{(4 + 2\text{i})(a + b\text{i})(2 - 4\text{i})}{(2 + 4\text{i})(2 - 4\text{i})} = \frac{(16 - 12\text{i})(a + b\text{i})}{20}$
$= \left(\frac{4a + 3b}{5}\right) + \left(\frac{4b - 3a}{5}\right)\text{i}$
c $a = 6, b = -1$ d −0.81

11 a $\sqrt{45} = 3\sqrt{5}$ b 0.46 c $\lambda = 2$

12 a $|-1 - \text{i}\sqrt{3}| = \sqrt{(-1)^2 + (-\sqrt{3})^2} = \sqrt{4} = 2$
b $\frac{z}{z^*} = \frac{(-1 - \text{i}\sqrt{3})(-1 - \text{i}\sqrt{3})}{(-1 + \text{i}\sqrt{3})(-1 - \text{i}\sqrt{3})} = -\frac{1}{2} + \frac{\sqrt{3}}{2}\text{i}$
$\left|\frac{z}{z^*}\right| = \left|-\frac{1}{2} + \frac{\sqrt{3}}{2}\text{i}\right| = \sqrt{\left(-\frac{1}{2}\right)^2 + \left(\frac{\sqrt{3}}{2}\right)^2} = 1$
c $\arg z = -\frac{2\pi}{3}$, $\arg z^* = \frac{2\pi}{3}$ and $\arg\frac{z}{z^*} = \frac{2\pi}{3}$,

13 $k = \frac{4\sqrt{3} - 1}{5 + \sqrt{3}} = \frac{-17 + 21\sqrt{3}}{22}$

14 Using sine rule, $|z| = \frac{5\sin\left(\frac{\pi}{10}\right)}{\sin\left(\frac{\pi}{5}\right)} \approx 2.63$

Exercise 1G

1 a $2\sqrt{2}\left(\cos\frac{\pi}{4} + \text{i}\sin\frac{\pi}{4}\right)$ b $3\left(\cos\frac{\pi}{2} + \text{i}\sin\frac{\pi}{2}\right)$
c $5(\cos 2.21 + \text{i}\sin 2.21)$ d $2\left(\cos\left(-\frac{\pi}{3}\right) + \text{i}\sin\left(-\frac{\pi}{3}\right)\right)$
e $\sqrt{29}(\cos(-1.95) + \text{i}\sin(-1.95))$
f $20(\cos\pi + \text{i}\sin\pi)$
g $25(\cos(-1.29) + \text{i}\sin(-1.29))$
h $5\sqrt{2}\left(\cos\frac{3\pi}{4} + \text{i}\sin\frac{3\pi}{4}\right)$

2 a $\frac{3}{2}\left(\cos\left(-\frac{\pi}{3}\right) + \text{i}\sin\left(-\frac{\pi}{3}\right)\right)$ b $\frac{\sqrt{5}}{5}(\cos 0.46 + \text{i}\sin 0.46)$
c $1\left(\cos\frac{\pi}{2} + \text{i}\sin\frac{\pi}{2}\right)$

3 a 5i b $\frac{\sqrt{3}}{4} + \frac{1}{4}\text{i}$ c $-3\sqrt{3} + 3\text{i}$
d $-\frac{3}{2} - \frac{3\sqrt{3}}{2}\text{i}$ e $2 - 2\text{i}$ f $2\sqrt{3} + 2\text{i}$

4 a $-2 + 2\text{i}\sqrt{3}$
b $-2 + 2\text{i}\sqrt{3}$ shown on an Argand diagram.

5 $p = \frac{7\sqrt{3}}{2}, q = -\frac{7}{2}$

6 $a = -\frac{5}{2}, b = \frac{5\sqrt{3}}{2}$

Exercise 1H

1 a $-1 + 5\text{i}, -1 - 5\text{i}$ b −2 c 26
2 a $4 + 3\text{i}, 4 - 3\text{i}$ b 8 c 25
3 a $2 - 3\text{i}$ b $z^2 - 4z + 13 = 0$
4 a $5 + \text{i}$
b $(z - (5 - \text{i}))(z - (5 + \text{i})) = 0$
$z^2 - (5 + \text{i})z - (5 - \text{i})z + (5 - \text{i})(5 + \text{i}) = 0$
$z^2 - 10z + 26 = 0 \Rightarrow p = -10, q = 26$
5 $z^2 + 10z + 41 = 0$
6 $z^2 - 2z + 5 = 0$
7 $z^2 - 6z + 34 = 0$
8 a $z = \frac{3}{2} + \frac{1}{2}\text{i}$
b $\left(z - \left(\frac{3}{2} + \frac{1}{2}\text{i}\right)\right)\left(z - \left(\frac{3}{2} - \frac{1}{2}\text{i}\right)\right)$
$= z^2 - z\left(\frac{3}{2} - \frac{1}{2}\text{i}\right) - z\left(\frac{3}{2} + \frac{1}{2}\text{i}\right) + \left(\frac{3}{2} - \frac{1}{2}\text{i}\right)\left(\frac{3}{2} + \frac{1}{2}\text{i}\right)$
$= z^2 - 3z + \frac{5}{2}$ so $p = -3, q = \frac{5}{2}$
9 $(z - (5 + q\text{i}))(z - (5 - q\text{i})) = z^2 - 10z + 25 + q^2$
$\Rightarrow 4p = 10 \Rightarrow p = \frac{5}{2} \Rightarrow 25 + q^2 = 34 \Rightarrow q = 3$

Exercise 1I

1 a $f(2) = 8 - 24 + 42 - 26 = 0$
b $z = 2, z = 2 + 3\text{i}$ or $z = 2 - 3\text{i}$

2 a Substitute $z = \frac{1}{2}$ into f(z).
b $b = 3, c = 6$
c $z = \frac{1}{2}$, or $z = -\frac{3}{2} \pm \frac{\sqrt{15}}{2}\mathrm{i}$

3 $3, -\frac{1}{2} + \frac{1}{2}\mathrm{i}$ and $-\frac{1}{2} - \frac{1}{2}\mathrm{i}$

4 a $(z - (-4 + \mathrm{i}))(z - (-4 - \mathrm{i})) = z^2 + 8z + 16 + 1$
$= z^2 + 8z + 17$
b $z = 4, z = -4 + \mathrm{i}$ or $z = -4 - \mathrm{i}$

5 a $a = 8, b = 25$ **b** $-1, -4 + 3\mathrm{i}, -4 - 3\mathrm{i}$ **c** -9

6 a $3 - \mathrm{i}$ **b** $c = 46, d = -60$

7 $-\frac{1}{2}, -\frac{1}{2} + \frac{\sqrt{3}}{2}\mathrm{i}$ and $-\frac{1}{2} - \frac{\sqrt{3}}{2}\mathrm{i}$

8 a $k = -40$ **b** $2 - 4\mathrm{i}, 2 + 4\mathrm{i}$

9 $2, -2, 2\mathrm{i}$ and $-2\mathrm{i}$

10 a $(z^2 - 9)(z^2 - 12z + 40)$ **b** $z = \pm 3, 6 \pm 2\mathrm{i}$

11 $-3 + \mathrm{i}, -3 - \mathrm{i}, 2 + 3\mathrm{i}$ and $2 - 3\mathrm{i}$
$(z - (2 - 3\mathrm{i}))(z - (2 + 3\mathrm{i})) = z^2 - 4z + 13$

12 a $(z^2 - 4z + 13)(z^2 + bz + c)$
$= z^4 - 10z^3 + 71z^2 + Qz + 442$
$b = -6, c = 34$
b $Q = -214$ **c** $z = 2 + 3\mathrm{i}, 2 - 3\mathrm{i}, 3 + 5\mathrm{i}$ or $3 - 5\mathrm{i}$

Challenge
$b = 0, c = 2, d = 4, e = -8, f = 16$

Chapter review 1

1 a $6 + \mathrm{i}$ **b** $-6 + 12\mathrm{i}$ **c** $50 - 22\mathrm{i}$

2 $-2\sqrt{14} < b < 2\sqrt{14}$

3 $3 + \mathrm{i}\sqrt{3}, 3 - \mathrm{i}\sqrt{3}$

4 $(1 + 2\mathrm{i})^5$
$= 1^5 + 5(1)^4(2\mathrm{i}) + 10(1)^3(2\mathrm{i})^2 + 10(1)^2(2\mathrm{i})^3 + 5(1)(2\mathrm{i})^4 + (2\mathrm{i})^5$
$= 1 + 10\mathrm{i} + 40\mathrm{i}^2 + 80\mathrm{i}^3 + 80\mathrm{i}^4 + 32\mathrm{i}^5$
$= 1 + 10\mathrm{i} - 40 - 80\mathrm{i} + 80 + 32\mathrm{i}$
$= 41 - 38\mathrm{i}$

5 Substitute $z = 3 + \mathrm{i}$ into f(z) to get f(z) = 0.

6 a $4 - 2\mathrm{i}$ **b** $-14 - 2\mathrm{i}$ **c** $-1 - \mathrm{i}$

7 $\frac{(45 - 28\mathrm{i})(1 - \mathrm{i}\sqrt{3})}{(1 + \mathrm{i}\sqrt{3})(1 - \mathrm{i}\sqrt{3})} = \frac{45 - 28\sqrt{3}}{4} + \left(\frac{-45\sqrt{3} - 28}{4}\right)\mathrm{i}$

8 $\frac{4 - 7\mathrm{i}}{3 + \mathrm{i}} = \frac{(4 - 7\mathrm{i})(3 - \mathrm{i})}{(3 + \mathrm{i})(3 - \mathrm{i})} = \frac{12 - 25\mathrm{i} + 7\mathrm{i}^2}{10} = \frac{1}{2} - \frac{5}{2}\mathrm{i}$

9 a $\frac{3}{25} - \frac{4}{25}\mathrm{i}$ **b** $\frac{-8}{5} - \frac{6}{5}\mathrm{i}$

10 $\frac{z}{z^*} = \frac{(a + b\mathrm{i})(a + b\mathrm{i})}{(a - b\mathrm{i})(a + b\mathrm{i})} = \frac{a^2 + 2ab\mathrm{i} + b^2\mathrm{i}^2}{a^2 - b^2\mathrm{i}^2}$
$= \frac{a^2 - b^2}{a^2 + b^2} + \left(\frac{2ab}{a^2 + b^2}\right)\mathrm{i}$

11 a $\frac{3 + q\mathrm{i}}{q - 5\mathrm{i}} \times \frac{q + 5\mathrm{i}}{q + 5\mathrm{i}} = \frac{3q - 5q}{q^2 + 25} + \frac{q^2 + 15}{q^2 + 25}\mathrm{i}$
$\frac{-2q}{q^2 + 25} = \frac{1}{13} \Rightarrow q^2 + 26q + 25 = 0 \Rightarrow q = -1, q = -25$
b $\frac{1}{13} + \frac{8}{13}\mathrm{i}, \frac{1}{13} + \frac{64}{65}\mathrm{i}$

12 $x + \mathrm{i}y + 4\mathrm{i}(x - \mathrm{i}y) = -3 + 18\mathrm{i}$
$(x + 4y) + \mathrm{i}(4x + y) = -3 + 18\mathrm{i}$
$x + 4y = -3, 4x + y = 18 \Rightarrow x = 5, y = -2$

13 $\frac{(9 + 6\mathrm{i})(2 + 3\mathrm{i})}{(2 - 3\mathrm{i})(2 + 3\mathrm{i})} = \frac{18 + 39\mathrm{i} + 18\mathrm{i}^2}{4 - 9\mathrm{i}^2} = 3\mathrm{i}$

14 $\frac{(q + 3\mathrm{i})(4 - q\mathrm{i})}{(4 + q\mathrm{i})(4 - q\mathrm{i})} = \frac{7q}{q^2 + 16} + \frac{12 - q^2}{q^2 + 16}\mathrm{i}$

15 a $z = -\frac{5}{2} + \frac{\sqrt{15}}{2}\mathrm{i}$ and $z = -\frac{5}{2} - \frac{\sqrt{15}}{2}\mathrm{i}$
b
Im
$z = -\frac{5}{2} + \mathrm{i}\frac{\sqrt{15}}{2}$
O Re
$z = -\frac{5}{2} - \mathrm{i}\frac{\sqrt{15}}{2}$

16 a $6 + 2\mathrm{i}$ **b** $z^2 - 12z + 40$

17 $k = 6, m = 4$

18 $z = 2 + \mathrm{i}, 2 - \mathrm{i}$ or -4

19 $z = -2, 1 + 3\mathrm{i}$ or $1 - 3\mathrm{i}$

20 a $k = 16$ **b** $-4\mathrm{i}$ and -3

21 a $-1 + 2\mathrm{i}, -1 - 2\mathrm{i}$ are two of the roots. These roots can be used to form the quadratic $z^2 + 2z + 5$.
$(z - 1)(z^2 + 2z + 5) = \mathrm{f}(z)$, so third root is 1.
b Argand diagram showing $-1 + 2\mathrm{i}, -1 - 2\mathrm{i}$ and 1.
c Sides of triangle are $\sqrt{8}, \sqrt{8}$ and 4. $(\sqrt{8})^2 + (\sqrt{8})^2 = 4^2$.

22 a $b = 4, c = 10$ **b** $z = 6, -1, -2 + \sqrt{6}\mathrm{i}$ or $-2 - \sqrt{6}\mathrm{i}$

23 $3 - 2\mathrm{i}, 3 + 2\mathrm{i}, \mathrm{i}\sqrt{6}$ and $-\mathrm{i}\sqrt{6}$

24 a $p = -18$ **b** $1, 4, -\frac{3}{2} + \frac{\sqrt{15}}{2}\mathrm{i}$ and $-\frac{3}{2} - \frac{\sqrt{15}}{2}\mathrm{i}$

25 a $-1 + 4\mathrm{i}, -1 - 4\mathrm{i}, 2, 1$
b Argand diagram showing above roots.

26 a $4x - y = 3$
$-3x - 6y = 0 \Rightarrow x = -2y$
$-9y = 3 \Rightarrow y = -\frac{1}{3} \Rightarrow x = \frac{2}{3}$
b Argand diagram showing the point $z = \frac{2}{3} - \frac{1}{3}i$
c $\frac{\sqrt{5}}{3}$
d -0.46 rad

27 a $z^2 = (a^2 - 16) + 8a\mathrm{i}$
$2z = 2a + 8\mathrm{i}$
$z^2 + 2z = (a^2 + 2a - 16) + (8 + 8a)\mathrm{i}$
b $a = -1$
c $z = -1 + 4\mathrm{i}$
$|z| = \sqrt{17} \approx 4.12$
$\arg z \approx 1.82$
d Show $z = -1 + 4\mathrm{i}, z^2 = -15 - 8\mathrm{i}$ and $z^2 + 2z = -17$ on a single Argand diagram.

28 a $z = \frac{(3 + 5\mathrm{i})(2 + \mathrm{i})}{(2 - \mathrm{i})(2 + \mathrm{i})} = \frac{1}{5} + \frac{13}{5}\mathrm{i}$
$|z| = \frac{1}{5}\sqrt{170}$
b $\arg z = 1.49$

29 a $z^2 = -3 + 4\mathrm{i}$
$z^2 - z = -4 + 2\mathrm{i}$
$|-4 + 2\mathrm{i}| = \sqrt{(-4)^2 + (2)^2} = \sqrt{20} = 2\sqrt{5}$
b 2.68
c

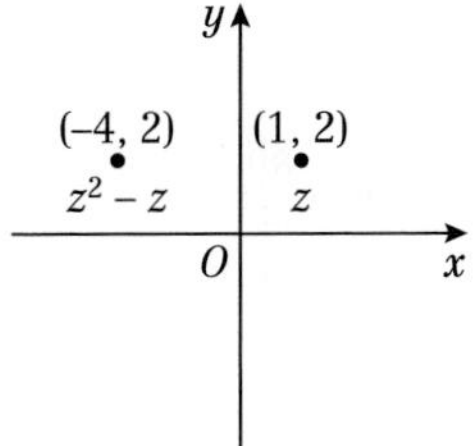

30 **a** **i** $\frac{3}{25} - \frac{4}{25}\text{i}$ **ii** $\frac{-8}{5} - \frac{6}{5}\text{i}$ **b** $\frac{1}{5}$ **c** -2.50

31 **a** $\frac{\sqrt{5}}{2}$

b $\frac{a + 3\text{i}}{2 + a\text{i}} = \frac{5a}{4 + a^2} + \frac{-a^2 + 6}{4 + a^2}\text{i}$,

for $\arg z = \frac{\pi}{4}$, the real and imaginary parts must be equal

$\Rightarrow a^2 + 5a - 6 = 0$

$\Rightarrow a = -6$ or 1

a cannot be negative otherwise $\arg z$ is negative

$\therefore a = 1$

32 $4\sqrt{2}\left(\cos\left(-\frac{\pi}{4}\right) + \text{i}\sin\left(-\frac{\pi}{4}\right)\right)$

Challenge

a If a root is not real, the other root must be its complex conjugate, but only real numbers are equal to their conjugate.

b $(z + \text{i})^2(z - \text{i})^2 = z^4 + 2z^2 + 1$

CHAPTER 2

Prior knowledge check

1 **a** $x = -2 \pm \text{i}$ **b** $x = \frac{7 \pm \text{i}\sqrt{15}}{4}$

2 **a** -1 and 3 **b** 4 and 8 **c** $-\frac{1}{2}$ and $\frac{3}{2}$

Exercise 2A

1 **a** $-\frac{7}{3}$ **b** $-\frac{4}{3}$ **c** $\frac{7}{4}$ **d** $\frac{73}{9}$

2 **a** $\frac{3}{7}$ **b** $\frac{1}{7}$ **c** 3 **d** $-\frac{5}{49}$

3 **a** $\frac{3}{2}$ **b** $\frac{1}{9}$ **c** $\frac{9}{2}$ **d** $\frac{15}{8}$

4 $a = 1, b = 1, c = -6$

5 $a = 6, b = 5, c = 1$

6 $a = 2, b = 2, c = 1$

7 **a** $-1 + 4\text{i}$ **b** $b = 2, c = 17$

8 $\frac{3}{5}$

9 -16

10 6

11 $k = \sqrt{2}$ and $m = -\frac{2\sqrt{2}}{3}$ or $k = -\sqrt{2}$ and $m = \frac{2\sqrt{2}}{3}$

Exercise 2B

1 **a** $x^2 + 8x - 1 = 0$ **b** $x^2 - 6x + 8 = 0$

2 **a** $9x^2 + 14x + 9 = 0$ **b** $27x^2 + 46x + 27 = 0$

3 **a** $27x^2 + 24x + 197 = 0$ **b** $27x^2 - 35x + 216 = 0$

4 **a** $64x^2 + 189x + 216 = 0$ **b** $27x^2 - 9x + 8 = 0$

Chapter review 2

1 **a** **i** $\frac{4}{3}$ **ii** 2 **c** $54x^2 + 76x + 27 = 0$

2 **a** $2x^2 + 5x - 12 = 0$ **b** **i** $\frac{73}{4}$ **ii** $-\frac{485}{8}$

c $288x^2 + 866x - 1235 = 0$

3 **a** $3x^2 + 7x - 6 = 0$ **c** $18x^2 + 85x + 77 = 0$

Challenge

1 **a** **i** $p = 5$, **ii** $q = -8$ **b** $\sqrt{\frac{89}{4}}$

2 **a** 24 **b** $q > 36$

CHAPTER 3

Prior knowledge check

1 **a** 3.25 **b** 11.24

2 **a** $f'(x) = 21x^2 - 4x$

b $f'(x) = \frac{3}{2\sqrt{x}} + 8x + \frac{15}{x^4}$

3 $u_1 = 2, u_2 = 2.5, u_3 = 2.9$

Exercise 3A

1 **a** $f(-2) = -1 < 0$, $f(-1) = 5 > 0$. Sign change implies root.

b $f(3) = -2.732 < 0$, $f(4) = 4 > 0$. Sign change implies root.

c $f(-0.5) = -0.125 < 0$, $f(-0.2) = 2.992 > 0$. Sign change implies root.

d $f(1.65) = -0.294 < 0$, $f(1.75) = 0.195 > 0$. Sign change implies root.

2 **a** $f(1.8) = 0.408 > 0$, $f(1.9) = -0.249$. Sign change implies root.

b $f(1.8635) = 0.00138... > 0$, $f(1.8645) = -0.00531... < 0$. Sign change implies root.

3 **a** $h(1.4) = -0.0512... < 0$, $h(1.5) = 0.0739.... > 0$. Sign change implies root.

b $h(1.4405) = -0.00055... < 0$, $h(1.4415) = 0.00069... > 0$. Sign change implies root.

4 **a** $f(1.5) = 16.10... > 0$, $f(1.6) = -32.2... < 0$. Sign change implies root.

b There is an asymptote in the graph of $y = f(x)$ at $x = \frac{\pi}{2} \approx 1.57$. So there is not a root in this interval.

5

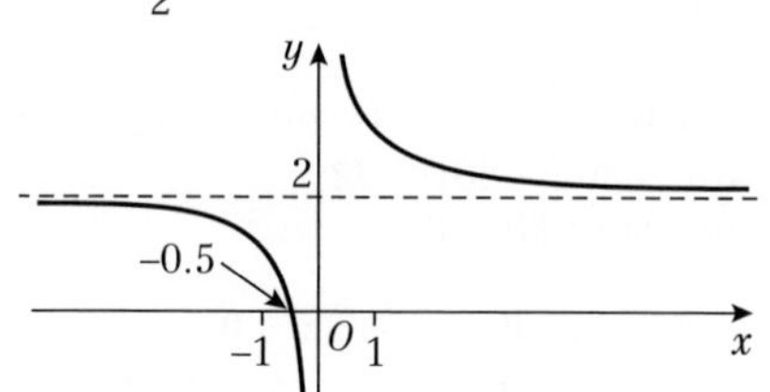

Alternatively: $\frac{1}{x} + 2 = 0 \Rightarrow \frac{1}{x} = -2 \Rightarrow x = -\frac{1}{2}$

6 **a**

b 1 point of intersection $\Rightarrow$ 1 root

c $f(1) = -1$, $f(2) = 0.414...$ **d** $p = 3, q = 4$ **e** $4^{\frac{1}{3}}$

7 **a** $f(-0.9) = 1.5561 > 0$, $f(-0.8) = -0.7904 < 0$. There is a change of sign in the interval $[-0.9, -0.8]$, so there is at least one root in this interval.

b $(1.74, -45.37)$ to 2 d.p. **c** $a = 3, b = 9$ and $c = 6$

d

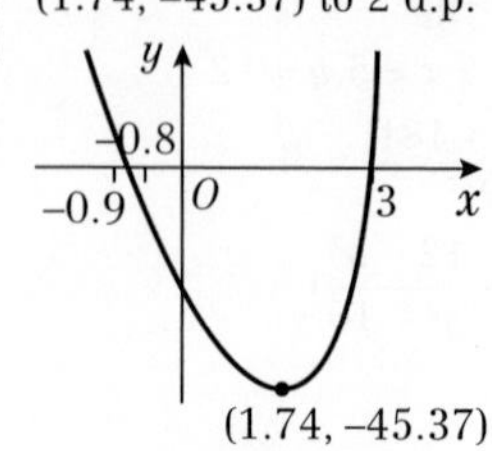

Exercise 3B

1 2.6

2 a $2^3 - 7 \times 2 + 2 = -4$ neg, $3^3 - 7 \times 3 + 2 = 8$ pos
There is a change of sign so there is a root in the interval [2, 3]

2 b 2.49

3 a $2^3 + 2 \times 2^2 - 8 \times 2 + 3 = 3$ pos,
$1^3 + 2 \times 1^2 - 8 \times 1 + 3 = -2$ neg
There is a change of sign so there is a root in the interval [1, 2]

3 b 1.7

4 a $\frac{1}{2} - \frac{1}{1} = -\frac{1}{2}$ neg $\frac{2}{2} - \frac{1}{2} = \frac{1}{2}$ pos
There is a change of sign so there is a root in the interval [1, 2]

4 b 1.4

5 2.4

Exercise 3C

1 a $2^3 - 3 \times 2 - 5 = -3$ neg $3^3 - 3 \times 3 - 5 = 13$ pos
There is a change of sign so there is a root in the interval [2, 3]
b 2.3

2 a $5 \times 1^3 - 8 \times 1 + 1 = -2$ neg $5 \times 2^3 - 8 \times 2 + 1 = 25$ pos
There is a change of sign so there is a root in the interval [1, 2]
b 1.5

3 a $\frac{3}{3} - 3 + 3 = 1$ pos $\frac{3}{4} - 4 + 3 = -\frac{1}{4}$ neg
There is a change of sign so there is a root in the interval [3, 4]
b 3.8

4 a $2 \times 1\cos(1) - 1 = 0.08...$ pos $2 \times 1.5\cos(1.5) - 1 = -0.787...$ neg
There is a change of sign so there is a root in the interval [1, 1.5]
b 1.1

5 a $2^3 - 2 \times 2^2 - 3 = -3$ neg $3^3 - 2 \times 3^2 - 3 = 6$ pos
There is a change of sign so there is a root in the interval [2, 3]
b 2.5

6 3.4

Exercise 3D

1 a f(1) = −2, f(2) = 3. There is a sign change in the interval $1 < \alpha < 2$, so there is a root in this interval.
b $x_1 = 1.632$

2 a $f'(x) = 2x + \frac{4}{x^2} + 6$ b −0.326

3 a f(1.3) = −0.085…, f(1.4) = 0.429… As there is a change of sign in the interval, there must be a root α in this interval.
b $f'(x) = 2x + \frac{6}{x^3}$ c 1.316

4 a

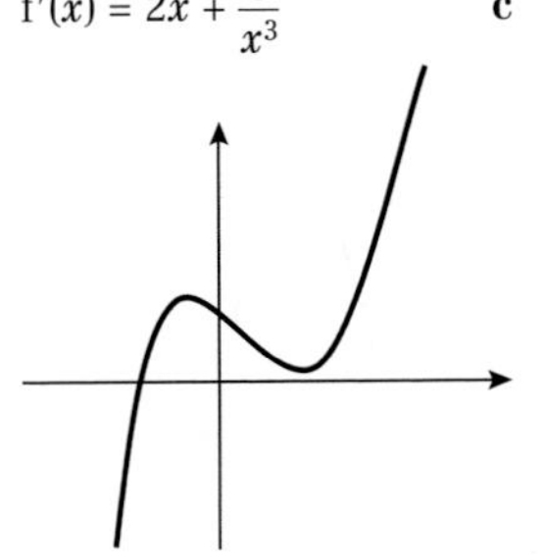

b −2.355

c f′(1) = 0 hence NR will not work.

5 a $0^4 - 7 \times 0^3 + 1 = 1$ pos $1^4 - 7 \times 1^3 + 1 = -5$ neg
There is a change of sign so there is a root in the interval [0, 1]
b 0.5368 c f′(0) = 0 hence NR will not work.

Chapter review 3

1 a f(3.9) = 13, f(4.1) = −7
b There is an asymptote at $x = 4$ which causes the change of sign, not a root.
c $\alpha = \frac{13}{3}$

2 −1.8

3 3.73

4 a

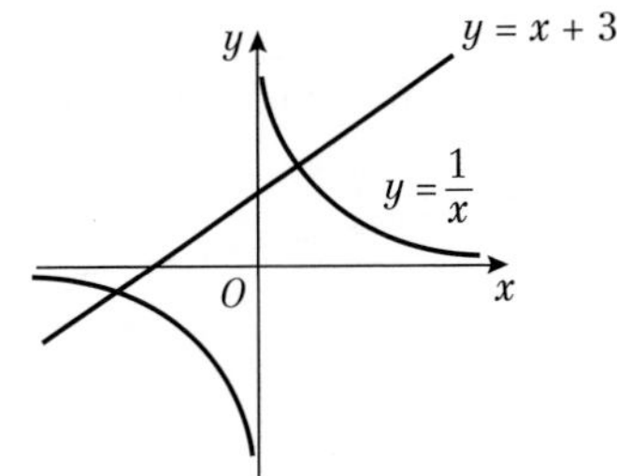

b 2
c $\frac{1}{x} = x + 3 \Rightarrow 0 = x + 3 - \frac{1}{x}$, let $f(x) = x + 3 - \frac{1}{x}$
f(0.30) = −0.0333…< 0, f(0.31) = 0.0841… > 0.
Sign change implies root.
d $\frac{1}{x} = x + 3 \Rightarrow 1 = x^2 + 3x \Rightarrow 0 = x^2 + 3x - 1$
e 0.303

5 a $g'(x) = 3x^2 - 14x + 2$ b 6.606
c $(x - 1)(x^2 - 6x - 4) \Rightarrow x^2 - 6x - 4 = 0 \Rightarrow x = 3 \pm \sqrt{13}$
d 0.007%

6 1.25

7 a

x	−1	0	1	2	3	4
$f(x)$	−3*	1*	−1	−3*	1	1.7

There are 3 changes of sign (*) so there are 3 roots.
b −0.532, 0.653, 2.879

Challenge

a Let $y = f(x)$
$f(y) = y^2 - 6y + 7$
$f(4) = 4^2 - 6 \times 4 + 7 = -1$
$f(5) = 5^2 - 6 \times 5 + 7 = 2$ Change of sign, so root
$f(1) = 1^2 - 6 \times 1 + 7 = 2$
$f(2) = 2^2 - 6 \times 2 + 7 = -1$ Change of sign, so root

b $y_0 = 5$, $y_1 = 4.5$, $y_2 = 4.416666$, $y_3 = 4.414242156$,
$y_4 = 4.414213562 \Rightarrow 4.4142$
Both y_4 and y_5 are 4.4142 correct to 5 significant figures.

CHAPTER 4

Prior knowledge check

1 a (1, 5) b $2\sqrt{13}$ or 7.211

2 $y = 3x$

3 $x + 4y - 27 = 0$

Exercise 4A

1

t	−4	−3	−2	−1	−0.5	0	0.5	1	2	3	4
$x = 2t^2$	32	18	8	2	0.5	0	0.5	2	8	18	32
$y = 4t$	−16	−12	−8	−4	−2	0	2	4	8	12	16

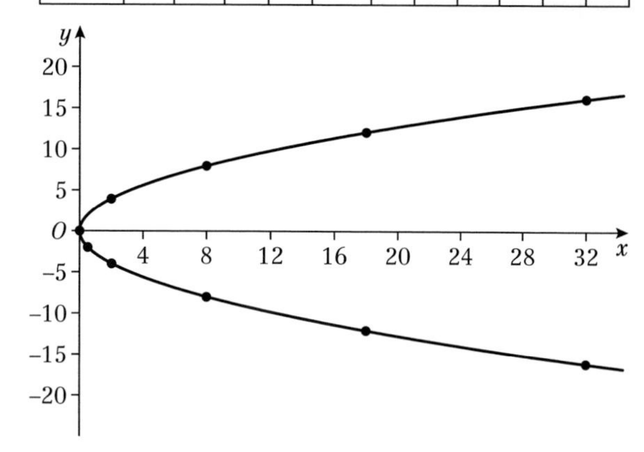

2

t	−3	−2	−1	−0.5	0	0.5	1	2	3
$x = 3t^2$	27	12	3	0.75	0	0.75	3	12	27
$y = 6t$	−18	−12	−6	−3	0	3	6	12	18

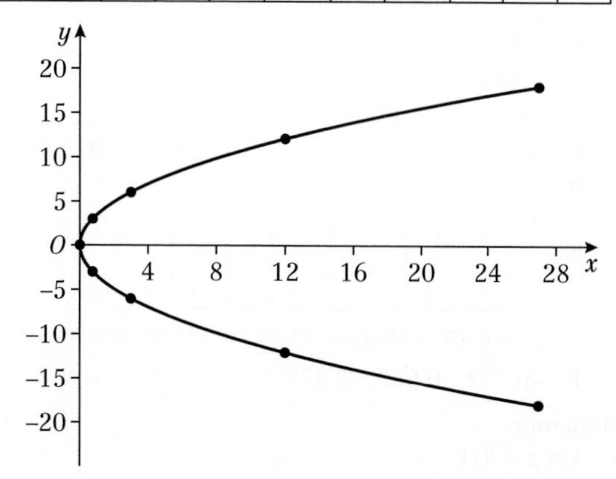

3

t	−4	−3	−2	−1	−0.5	0.5	1	2	3	4
$x = 4t$	−16	−12	−8	−4	−2	2	4	8	12	16
$y = \frac{4}{t}$	−1	$-\frac{4}{3}$	−2	−4	−8	8	4	2	$\frac{4}{3}$	1

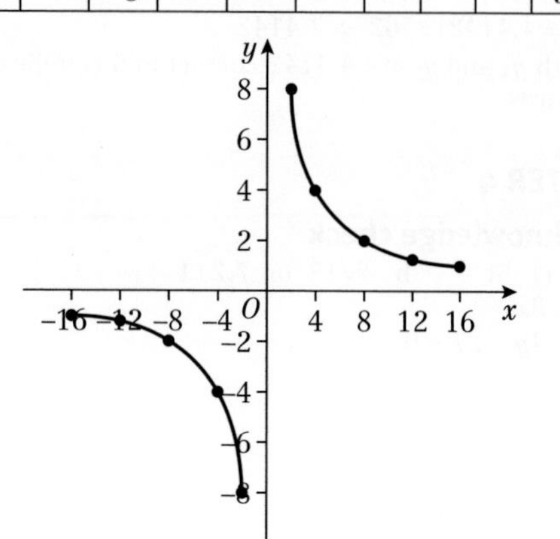

4 **a** $y^2 = 20x$ **b** $y^2 = 2x$
c $y^2 = 200x$ **d** $y^2 = \frac{4}{5}x$
e $y^2 = 10x$ **f** $y^2 = 4\sqrt{3}x$
g $x^2 = 8y$ **h** $x^2 = 12y$

5 **a** $xy = 1$ **b** $xy = 49$
c $xy = 45$ **d** $xy = \frac{1}{25}$

6 **a** $xy = 9$
b

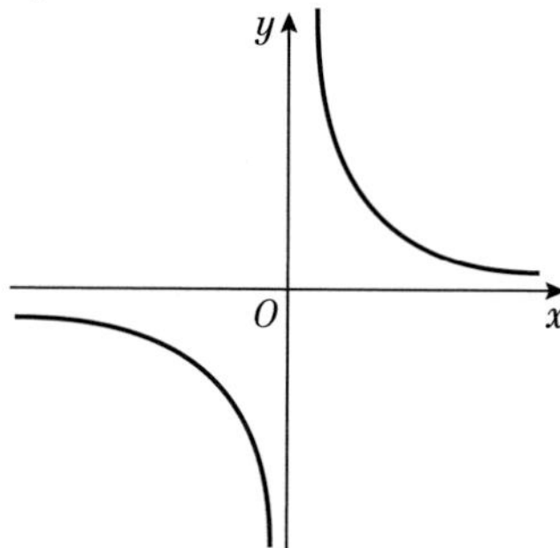

7 **a** $xy = 2$
b

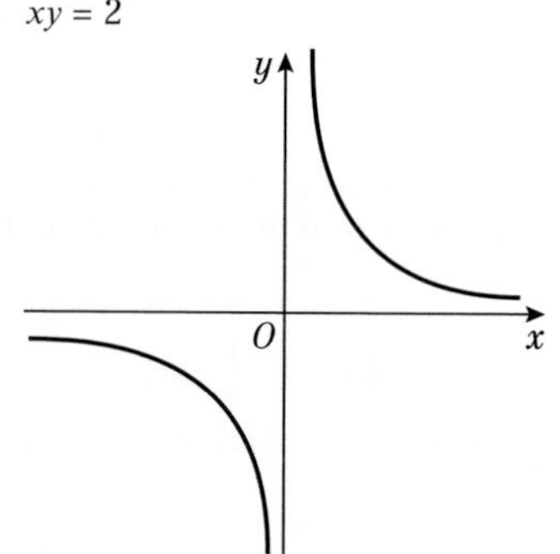

Exercise 4B

1 **a** $y^2 = 20x$ **b** $y^2 = 32x$
c $y^2 = 4x$ **d** $y^2 = 6x$
e $y^2 = 2\sqrt{3}x$

2 **a** $(3, 0)$; $x + 3 = 0$ **b** $(5, 0)$; $x + 5 = 0$
c $\left(\frac{5}{2}, 0\right)$; $x + \frac{5}{2} = 0$ **d** $(\sqrt{3}, 0)$; $x + \sqrt{3} = 0$
e $\left(\frac{\sqrt{2}}{4}, 0\right)$; $x + \frac{\sqrt{2}}{4} = 0$ **f** $\left(\frac{5\sqrt{2}}{4}, 0\right)$; $x + \frac{5\sqrt{2}}{4} = 0$

3 Distance from P to point $(3, 0)$ is the same as the distance from P to the directrix:
$(x - 3)^2 + (y - 0)^2 = (x + 3)^2$
$x^2 - 6x + 9 + y^2 = x^2 + 6x + 9$
$-6x + y^2 = 6x$
$y^2 = 12x$
$\Rightarrow a = 3$

4 Distance from P to point $(2\sqrt{5}, 0)$ is the same as the distance from P to the directrix:
$(x - 2\sqrt{5})^2 + (y - 0)^2 = (x + 2\sqrt{5})^2$
$x^2 - 4\sqrt{5}x + 20 + y^2 = x^2 + 4\sqrt{5}x + 20$
$-4\sqrt{5}x + y^2 = 4\sqrt{5}x$
$y^2 = 8\sqrt{5}x$
$\Rightarrow a = 2\sqrt{5}$

5 **a**

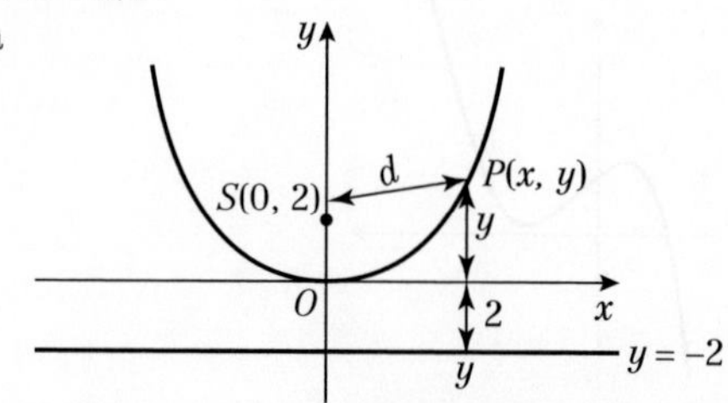

From the sketch, the locus satisfies $SP = YP$

So, $SP^2 = YP^2$

and, $(x-0)^2 + (y-2)^2 = (y-[-2])^2$

$\Rightarrow x^2 + y^2 - 4y + 4 = y^2 + 4y + 4$

$\Rightarrow x^2 = 8y$

$\Rightarrow y = \frac{x^2}{8}$

So the locus of P has an equation of the form

$y = \frac{x^2}{8}$ where $k = \frac{1}{8}$

Hence shown.

b $(0, 2)$; $y + 2 = 0$

c

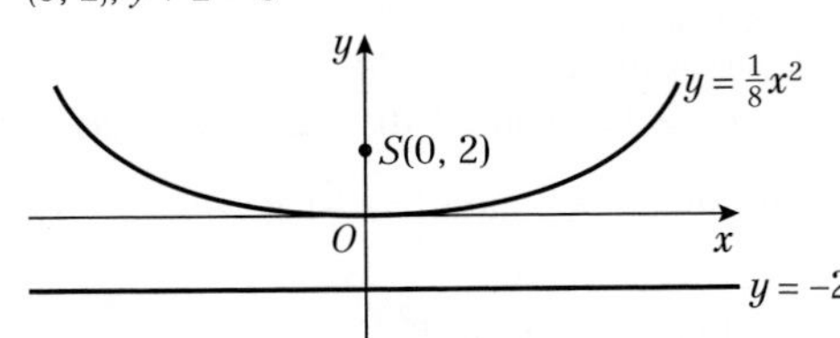

Exercise 4C

1 $(3, 3)$ and $(0.75, -1.5)$

2 $16\sqrt{2}$

3 $A = (10, -10)$, $B = (40, 20)$, $M = (25, 5)$

4 **a** $y^2 = 24x$ **b** $(6, 0)$; $x + 6 = 0$

c

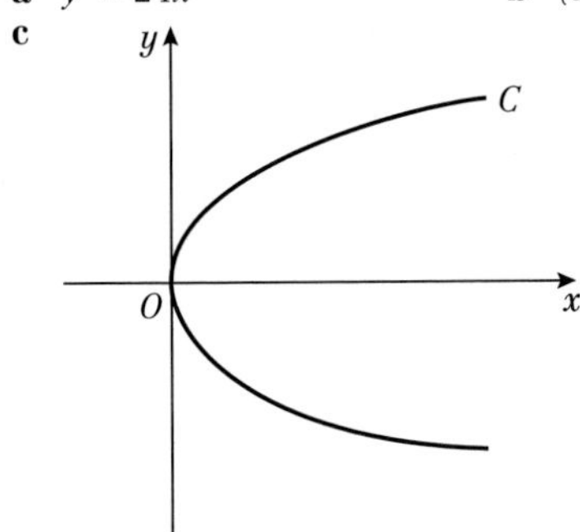

d 9 **e** $12\sqrt{2}$

f $18\sqrt{2}$

5 **a** $y^2 = 5x$ **b** 5

c $\left(-\frac{5}{4}, 3\right)$ **d** $8x - 25y + 85 = 0$

6 **a** $(1, 0)$ **b** 4

c $4x - 3y - 4 = 0$ **d** $\left(\frac{1}{4}, -1\right)$

e $\frac{5}{4}$

7 **a** $R(-3, 0)$, $S(3, 0)$ **b** $P(9, 6\sqrt{3})$, $Q(-3, 6\sqrt{3})$

c $54\sqrt{3}$

8 **a** $a = 1, b = -4$ **b** $y = x - 8$

c $(10, 2)$ **d** $y = -x + 12$

e $x = 14 \pm 2\sqrt{13}$

Exercise 4D

1 **a** $x - 4y + 16 = 0$ **b** $\sqrt{2}x - 2y + 4\sqrt{2} = 0$

c $x + y - 10 = 0$ **d** $16x + y - 16 = 0$

e $x + 2y + 7 = 0$ **f** $2x + y - 8\sqrt{2} = 0$

2 **a** $x + y - 15 = 0$ **b** $2x - 8y - 45 = 0$

3 **a** 4 **b** $y = -x + 12$

c $(36, -24)$ **d** $32\sqrt{2}$

4 **a** $x - 8y - 126 = 0$ **b** $\left(128, \frac{1}{4}\right)$

5 **a** The points P and Q have coordinates $(4, 12)$ and $(-8, -6)$ respectively

Gradient of PQ $m_{PQ} = \frac{-6-12}{-8-4} = \frac{-18}{-12} = \frac{3}{2}$

$y - 12 = \frac{3}{2}(x-4) \Rightarrow 3x - 2y + 12 = 0$

b $(6\sqrt{2}, 4\sqrt{2})$ or $(-6\sqrt{2}, -4\sqrt{2})$

6 **a** $xy = 3$ **b** $8x - 2y - 15\sqrt{3} = 0$

c $\left(-\frac{1}{8}\sqrt{3}, -8\sqrt{3}\right)$

7 **a** $\frac{1}{2}$; $(1, 4)$ **b** $(-15, 0)$

c $(-1, 0)$ **d** 28

Exercise 4E

1 **a** $y^2 = 12x \Rightarrow y = \pm\sqrt{12}x^{\frac{1}{2}} = \pm 2\sqrt{3}x^{\frac{1}{2}}$

$\frac{dy}{dx} = \pm\left(\frac{1}{2}\right)2\sqrt{3}x^{-\frac{1}{2}} = \pm\frac{\sqrt{3}}{\sqrt{x}}$

At $P(3t^2, 6t)$, $m_T = \pm\frac{\sqrt{3}}{\sqrt{3t^2}} = \frac{1}{t}$

$y - 6t = \frac{1}{t}(x - 3t^2) \Rightarrow yt = x + 3t^2$

Hence shown.

b Gradient of tangent is $\frac{1}{t}$, so gradient of normal is $-t$

$y - 6t = -t(x - 3t^2) \Rightarrow y + tx = 3t^3 + 6t$

Hence shown.

2 **a** $xy = 36 \Rightarrow y = 36x^{-1}$

$\frac{dy}{dx} = -36x^{-2}$

At $P\left(6t, \frac{6}{t}\right)$, $m_T = -36(6t)^{-2} = -\frac{1}{t^2}$

$y - \frac{6}{t} = -\frac{1}{t^2}(x + 6t) \Rightarrow yt^2 - 6t = x + 6t \Rightarrow t^2y + x = 12t$

b Gradient of tangent is $-\frac{1}{t^2}$, so gradient of normal is t^2

$y - \frac{6}{t} = t^2(x - 6t) \Rightarrow yt - 6 = xt^3 - 6t^4$

$\Rightarrow t^3x - ty = 6(t^4 - 1)$

Hence shown.

3 **a** 5 **c** $\frac{25}{2}t^3$

b The equation of C is $y^2 = 20x$ from part (a) where $a = 5$

$y^2 = 20x \Rightarrow y = \sqrt{20}\sqrt{x} = 2\sqrt{5}x^{\frac{1}{2}}$

So $y = \pm 2\sqrt{5}x^{\frac{1}{2}}$

$\frac{dy}{dx} = \pm\frac{1}{2} \times 2\sqrt{5}x^{-\frac{1}{2}} = \pm\sqrt{5}x^{-\frac{1}{2}} = \pm\frac{\sqrt{5}}{\sqrt{x}}$

At P $(5t^2, 10t)\frac{dy}{dx} = \frac{\sqrt{5}}{\sqrt{5t^2}} = \frac{\sqrt{5}}{\sqrt{5}t} = \frac{1}{t}$

The equation of the line is therefore:

$y - 10t = \frac{1}{t}(x - 5t^2) \Rightarrow ty - 10t^2 = x - 5t^2 \Rightarrow ty = x + 5t^2$

So the equation of the tangent to C at P is $ty = x + 5t^2$ as required

4 **a** $y^2 = 4ax \Rightarrow y = 2\sqrt{a}x^{\frac{1}{2}}$

$\frac{dy}{dx} = \frac{1}{2}(2\sqrt{a})x^{-\frac{1}{2}} = \frac{\sqrt{a}}{\sqrt{x}}$

At P $(at^2, 2at)$ $\frac{dy}{dx} = \frac{\sqrt{a}}{\sqrt{at^2}} = \frac{1}{t}$

$y - 2at = \frac{1}{t}(x - at^2) \Rightarrow ty = x + at^2$

Hence shown.

b $(a, -2a)$ and $(16a, 8a)$

5 **a** $xy = 16 \Rightarrow y = 16x^{-1}$

$\frac{dy}{dx} = -16x^{-2}$

At $P\left(4t, \frac{4}{t}\right)$, $m_T = -16(4t)^{-2} = -\frac{1}{t^2}$

$y - \frac{4}{t} = -\frac{1}{t^2}(x + 4t) \Rightarrow t^2y - 4t = x + 4t \Rightarrow t^2y + x = 8t$

Hence shown.

b $(-4, 5)$ **c** $(8, 2)$ and $\left(-\frac{8}{5}, -10\right)$

d $x + 4y - 16 = 0$; $25x + 4y + 80 = 0$

6 **a** $(-at^2, 0)$ **b** $(2a + at^2, 0)$ **c** $2at^2t(1 + t^2)$

7 a $y^2 = 8x \Rightarrow y = 2\sqrt{2}x^{\frac{1}{2}}$

$\frac{dy}{dx} = \frac{1}{2}(2\sqrt{2})x^{-\frac{1}{2}} = \frac{\sqrt{2}}{\sqrt{x}}$

At $P\ (2t^2, 4t)$ $\frac{dy}{dx} = \frac{\sqrt{2}}{\sqrt{2t^2}} = \frac{1}{t}$

So gradient of normal $m_N = -\frac{1}{\frac{1}{t}} = -t$

$y - 4t = -t(x - 2t^2) \Rightarrow x + ty = 2t^3 + 4t$

Hence shown.

b (0, 0), (8, 8) and (8, −8)

c $y = 0$, $2x + y - 24 = 0$ and $2x - y - 24 = 0$

8 a $(0, at)$ b $(a, 0)$

c $P(at^2, 2at)$, $Q(0, at)$ and $S(a, 0)$

$m_{pq} = \frac{at - 2at}{0 - at^2} = \frac{-at}{-at^2} = \frac{1}{t}$

$m_{SQ} = \frac{0 - at}{a - 0} = \frac{-at}{a} = -t$

$m_{pq} \times m_{SQ} = \frac{1}{t} \times -t = -1$ Hence perpendicular

Hence shown.

9 a $y^2 = 24x \Rightarrow y = 2\sqrt{6}x^{\frac{1}{2}}$

$\frac{dy}{dx} = \frac{1}{2}(2\sqrt{6})x^{-\frac{1}{2}} = \frac{\sqrt{6}}{\sqrt{x}}$

At $P\ (6t^2, 12t)$ $\frac{dy}{dx} = \frac{\sqrt{6}}{\sqrt{6t^2}} = \frac{1}{t}$

$y - 12t = \frac{1}{t}(x - 6t^2) \Rightarrow ty = x + 6t^2$

Hence shown.

b −6 c (24, 24) and $(\frac{3}{2}, -6)$

Chapter review 4

1 a (3, 0) b $(\frac{4}{3}, 4)$ c 6

2 a $\frac{3}{2}$ b (6, 0)

c The point P and S have coordinates

$\left(\frac{3}{2}, 6\right)$ and $(6, 0)$ respectively

$m_l = m_{PS} = \frac{0 - 6}{6 - \frac{3}{2}} = -\frac{4}{3}$

$y - 0 = -\frac{4}{3}(x - 6) \Rightarrow 3y = -4x + 24 \Rightarrow 3y + 4x - 24 = 0$

Hence shown.

d 30

3 a $y^2 = 48x$ b $x + 12 = 0$

c $(16, 16\sqrt{3})$ d $96\sqrt{3}$

4 a (1, 4) and (64, 32)

c $x + 2y - 9 = 0$ and $4x + y - 288 = 0$

d (81, −36)

e $9\sqrt{97}$

5 a focus of $C(a, 0)$, $Q(-a, 0)$

b $(a, 2a)$ or $(a, -2a)$

6 a $xy = c^2 \Rightarrow y = c^2x^{-1}$

$m_T = \frac{dy}{dx} = -\frac{c^2}{x^2}$

At $P\left(ct, \frac{c}{t}\right)$ $\frac{dy}{dx} = -\frac{c^2}{ct^2} = -\frac{1}{t^2}$

Gradient of normal $m_N = -\frac{1}{-\frac{1}{t^2}} = t^2$

$y - \frac{c}{t} = t^2(x - ct) \Rightarrow ty - c = t^3x - ct^4 \Rightarrow t^3x - ty = c(1 - t^4)$

Hence shown.

b $4x - y = 45$

c $\left(-\frac{3}{4}, -48\right)$

7 $x + 4y - 12 = 0$ and $x + 4y + 12 = 0$

8 a $X(2ct, 0)$ and $Y(0, \frac{2c}{t})$

b $6\sqrt{2}$

9 a $xy = c^2 \Rightarrow y = c^2x^{-1}$

$m_T = \frac{dy}{dx} = -\frac{c^2}{x^2}$

At $P\left(ct, \frac{c}{t}\right)$ $\frac{dy}{dx} = -\frac{c^2}{ct^2} = -\frac{1}{t^2}$

$y - \frac{c}{t} = -\frac{1}{t^2}(x - ct) \Rightarrow t^2y - ct = -x + ct \Rightarrow x + t^2y = 2ct$

Hence shown.

b T passes through X (2a, 0)

$x + t^2y = 2ct \Rightarrow 2a + t^2(0) = 2ct \Rightarrow 2a = 2ct \Rightarrow t = \frac{a}{c}$

Substitute $t = \frac{a}{c}$ into $\left(ct, \frac{c}{t}\right)$ gives

$P\left(c\left(\frac{a}{c}\right), \frac{c}{\frac{a}{c}}\right) = P\left(a, \frac{c^2}{a}\right)$

Hence shown.

c $\frac{c^2}{2a}$ d $y = \frac{c^2x}{4a^2}$

e $\frac{8a}{5}$

f From earlier parts;

$m_{OQ} = \frac{c^2}{4a^2}$ and $m_{XP} = -\frac{c^2}{a^2}$

OP is perpendicular to $XP \Rightarrow m_{OQ} \times m_{XP} = -1$

$\left(\frac{c^2}{4a^2}\right) \times \left(-\frac{c^2}{a^2}\right) = -\frac{c^4}{4a^4} = -1$

$-c^4 = -4a^4 \Rightarrow (c^2)^2 = (2a^2)^2 \Rightarrow c^2 = 2a^2$

Hence shown.

g $\frac{4a}{5}$

Challenge

$x + y + a = 0$

Review exercise 1

1 a $4 - (5 + p)i$ b $5p + 4pi$ c $-\frac{5}{p} - \frac{4}{p}i$

2 a $-2\sqrt{3} < k < 2\sqrt{3}$ b $0, 1 \pm \sqrt{2}i$

3 $\frac{5}{2} + i\frac{3\sqrt{3}}{2}, \frac{5}{2} - i\frac{3\sqrt{3}}{2}$

4 $x = 3, y = -1$

5 a $\frac{1}{2}$ b $-\frac{1}{4}$

6 $-1 + i, -1 - i, -3$

7 a 21 b $2 + 3i, 2$

8 a $(z^2 - 3z - 4)(z^2 + 2z + 4)$

b $-1, 4, -1 + i\sqrt{3}, -1 - i\sqrt{3}$

9 c $3 + i, 3 - i, 1 - 2i, 1 + 2i$

10 a $1 + i\sqrt{3}, 1 - i\sqrt{3}, 3$ b −5, 10

11 a $\frac{1}{3} - \frac{4}{3}i$

b

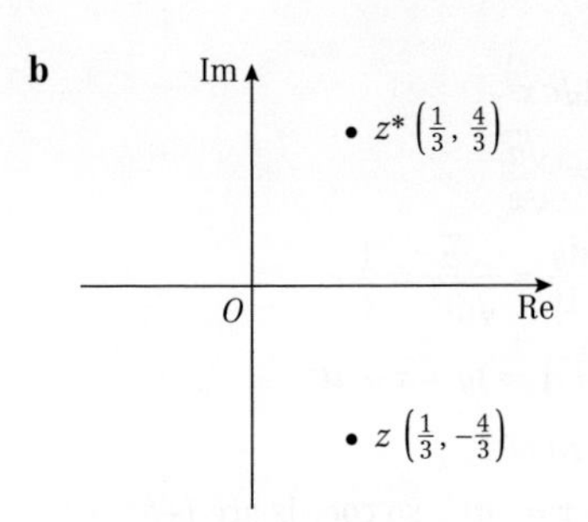

c $z = \frac{\sqrt{17}}{3}(\cos(-76°) + i\sin(-76°))$

$z^* = \frac{\sqrt{17}}{3}(\cos 76° + i\sin 76°)$

12 a

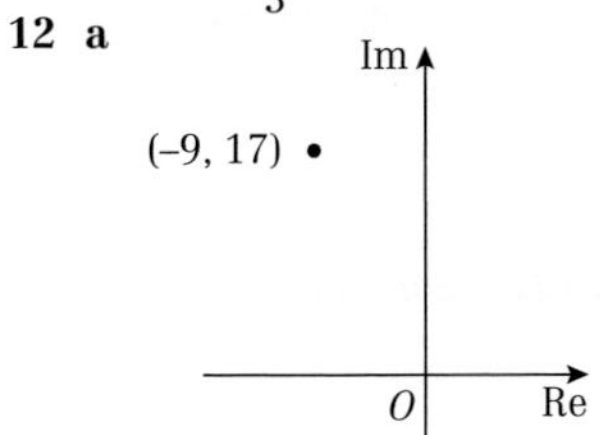

b 2.06 **c** $1 - 2i$

13 a $z^2 = (2 - i)^2 = 4 - 4i + i^2 = 4 - 4i - 1 = 3 - 4i$

b $2 - 2i$ and -2

c

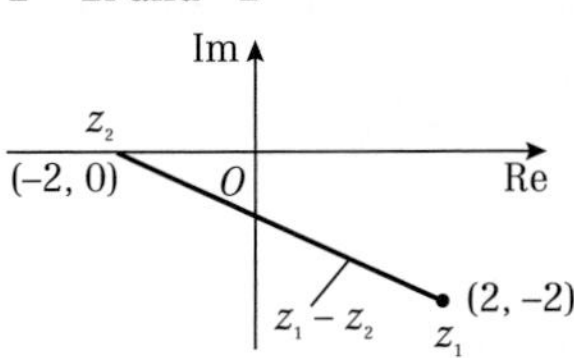

d $|z_1 - z_2|^2 = \sqrt{(-2-2)^2 + (-2)^2} = 2\sqrt{5}$ **e** $-\frac{\pi}{2}$

14 a $g(1.4) = -0.216 < 0$, $g(1.5) = 0.125 > 0$. Sign change implies root.

b $g(1.4655) = -0.00025\ldots < 0$, $g(1.4665) = 0.00326\ldots > 0$. Sign change implies root.

15 a i $\frac{22}{9}$ **ii** $-\frac{100}{27}$

b $3x^2 - 100x - 9 = 0$

16 a $-\frac{5}{2}, -2$ **b** $4x^2 - 5x - 2 = 0$

17 a i 7

ii $\alpha + \beta = 3,\ \alpha\beta = 1$

$\alpha^3 + \beta^3 = (\alpha + \beta)^3 - 3\alpha\beta(\alpha + \beta)$

$= 3^3 - 3 \times 1 \times 3 = 18$

Hence shown.

iii $(\alpha^2 + \beta^2)^2 = \alpha^4 + 2\alpha^2\beta^2 + \beta^4 \Rightarrow \alpha^4 + \beta^4$

$= (\alpha^2 + \beta^2)^2 - 2(\alpha\beta)^2$

Hence shown.

b $x^2 - 15x - 45 = 0$

18 a

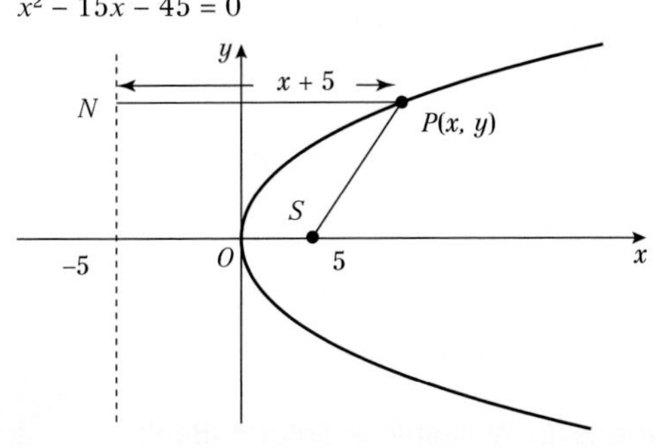

By the definition of a parabola

$SP = PN \Rightarrow SP^2 = PN^2$

$S(5,0),\ P(x,y)$

$SP^2 = (x_1 - x_2)^2 + (y_1 - y_2)^2$

$= (x - 5)^2 + y^2$

$PN = x + 5$

$(x - 5)^2 + y^2 = (x + 5)^2$

$\Rightarrow x^2 - 10x + 25 + y^2 = x^2 + 10x + 25$

$y^2 = 20x$

Comparing with $y^2 = 4ax$ this is the required form with $a = 5$

19 a (4,0) **b** $4x - 3y - 16 = 0$ **c** (1, –4)

20 a

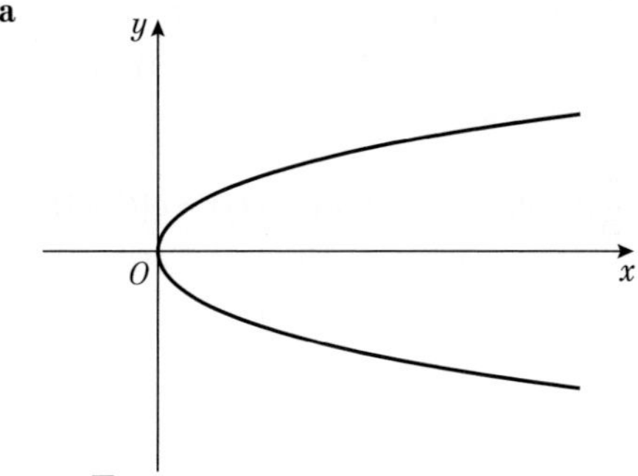

b $60\sqrt{2}$

21 $4\sqrt{15}$

22 a 8 **b** $y = 2x + 4$ **c** 4

23 a $y = \frac{4}{5}x + \frac{8}{5}$ **b** (–5, –2.4)

24 a $x + 4y = 24$ **b** $6\sqrt{17}$

25 (–8, 2)

26 a $t = \frac{1}{2}$, $P(6, 24)$ **b** $y = 2x + 12$

27 a Using $\frac{y - y_1}{y_2 - y_1} = \frac{x - x_1}{x_2 - x_1}$ with

$P(9,8),\ Q(6,12)$ an equation of the chord is

$\frac{y - 8}{12 - 8} = \frac{x - 9}{6 - 9} \Rightarrow 4x + 3y = 60$

Hence shown.

b $(3\sqrt{6}, 4\sqrt{6})$ and $(-3\sqrt{6}, -4\sqrt{6})$

28 a

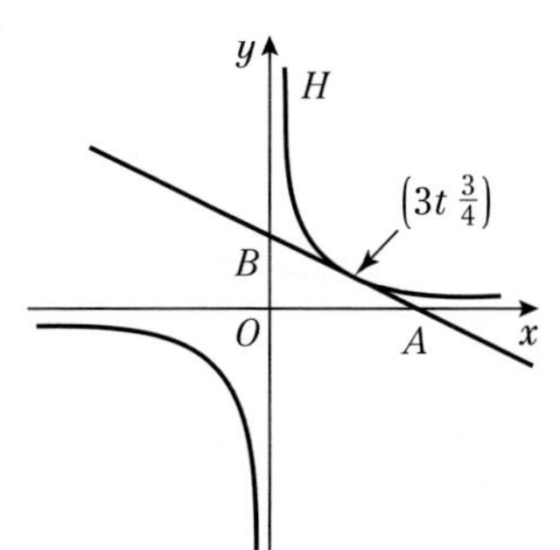

$y = \frac{9}{x} = 9x^{-1}$

$\frac{dy}{dx} = -9x^{-2} = -\frac{9}{x^2}$

When $x = 3t$, $\frac{dy}{dx} = -\frac{9}{9t^2} = -\frac{1}{t^2}$

At $\left(3t, \frac{3}{t}\right)$ $y - \frac{3}{t} = -\frac{1}{t^2}(x - 3t) \Rightarrow t^2y - 3t = -x + 3t$

$\Rightarrow x + t^2y = 6t$

Hence shown.

b At A, when $y = 0$ $x = 6t$ $OA = 6t$
At B, when $x = 0$ $0 + t^2y = 6t \Rightarrow y = \frac{6t}{t^2} = \frac{6}{t}$

Area $\Delta OAB = \frac{1}{2} \times OA \times OB = \frac{1}{2} \times 6t \times \frac{6}{t} = 18$

The area of 18 is constant and independent of t

29 a $xy = c^2 \Rightarrow y = c^2x^{-1}$

$m_T = \frac{dy}{dx} = -\frac{c^2}{x^2}$

At $P\left(ct, \frac{c}{t}\right)$ $\frac{dy}{dx} = -\frac{c^2}{ct^2} = -\frac{1}{t^2}$

Gradient of normal $m_N = -\frac{1}{-\frac{1}{t^2}} = t^2$

$y - \frac{c}{t} = t^2(x - ct) \Rightarrow ty - c = t^3x + ct^4$

$\Rightarrow t^3x - ty - c(t^4 - 1) = 0$

Hence shown.

b At G $y = x$

$t^3x - tx - c(t^4 - 1) = 0 \Rightarrow (t^3 - t)x = c(t^4 - 1)$

$\Rightarrow x = \frac{c(t^4 - 1)}{t^3 - t} = \frac{c(t^2 - 1)(t^2 + 1)}{t(t^2 - 1)} = \frac{c(t^2 + 1)}{t} = ct + \frac{c}{t}$

The co'ords of G are $\left(ct + \frac{c}{t},\ ct + \frac{c}{t}\right)$

$PG^2 = (x_1 - x_2)^2 + (y_1 - y_2)^2$

$PG^2 = \left(ct + \frac{c}{t} - ct\right)^2 + \left(ct + \frac{c}{t} - \frac{c}{t}\right)^2$

$PG^2 = \frac{c^2}{t^2} + c^2t^2 = c^2\left(t^2 + \frac{1}{t^2}\right)$

30 a $xy = c^2 \Rightarrow y = c^2x^{-1}$

$m_T = \frac{dy}{dx} = -\frac{c^2}{x^2}$

At $P\left(ct, \frac{c}{t}\right)$ $\frac{dy}{dx} = -\frac{c^2}{ct^2} = -\frac{1}{t^2}$

Gradient of normal $m_N = -\frac{1}{-\frac{1}{t^2}} = t^2$

$y - \frac{c}{t} = t^2(x - ct) \Rightarrow ty - c = t^3x - ct^4 \Rightarrow y = t^2x + \frac{c}{t} - ct^3$

Hence shown

b $\left(-\frac{4}{3}, -12\right)$ and $\left(12, \frac{4}{3}\right)$

31

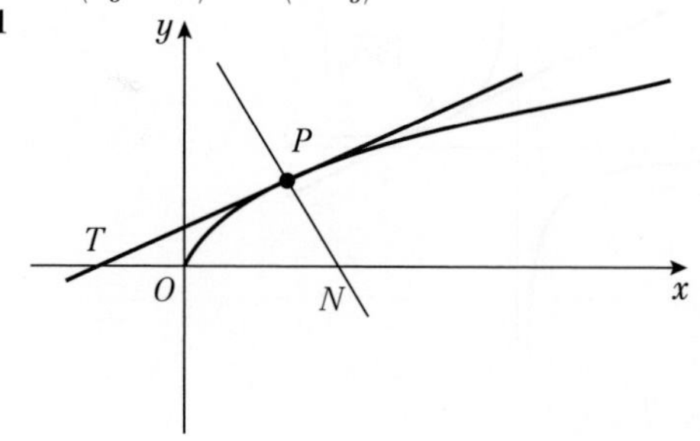

$y^2 = 4ax \Rightarrow y = 2\sqrt{a}x^{\frac{1}{2}}$

$\frac{dy}{dx} = \frac{1}{2}(2\sqrt{a})x^{-\frac{1}{2}} = \frac{\sqrt{a}}{\sqrt{x}}$

At $P\ (at^2, 2at)$ $\frac{dy}{dx} = \frac{\sqrt{a}}{\sqrt{at^2}} = \frac{1}{t}$

$y - 2at = \frac{1}{t}(x - at^2) \Rightarrow ty = x + at^2$

Finding the x coord of T,

$t \times 0 = x + at^2 \Rightarrow x = -at^2$ so coords are $(-at^2, 0)$

Length of PT

$PT^2 = (at^2 - [-at^2])^2 + (2at - 0)^2 = 4a^2t^4 + 4a^2t^2$

$= 4a^2t^2(t^2 + 1)$

The normal PN

$m_T = \frac{1}{t} \Rightarrow m_n = -t$

$y - 2at = -t(x - at^2) \Rightarrow y + tx = 2at + at^3$

Finding the x coord of N

$0 + tx = 2at + at^3 \Rightarrow x = \frac{2at + at^3}{t} = 2a + at^2$

Length of PN

$PN^2 = (at^2 - (2a + at^2)) + (2at - 0)^2 = 4a^2 + 4a^2t^2 = 4a^2(1 + t^2)$

$\frac{PT^2}{PN^2} = \frac{4a^2t^2(t^2 + 1)}{4a^2(t^2 + 1)} = t^2 \Rightarrow \frac{PT}{PN} = t$

Hence shown.

32 a $y^2 = 4ax \Rightarrow y = 2\sqrt{a}x^{\frac{1}{2}}$

$\frac{dy}{dx} = \frac{1}{2}(2\sqrt{a})x^{-\frac{1}{2}} = \frac{\sqrt{a}}{\sqrt{x}}$

At $P\ (ap^2, 2ap)$ $\frac{dy}{dx} = \frac{\sqrt{a}}{\sqrt{ap^2}} = \frac{1}{p}$

$y - 2ap = \frac{1}{p}(x - ap^2) \Rightarrow py = x + ap^2$

Hence shown.

b $(apq, a(p + q))$ **c** $p = 4 - q$

33 a $y^2 = 4ax \Rightarrow y = 2\sqrt{a}x^{\frac{1}{2}}$

$\frac{dy}{dx} = \frac{1}{2}(2\sqrt{a})x^{-\frac{1}{2}} = \frac{\sqrt{a}}{\sqrt{x}}$

At $P\ (at^2, 2at)$ $\frac{dy}{dx} = \frac{\sqrt{a}}{\sqrt{at^2}} = \frac{1}{t}$

$m_T = \frac{1}{t} \Rightarrow m_n = -t$

$y - 2at = -t(x - at^2) \Rightarrow y + tx = 2at + at^3$

Hence shown.

b $\left(a\left[\frac{t^2 + 2}{t}\right]^2, -2a\left[\frac{t^2 + 2}{t}\right]\right)$

34 a $xy = c^2 \Rightarrow y = c^2x^{-1}$

$\frac{dy}{dx} = -c^2x^{-2}$

At $P\left(ct, \frac{c}{t}\right)$, $m_T = -c^2(ct)^{-2} = -\frac{1}{t^2}$

$m_n = t^2$

$y - \frac{c}{t} = t^2(x - ct) \Rightarrow y = t^2x - ct^3 \Rightarrow y = t^2x + \frac{c}{t} - ct^3$

Hence shown.

b $\left(-\frac{c}{t^3}, -ct^3\right)$

c $X = \dfrac{ct + \left(-\frac{c}{t^3}\right)}{2} = \dfrac{c(t^4 - 1)}{2t^3}$

$Y = \dfrac{\frac{c}{t} + (-ct^3)}{2} = \dfrac{c(1 - t^4)}{2t}$

$\dfrac{X}{Y} = \dfrac{\frac{c(t^4 - 1)}{2t^3}}{\frac{c(1 - t^4)}{2t}} = -\dfrac{1}{t^2}$

Hence shown.

35 a $xy = c^2 \Rightarrow y = c^2x^{-1}$

$\dfrac{dy}{dx} = -c^2x^{-2}$

At $P\left(cp, \dfrac{c}{p}\right)$, $m_T = -c^2(cp)^{-2} = -\dfrac{1}{p^2}$

$y - \dfrac{c}{p} = -\dfrac{1}{p^2}(x - cp) \Rightarrow y = -\dfrac{x}{p^2} + \dfrac{2c}{p} \Rightarrow p^2y = -x + 2cp$

Hence shown.

b The tangent at Q is

$q^2y = -x + 2cq$ **1**

The tangent at P is

$p^2y = -x + 2cp$ **2**

2 - 1

$(p^2 - q^2)y = 2c(p - q) \Rightarrow y = \dfrac{2c(p - q)}{(p^2 - q^2)}$

$\Rightarrow y = \dfrac{2c(p - q)}{(p - q)(p + q)} = \dfrac{2c}{(p + q)}$

Hence shown.

c 1

36 a $y^2 = 4ax \Rightarrow y = 2\sqrt{a}x^{\frac{1}{2}}$

$\dfrac{dy}{dx} = \dfrac{1}{2}(2\sqrt{a})x^{-\frac{1}{2}} = \dfrac{\sqrt{a}}{\sqrt{x}}$

At $P\ (ap^2, 2ap)$ $\dfrac{dy}{dx} = \dfrac{\sqrt{a}}{\sqrt{ap^2}} = \dfrac{1}{p}$

$m_T = \dfrac{1}{p} \Rightarrow m_n = -p$

$y - 2ap = -p(x - ap^2) \Rightarrow y + px = 2ap + ap^3$

Hence shown.

b $q = -p - \dfrac{2}{p}$

c $p = \dfrac{2}{3}$

37 a $(8, 0)$ **b** $x = -8$

c $P(2, 8)$ $Q(32, -32)$

$\dfrac{y - 8}{-32 - 8} = \dfrac{x - 2}{32 - 2} \Rightarrow 3y + 4x = 32$ Equation of PQ

When $y = 0$

$0 + 4x = 32 \Rightarrow x = 8$

The coordinates of S satisfy the equation of PQ

Hence S lies on the line joining P and Q

d $y^2 = 32x \Rightarrow y = 4\sqrt{2}x^{\frac{1}{2}}$

P is on the upper half of the parabola where $y = +4\sqrt{2}x^{\frac{1}{2}}$

$\dfrac{dy}{dx} = \dfrac{1}{2}(4\sqrt{2})x^{-\frac{1}{2}} = \dfrac{2\sqrt{2}}{\sqrt{x}}$

At $x = 2$ $\dfrac{dy}{dx} = \dfrac{2\sqrt{2}}{\sqrt{2}} = 2$

$y - 8 = 2(x - 2) \Rightarrow y = 2x + 4$ **1**

Q is on the upper half of the parabola where $y = -4\sqrt{2}x^{\frac{1}{2}}$

$\dfrac{dy}{dx} = -\dfrac{1}{2}(4\sqrt{2})x^{-\frac{1}{2}} = -\dfrac{2\sqrt{2}}{\sqrt{x}}$

At $x = 32$ $\dfrac{dy}{dx} = -\dfrac{2\sqrt{2}}{\sqrt{32}} = -\dfrac{1}{2}$

$y - (-32) = -\dfrac{1}{2}(x - 32) \Rightarrow y = -\dfrac{1}{2}x - 16$ **2**

Finding the x-coordinate of the intersection of the tangents;

$2x + 4 = -\dfrac{1}{2}x - 16 \Rightarrow x = -8$

The equation of the directrix is $x = -8$ and hence, the intersection of the tangents lies on the directrix.

Challenge

1 $x^2 + 2ix + 5 = 0 \Rightarrow (x + i)^2 - i^2 + 5 = 0$

$\Rightarrow (x + i)^2 = -6 \Rightarrow x = -i \pm i\sqrt{6}$

2 $\alpha^4 + \beta^4 = (\alpha^2 + \beta^2) - 2(\alpha\beta)^2 \Rightarrow \alpha\beta = \dfrac{\left(-\frac{7}{4}\right)^2 + \frac{79}{16}}{2} = 2\ [\alpha\beta > 0]$

$(\alpha + \beta)^2 = \alpha^2 + \beta^2 + 2\alpha\beta = -\dfrac{7}{4} + 2 \times 2 = \dfrac{9}{4}$

$\Rightarrow \alpha + \beta = \dfrac{3}{2}\ [\alpha > \beta]$

Equation is:

$x^2 - \dfrac{3}{2}x + 2 = 0 \Rightarrow 2x^2 - 3x + 4 = 0$ $(a = 2, b = -3, c = 4)$

CHAPTER 5

Prior knowledge check

1 a $\begin{pmatrix}6\\2\end{pmatrix}$ **b** $\begin{pmatrix}-2\\11\end{pmatrix}$ **c** $\begin{pmatrix}8\\-16\end{pmatrix}$

Exercise 5A

1 a 2×2 **b** 2×1 **c** 2×3
d 1×3 **e** 1×2 **f** 3×3

2 $a = 6, b = 2$

3 a $\begin{pmatrix}8 & -1\\1 & 4\end{pmatrix}$ **b** $\begin{pmatrix}2 & 2\\-2 & -5\end{pmatrix}$ **c** $\begin{pmatrix}0 & 0\\0 & 0\end{pmatrix}$

4 a Not possible **b** $\begin{pmatrix}-2\\3\end{pmatrix}$ **c** $(1\ \ 1\ \ 4)$
d Not possible **e** $(3\ \ -1\ \ 4)$
f Not possible **g** $(-3\ \ 1\ \ -4)$

5 $a = 6, b = 3, c = 2, d = -1$

6 $a = 4, b = 3, c = 5$

7 $a = 2, b = -2, c = 2, d = 1, e = -1, f = 3$

8 a $\begin{pmatrix}6 & 0\\12 & -18\end{pmatrix}$ **b** $\begin{pmatrix}1 & 0\\2 & -3\end{pmatrix}$ **c** $\begin{pmatrix}2\\-2\end{pmatrix}$

d **A** and **B** are not the same size, so you can't subtract them.

9 a $\begin{pmatrix}13 & -4\\-1 & 6\end{pmatrix}$ **b** $\begin{pmatrix}-2 & -8\\10 & -12\end{pmatrix}$

c $\begin{pmatrix} 11 & -12 \\ 9 & -6 \end{pmatrix}$ **d** $\begin{pmatrix} \frac{9}{2} & \frac{1}{2} \\ -\frac{5}{2} & \frac{9}{2} \end{pmatrix}$

10 $k = 3, x = -1$

11 $a = 3, b = -3.5, c = -1, d = 2$

12 $a = 5, b = 5, c = -2, d = 2$

13 $k = \frac{3}{2}$

Exercise 5B

1 **a** 1×2 **b** 3×3 **c** 1×2 **d** 2×2 **e** 2×3 **f** 3×2

2 **a** $\begin{pmatrix} 3 \\ 6 \end{pmatrix}$ **b** $\begin{pmatrix} -2 & 1 \\ -4 & 7 \end{pmatrix}$

3 **a** $\begin{pmatrix} -3 & -2 & -1 \\ 3 & 3 & 0 \end{pmatrix}$ **b** $\begin{pmatrix} 1 & -4 \\ 0 & 9 \end{pmatrix}$

4 **a** Not possible **b** $\begin{pmatrix} -6 & -4 \\ -3 & -2 \end{pmatrix}$

c Not possible **d** $\begin{pmatrix} 7 \\ 0 \end{pmatrix}$

e (-8) **f** $(-7 \quad -7)$

5 $\begin{pmatrix} 2 & 6-a & 2a \\ 1 & 4 & -2 \end{pmatrix}$

6 $\begin{pmatrix} 3x+2 & 0 \\ 0 & 3x+2 \end{pmatrix}$

7 **a** $\begin{pmatrix} 8 & -3 \\ -10 & 6 \end{pmatrix}$ **b** $\begin{pmatrix} 3 & -2 \\ 20 & 13 \end{pmatrix}$ **c** $\begin{pmatrix} 13 & -21 \\ -36 & -13 \end{pmatrix}$

8 **a** $\begin{pmatrix} 3+k^2 & 2k \\ 2k & k^2-1 \end{pmatrix}$ **b** $\begin{pmatrix} 3+k^2 & 2k \\ 2k & k^2-1 \end{pmatrix}$

c $\begin{pmatrix} 7 & k \\ k & 5 \end{pmatrix}$ **d** $\begin{pmatrix} 9+2k^2 & 7k \\ 7k & 2k^2-5 \end{pmatrix}$

9 **a** $\begin{pmatrix} 1 & 4 \\ 0 & 1 \end{pmatrix}$ **b** $\begin{pmatrix} 1 & 6 \\ 0 & 1 \end{pmatrix}$ **c** $\begin{pmatrix} 1 & 2 \times k \\ 0 & 1 \end{pmatrix}$

10 **a** $\begin{pmatrix} a^2 & 0 \\ ab & 0 \end{pmatrix}$ **b** 3

11 **a** $\begin{pmatrix} -8 & -14 \\ -4 & -7 \\ 0 & 0 \end{pmatrix}$ **b** $(-16 \quad 29)$

12 **a** $\begin{pmatrix} -1 \\ 1 \\ -2 \end{pmatrix}$ **b** $(-3 \quad 2 \quad 3)$

13 **a** $\begin{pmatrix} 1 & 0 \\ 0 & 1 \end{pmatrix}$

b $\mathbf{AI} = \begin{pmatrix} 2 & -2 \\ 1 & 3 \end{pmatrix}\begin{pmatrix} 1 & 0 \\ 0 & 1 \end{pmatrix} = \begin{pmatrix} 2+0 & 0-2 \\ 1+0 & 0+3 \end{pmatrix} = \begin{pmatrix} 2 & -2 \\ 1 & 3 \end{pmatrix} = \mathbf{A}$

$\mathbf{IA} = \begin{pmatrix} 1 & 0 \\ 0 & 1 \end{pmatrix}\begin{pmatrix} 2 & -2 \\ 1 & 3 \end{pmatrix} = \begin{pmatrix} 2+0 & -2+0 \\ 0+1 & 0+3 \end{pmatrix} = \begin{pmatrix} 2 & -2 \\ 1 & 3 \end{pmatrix} = \mathbf{A}$

14 $\mathbf{AB} = \begin{pmatrix} 9 & 4 \\ 10 & 6 \end{pmatrix}$ and $\mathbf{AC} = \begin{pmatrix} 2 & 5 \\ 3 & 4 \end{pmatrix}$, so $\mathbf{AB} + \mathbf{AC} = \begin{pmatrix} 11 & 9 \\ 13 & 10 \end{pmatrix}$

$\mathbf{B} + \mathbf{C} = \begin{pmatrix} 5 & 4 \\ -1 & -1 \end{pmatrix}$

so $\mathbf{A}(\mathbf{B} + \mathbf{C}) = \begin{pmatrix} 2 & -1 \\ 3 & 2 \end{pmatrix}\begin{pmatrix} 5 & 4 \\ -1 & -1 \end{pmatrix} = \begin{pmatrix} 11 & 9 \\ 13 & 10 \end{pmatrix} = \mathbf{AB} + \mathbf{AC}$

15 $\mathbf{A}^2 = \begin{pmatrix} 1 & 2 \\ 3 & 1 \end{pmatrix}\begin{pmatrix} 1 & 2 \\ 3 & 1 \end{pmatrix} = \begin{pmatrix} 7 & 4 \\ 6 & 7 \end{pmatrix}$

$2\mathbf{A} + 5\mathbf{I} = \begin{pmatrix} 2 & 4 \\ 6 & 2 \end{pmatrix} + \begin{pmatrix} 5 & 0 \\ 0 & 5 \end{pmatrix} = \begin{pmatrix} 7 & 4 \\ 6 & 7 \end{pmatrix}$

16 $\mathbf{AB} = \begin{pmatrix} p & 3 \\ 6 & p \end{pmatrix}\begin{pmatrix} q & 2 \\ 4 & q \end{pmatrix} = \begin{pmatrix} pq+12 & 2p+3q \\ 6q+4p & 12+pq \end{pmatrix}$

$\mathbf{BA} = \begin{pmatrix} q & 2 \\ 4 & q \end{pmatrix}\begin{pmatrix} p & 3 \\ 6 & p \end{pmatrix} = \begin{pmatrix} pq+12 & 2p+3q \\ 6q+4p & 12+pq \end{pmatrix}$

17 $p = 2, q = -3$

Challenge

a e.g. $\begin{pmatrix} 0 & 1 \\ 0 & 0 \end{pmatrix}$ **b** $\begin{pmatrix} 1 & 1 \\ -1 & -1 \end{pmatrix}$

Exercise 5C

1 **a** 10 **b** 6 **c** -3 **d** 0 **e** 21 **f** 4

2 **a** -3 **b** -5 **c** $\frac{1}{4}$

3 $k = 2 - \sqrt{3}, k = 2 + \sqrt{3}$

4 $k = -4, k = 1$

5 **a** $\det\mathbf{A} = 0, \det\mathbf{B} = 0$ **b** $\begin{pmatrix} 0 & 0 \\ 0 & 0 \end{pmatrix}$

6 **a** 7 **b** $k = -2.5$ **c** $\begin{pmatrix} -13 & -11.5 \\ 2 & -2 \end{pmatrix}$

d $\det\mathbf{MN} = 26 + 23 = 49, \det\mathbf{M}\det\mathbf{N} = 7 \times 7 = 49$

Challenge

a $\begin{pmatrix} 1 & 1 \\ 1 & 1 \end{pmatrix}, \begin{pmatrix} -1 & -1 \\ 1 & 1 \end{pmatrix}, \begin{pmatrix} 1 & 1 \\ -1 & -1 \end{pmatrix}, \begin{pmatrix} -1 & 1 \\ -1 & 1 \end{pmatrix},$

$\begin{pmatrix} 1 & -1 \\ 1 & -1 \end{pmatrix}, \begin{pmatrix} 1 & -1 \\ -1 & 1 \end{pmatrix}, \begin{pmatrix} -1 & 1 \\ 1 & -1 \end{pmatrix}, \begin{pmatrix} -1 & -1 \\ -1 & -1 \end{pmatrix}$

b $\begin{pmatrix} 0 & 0 \\ 0 & 0 \end{pmatrix}, \begin{pmatrix} 1 & 0 \\ 0 & 0 \end{pmatrix}, \begin{pmatrix} 0 & 1 \\ 0 & 0 \end{pmatrix}, \begin{pmatrix} 0 & 0 \\ 1 & 0 \end{pmatrix}, \begin{pmatrix} 0 & 0 \\ 0 & 1 \end{pmatrix},$

$\begin{pmatrix} 1 & 1 \\ 0 & 0 \end{pmatrix}, \begin{pmatrix} 1 & 0 \\ 1 & 0 \end{pmatrix}, \begin{pmatrix} 0 & 1 \\ 0 & 1 \end{pmatrix}, \begin{pmatrix} 0 & 0 \\ 1 & 1 \end{pmatrix}, \begin{pmatrix} 1 & 1 \\ 1 & 1 \end{pmatrix}$

Exercise 5D

1 **a** Non-singular. Inverse = $\begin{pmatrix} 1 & 0.5 \\ 2 & 1.5 \end{pmatrix}$

b Singular **c** Singular

d Non-singular. Inverse = $\begin{pmatrix} -5 & 2 \\ 3 & -1 \end{pmatrix}$ **e** Singular

f Non-singular. Inverse = $\begin{pmatrix} -0.2 & 0.3 \\ 0.6 & -0.4 \end{pmatrix}$

2 **a** $\begin{pmatrix} -(2+a) & 1+a \\ 1+a & -a \end{pmatrix}$

b $\begin{pmatrix} \frac{-1}{a} & \frac{-3}{a} \\ \frac{1}{b} & \frac{2}{b} \end{pmatrix}$ (provided $a \neq 0, b \neq 0$)

3 **a** $(\mathbf{A}^{-1}\mathbf{A})\mathbf{BC} = \mathbf{A}^{-1} \Rightarrow (\mathbf{B}^{-1}\mathbf{B})\mathbf{C} = \mathbf{B}^{-1}\mathbf{A}^{-1}$
$\Rightarrow \mathbf{CA} = \mathbf{B}^{-1}(\mathbf{A}^{-1}\mathbf{A}) = \mathbf{B}^{-1}$

b $\begin{pmatrix} 3 & 4 \\ -1 & -1 \end{pmatrix}$

4 **a** $\mathbf{B} = \mathbf{A}^{-1}\mathbf{C}$ **b** $\begin{pmatrix} 1 & 4 \\ -1 & 2 \end{pmatrix}$

5 **a** $\mathbf{A} = \mathbf{C}^{-1}$ **b** $\begin{pmatrix} 2 & -3 \\ -3 & 5 \end{pmatrix}$

6 **a** $\frac{1}{2ab}\begin{pmatrix} 2b & -b \\ -4a & 3a \end{pmatrix}$ **b** $\begin{pmatrix} -3 & 2 \\ -1 & \frac{3}{2} \end{pmatrix}$

Chapter review 5

1 $\begin{pmatrix} 1 & 4 & 3 \\ -1 & 1 & -2 \end{pmatrix}$

2 **a** $\begin{pmatrix} \frac{3}{a} & -\frac{1}{a} \\ -\frac{2}{b} & \frac{1}{b} \end{pmatrix}$ **b** $\begin{pmatrix} -1 & 1 \\ 4 & -1 \end{pmatrix}$

3 **a** $k \neq \pm 2\sqrt{2}$ **b** $\frac{1}{k^2 - 8}\begin{pmatrix} k & 2 \\ 4 & k \end{pmatrix}$

4 **a** $m = \sqrt{2}, m = -\sqrt{2}$ **b** $\frac{1}{m^2 - 2}\begin{pmatrix} -1 & m \\ -m & 2 \end{pmatrix}$

5 **a** $\mathbf{A}^2\mathbf{B} = \mathbf{ABA} \Rightarrow \mathbf{A}^2\mathbf{B} = \mathbf{B} \Rightarrow \mathbf{A}^2 = \mathbf{BB}^{-1} \Rightarrow \mathbf{A}^2 = \mathbf{I}$

b $\mathbf{AB} = \begin{pmatrix} c & d \\ a & b \end{pmatrix}, \mathbf{BA} = \begin{pmatrix} b & a \\ d & c \end{pmatrix} \Rightarrow \begin{pmatrix} c & d \\ a & b \end{pmatrix} = \begin{pmatrix} b & a \\ d & c \end{pmatrix}$

Hence $a = d$ and $b = c$.

6 **a** $\frac{1}{p-4}\begin{pmatrix} -1 & -\frac{p}{2} \\ 1 & 2 \end{pmatrix}$ **b** $p = 3$

7 **a** $k^2 + 3k + 12$

b Determinant $= \left(k + \frac{3}{2}\right)^2 + 9.75 \geqslant 9.75$, so non-singular.

c $k = -1$

8 **a** $k > -\frac{2}{3}$ or $k < -\frac{2}{3}$

b $\frac{1}{-2-3k}\begin{pmatrix} -1 & -3 \\ -k & 2 \end{pmatrix}$

9 **a** $\frac{1}{2a^2-6}\begin{pmatrix} 2a & -2 \\ -3 & a \end{pmatrix}$

b $a = \pm\sqrt{3}$

Challenge

Let $\mathbf{A} = \begin{pmatrix} a & b \\ c & d \end{pmatrix}$, $\mathbf{B} = \begin{pmatrix} h & j \\ k & l \end{pmatrix}$. $\det\mathbf{A} = ad - bc$, $\det\mathbf{B} = hl - jk$

So $\det\mathbf{A}\det\mathbf{B} = (ad - bc)(hl - jk) = adhl - adjk - bchl + bcjk$

$\mathbf{AB} = \begin{pmatrix} ah + bk & aj + bl \\ ch + dk & cj + dl \end{pmatrix}$

$\det(\mathbf{AB}) = (ah + bk)(cj + dl) - (aj + bl)(ch + dk)$
$= adhl - adjk - bchl + bcjk = \det\mathbf{A}\det\mathbf{B}$

CHAPTER 6

Prior knowledge check

1 **a** $\begin{pmatrix} 0 & 1 \\ 1 & 3 \end{pmatrix}$ **b** $\begin{pmatrix} -5 & -3 \\ 13 & 8 \end{pmatrix}$

2 **a** -10 **b** $\frac{-1}{10}\begin{pmatrix} -2 & -1 \\ -4 & 3 \end{pmatrix}$

Exercise 6A

1 **a** Not linear **b** Not linear
c Not linear **d** Linear
e Not linear **f** Linear

2 **a** Linear $\begin{pmatrix} 2 & -1 \\ 3 & 0 \end{pmatrix}$

b Not linear ($2y + 1$ and $x - 1$ cannot be written as $ax + by$)

c Not linear (xy cannot be written as $ax + by$)

d Linear $\begin{pmatrix} 0 & 2 \\ -1 & 0 \end{pmatrix}$ **e** Linear $\begin{pmatrix} 0 & 1 \\ 1 & 0 \end{pmatrix}$

3 **a** Not linear (x^2 and y^2 cannot be written as $ax + by$)

b Linear $\begin{pmatrix} 0 & -1 \\ 1 & 0 \end{pmatrix}$ **c** Linear $\begin{pmatrix} 1 & -1 \\ 1 & -1 \end{pmatrix}$

d Linear $\begin{pmatrix} 0 & 0 \\ 0 & 0 \end{pmatrix}$ **e** Linear $\begin{pmatrix} 1 & 0 \\ 0 & 1 \end{pmatrix}$

4 **a** Linear $\begin{pmatrix} 2 & 1 \\ 0 & -1 \end{pmatrix}$ **b** Linear $\begin{pmatrix} 0 & -1 \\ 1 & 2 \end{pmatrix}$

5 **a** (1, 1), (−2, 3), (−5, 1)
b (3, −2), (14, −6), (9, −2)
c (−2, −2), (−6, 4), (−2, 10)

6 **a** (−2, 0), (0, 3), (2, 0), (0, −3)
b (−1, −1), (−1, 1), (1, 1), (1, −1)
c (−1, −1), (1, −1), (1, 1), (−1, 1)

7 **a** (−2, −1), (−4, −1), (−4, −2), (−2, −2)

b

c Rotation through 180° about (0, 0)

8 **a** (2, 0), (8, 4), (6, 8), (0, 4)

b

c Enlargement, centre (0, 0), scale factor 2

9 **a** (0, −2), (0, −6), (4, −6), (4, −2)

b

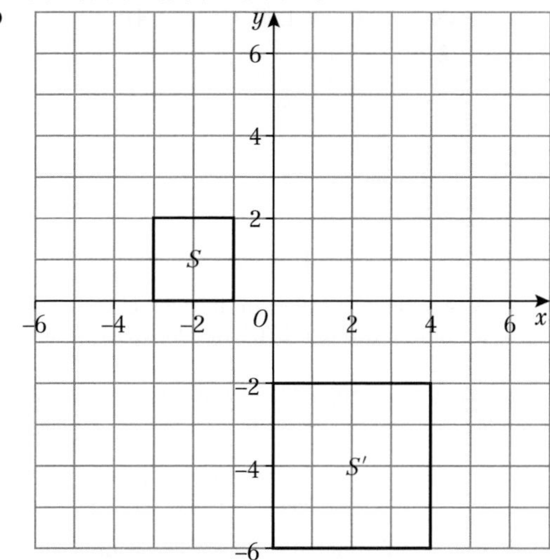

c Reflection in $y = x$ and enlargement, centre (0, 0), scale factor 2

10 **a** (4, 1), (4, 3), (1, 3)

b The transformation represented by the identity matrix leaves T unchanged.

Challenge

a $T = \begin{pmatrix} 2 & -3 \\ 1 & 1 \end{pmatrix}$ so $T\begin{pmatrix} kx \\ ky \end{pmatrix} = \begin{pmatrix} 2 & -3 \\ 1 & 1 \end{pmatrix}\begin{pmatrix} kx \\ ky \end{pmatrix} = \begin{pmatrix} 2kx - 3ky \\ kx + ky \end{pmatrix}$

$= k\begin{pmatrix} 2x - 3y \\ x + y \end{pmatrix} = kT\begin{pmatrix} x \\ y \end{pmatrix}$

b $T\left(\begin{pmatrix}x_1\\y_1\end{pmatrix}+\begin{pmatrix}x_2\\y_2\end{pmatrix}\right) = T\begin{pmatrix}x_1+x_2\\y_1+y_2\end{pmatrix}$

$= \begin{pmatrix}2&-3\\1&1\end{pmatrix}\begin{pmatrix}x_1+x_2\\y_1+y_2\end{pmatrix} = \begin{pmatrix}2(x_1+x_2)-3(y_1+y_2)\\x_1+x_2+y_1+y_2\end{pmatrix}$

$= \begin{pmatrix}2x_1-3y_1\\x_1+y_1\end{pmatrix}+\begin{pmatrix}2x_2-3y_2\\x_2+y_2\end{pmatrix} = T\begin{pmatrix}x_1\\y_1\end{pmatrix}+T\begin{pmatrix}x_2\\y_2\end{pmatrix}$

Exercise 6B

1 a $\begin{pmatrix}1&0\\0&-1\end{pmatrix}$

b $\begin{pmatrix}1&0\\0&-1\end{pmatrix}\begin{pmatrix}1\\3\end{pmatrix}=\begin{pmatrix}1\\-3\end{pmatrix}$ so $A' = (1, -3)$

$\begin{pmatrix}1&0\\0&-1\end{pmatrix}\begin{pmatrix}3\\3\end{pmatrix}=\begin{pmatrix}3\\-3\end{pmatrix}$ so $B' = (3, -3)$

$\begin{pmatrix}1&0\\0&-1\end{pmatrix}\begin{pmatrix}3\\2\end{pmatrix}=\begin{pmatrix}3\\-2\end{pmatrix}$ so $C' = (3, -2)$

2 a $\begin{pmatrix}0&-1\\-1&0\end{pmatrix}$

b $\begin{pmatrix}0&-1\\-1&0\end{pmatrix}\begin{pmatrix}1\\1\end{pmatrix}=\begin{pmatrix}-1\\-1\end{pmatrix}$ so $P' = (-1, -1)$

$\begin{pmatrix}0&-1\\-1&0\end{pmatrix}\begin{pmatrix}1\\3\end{pmatrix}=\begin{pmatrix}-3\\-1\end{pmatrix}$ so $Q' = (-3, -1)$

$\begin{pmatrix}0&-1\\-1&0\end{pmatrix}\begin{pmatrix}2\\3\end{pmatrix}=\begin{pmatrix}-3\\-2\end{pmatrix}$ so $R' = (-3, -2)$

$\begin{pmatrix}0&-1\\-1&0\end{pmatrix}\begin{pmatrix}2\\1\end{pmatrix}=\begin{pmatrix}-1\\-2\end{pmatrix}$ so $S' = (-1, -2)$

3 a $\begin{pmatrix}0&-1\\1&0\end{pmatrix}$ **b** $\begin{pmatrix}0&1\\-1&0\end{pmatrix}$ **c** $\frac{1}{2}\begin{pmatrix}\sqrt2&-\sqrt2\\\sqrt2&\sqrt2\end{pmatrix}$

d $\frac{1}{2}\begin{pmatrix}-\sqrt3&1\\-1&-\sqrt3\end{pmatrix}$ **e** $\frac{1}{2}\begin{pmatrix}-\sqrt2&\sqrt2\\-\sqrt2&-\sqrt2\end{pmatrix}$

4 a $A' = (-1, 1)$, $B' = (-1, 4)$, $C' = (-2, 4)$

b $A' = \left(-\frac{\sqrt3}{2}-\frac{1}{2}, \frac{1}{2}-\frac{\sqrt3}{2}\right)$, $B' = \left(-2\sqrt3-\frac{1}{2}, 2-\frac{\sqrt3}{2}\right)$,

$C' = \left(-2\sqrt3-1, 2-\sqrt3\right)$

5 a $P' = (2, -2)$, $Q' = (3, -2)$, $R' = (3, -4)$, $S' = (2, -4)$

b $P' = \left(0, -2\sqrt2\right)$, $Q' = \left(\frac{\sqrt2}{2}, -\frac{5\sqrt2}{2}\right)$,

$R' = \left(-\frac{\sqrt2}{2}, -\frac{7\sqrt2}{2}\right)$, $S' = \left(-\sqrt2, -3\sqrt2\right)$

6 a **A** represents a reflection in the x-axis. **B** represents a rotation through 270° anticlockwise about (0, 0).

b (3, −2) **c** $a = 0, b = 0$

7 a Rotation through 225° anticlockwise about (0, 0)

b $p = 3, q = 1$ **c** $\begin{pmatrix}\frac{1}{\sqrt2}&\frac{1}{\sqrt2}\\-\frac{1}{\sqrt2}&\frac{1}{\sqrt2}\end{pmatrix}$

d Rotation through 45° clockwise about (0, 0); (−3, −1)

8 a Reflection in the line $y = x$

b e.g. $y = x$, $y = -x$, $y = -x + 1$ **c** $\begin{pmatrix}1&0\\0&1\end{pmatrix}$

9 a $a = -0.5$

b $\theta = 120°$; $\begin{pmatrix}-0.5&-0.866\\0.866&-0.5\end{pmatrix}$

$\theta = 240°$; $\begin{pmatrix}-0.5&0.866\\-0.866&-0.5\end{pmatrix}$

10 a $\begin{pmatrix}0&-1\\1&0\end{pmatrix}$ **b** $a = 0, b = 0$

Challenge

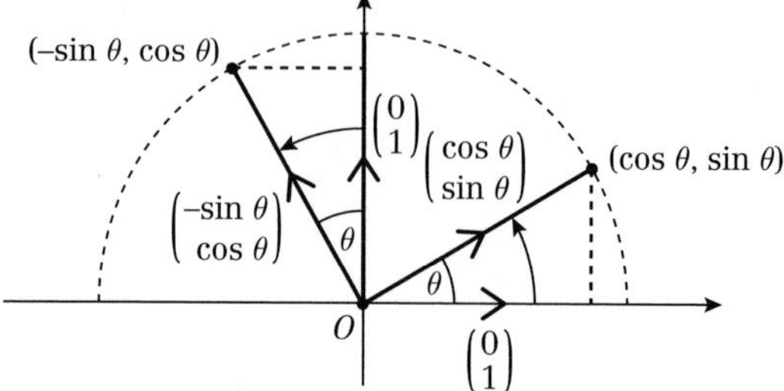

Rotating $\begin{pmatrix}1\\0\end{pmatrix}$ by θ takes it to $\begin{pmatrix}\cos\theta\\\sin\theta\end{pmatrix}$,

and rotating $\begin{pmatrix}0\\1\end{pmatrix}$ by θ takes it to $\begin{pmatrix}-\sin\theta\\\cos\theta\end{pmatrix}$,

so the matrix for the rotation is $\begin{pmatrix}\cos\theta&-\sin\theta\\\sin\theta&\cos\theta\end{pmatrix}$.

Exercise 6C

1 a $\begin{pmatrix}4&0\\0&1\end{pmatrix}$ **b** $\begin{pmatrix}1&0\\0&3\end{pmatrix}$ **c** $\begin{pmatrix}2&0\\0&2\end{pmatrix}$ **d** $\begin{pmatrix}5&0\\0&0.5\end{pmatrix}$

2 a 4 **b** 3 **c** 4 **d** 2.5

3 a (0, 0) **b** 12

4 a $\begin{pmatrix}-2&0\\0&1\end{pmatrix}$ **b** $\begin{pmatrix}-3&0\\0&4\end{pmatrix}$ **c** $\begin{pmatrix}-\frac{1}{2}&0\\0&-\frac{1}{2}\end{pmatrix}$

5 a Stretch parallel of the x-axis, scale factor 2 and stretch parallel to the y-axis, scale factor −3

b $k = 4$

6 a (3, 9), (15, 9), (15, 6) **b** 18

7 a (4, 0), (8, 0), (8, −15), (4, −15)

b 60

8 a Enlargement, centre (0, 0), scale factor $2\sqrt5$

b $a = 1$ or $a = 7$

9 a $\begin{pmatrix}p^2+p&p+q\\p^2+qp&p+q^2\end{pmatrix}$ **b** $p = -3, q = 3$

10 a $\begin{pmatrix}8&0\\0&-8\end{pmatrix}$

b Stretch parallel to the x-axis, scale factor 8; and stretch parallel to the y-axis, scale factor −8. Or enlargement scale factor 8 and centre (0, 0) and reflection in the x-axis.

c $k = \frac{7}{2}$ or $k = -\frac{3}{2}$

11 36

12 $k = 2$

13 a $\begin{pmatrix}\frac{1}{\sqrt2}&-\frac{1}{\sqrt2}\\\frac{1}{\sqrt2}&\frac{1}{\sqrt2}\end{pmatrix}$ **b** $(0,0)$, $(0,7\sqrt2)$, $\left(\frac{5}{\sqrt2},\frac{1}{\sqrt2}\right)$

c $\det \mathbf{M} = \frac{1}{\sqrt2}\times\frac{1}{\sqrt2}-\left(-\frac{1}{\sqrt2}\times\frac{1}{\sqrt2}\right) = \frac{1}{2}-\left(-\frac{1}{2}\right) = 1$

d 17.5

Challenge

a 0

b For any x, y, $\begin{pmatrix}x'\\y'\end{pmatrix}=\begin{pmatrix}7&0\\0&0\end{pmatrix}\begin{pmatrix}x\\y\end{pmatrix}=\begin{pmatrix}7x\\0\end{pmatrix}$

The y-coordinate is 0, so all points (x, y) map onto the x-axis.

Exercise 6D

1 a $\begin{pmatrix}0&1\\1&0\end{pmatrix}$; Reflection in $y = x$

b $\begin{pmatrix}0 & 1\\1 & 0\end{pmatrix}$; Reflection in $y = x$

c $\begin{pmatrix}-2 & 0\\0 & -2\end{pmatrix}$; Enlargement scale factor -2, centre (0, 0)

d $\begin{pmatrix}1 & 0\\0 & 1\end{pmatrix}$; Identity (no transformation)

e $\begin{pmatrix}4 & 0\\0 & 4\end{pmatrix}$; Enlargement scale factor 4, centre (0, 0)

2 a A: $\begin{pmatrix}0 & -1\\1 & 0\end{pmatrix}$, B: $\begin{pmatrix}-1 & 0\\0 & -1\end{pmatrix}$ C: $\begin{pmatrix}1 & 0\\0 & -1\end{pmatrix}$, D: $\begin{pmatrix}-1 & 0\\0 & 1\end{pmatrix}$

b i Reflection in y-axis
ii Reflection in y-axis
iii Rotation of 180° about (0, 0)
iv Reflection in $y = -x$
v No transformataion (Identity)
vi Rotation of 90° anticlockwise about (0, 0)
vii No transformataion (Identity)

3 a $\begin{pmatrix}0 & 3\\2 & 0\end{pmatrix}$; reflection in $y = x$ with a stretch by scale factor 3 parallel to the x-axis and by scale factor 2 parallel to the y-axis.

b $\begin{pmatrix}15 & 0\\0 & -10\end{pmatrix}$; stretch by scale factor 15 parallel to the x-axis and by scale factor -10 parallel to the y-axis

c $\begin{pmatrix}0 & 5\\-5 & 0\end{pmatrix}$; enlargement by scale factor 5 about (0, 0) and rotation through 270° anticlockwise.

d $\begin{pmatrix}15 & 0\\0 & -10\end{pmatrix}$; stretch by scale factor 15 parallel to the x-axis and by scale factor -10 parallel to the y-axis

e $\begin{pmatrix}0 & 5\\-5 & 0\end{pmatrix}$; enlargement by scale factor 5 about (0, 0) and rotation through 270° anticlockwise.

f $\begin{pmatrix}0 & 15\\10 & 0\end{pmatrix}$; reflection in $y = x$ with a stretch by scale factor 15 parallel to the x-axis and by scale factor 10 parallel to the y-axis.

4 a A: $\begin{pmatrix}2 & 0\\0 & 3\end{pmatrix}$, B: $\begin{pmatrix}-2 & 0\\0 & -2\end{pmatrix}$, C: $\begin{pmatrix}4 & 0\\0 & 4\end{pmatrix}$

b i $\begin{pmatrix}-4 & 0\\0 & -6\end{pmatrix}$ **ii** $\begin{pmatrix}8 & 0\\0 & 12\end{pmatrix}$ **iii** $\begin{pmatrix}-8 & 0\\0 & -8\end{pmatrix}$

iv $\begin{pmatrix}16 & 0\\0 & 16\end{pmatrix}$ **v** $\begin{pmatrix}-16 & 0\\0 & -24\end{pmatrix}$

5 Reflection in y-axis = $\mathbf{M} = \begin{pmatrix}-1 & 0\\0 & 1\end{pmatrix}$

Reflection in line $y = -x = \mathbf{N} = \begin{pmatrix}0 & -1\\-1 & 0\end{pmatrix}$

Combined transformation = **NM**

$= \begin{pmatrix}0 & -1\\-1 & 0\end{pmatrix}\begin{pmatrix}-1 & 0\\0 & 1\end{pmatrix} = \begin{pmatrix}0 & -1\\1 & 0\end{pmatrix}$

$\begin{pmatrix}0 & -1\\1 & 0\end{pmatrix} = \begin{pmatrix}\cos 90° & -\sin 90°\\\sin 90° & \cos 90°\end{pmatrix}$

so it represents a rotation through 90° anticlockwise about (0, 0).

6 $\mathbf{T} = \begin{pmatrix}1 & 0\\0 & -1\end{pmatrix}$ and $\mathbf{U} = \begin{pmatrix}0 & -1\\1 & 0\end{pmatrix}$

$\mathbf{UT} = \begin{pmatrix}0 & -1\\1 & 0\end{pmatrix}\begin{pmatrix}1 & 0\\0 & -1\end{pmatrix} = \begin{pmatrix}0 & 1\\1 & 0\end{pmatrix}$

$\mathbf{TU} = \begin{pmatrix}1 & 0\\0 & -1\end{pmatrix}\begin{pmatrix}0 & -1\\1 & 0\end{pmatrix} = \begin{pmatrix}0 & -1\\-1 & 0\end{pmatrix} \neq \mathbf{UT}$

7 a $\begin{pmatrix}-4k & 0\\0 & 2k\end{pmatrix}$

b Stretch by scale factor $-4k$ parallel to the x-axis and by scale factor $2k$ parallel to the y-axis.

c $\mathbf{QP} = \begin{pmatrix}k & 0\\0 & k\end{pmatrix}\begin{pmatrix}-4 & 0\\0 & 2\end{pmatrix} = \begin{pmatrix}-4k & 0\\0 & 2k\end{pmatrix} = \mathbf{PQ}$ (from part a)

8 a $\begin{pmatrix}9 & 0\\0 & 16\end{pmatrix}$

b Stretch by scale factor 9 parallel to the x-axis and by scale factor 16 parallel to the y-axis.

c $\begin{pmatrix}a^2 & 0\\0 & b^2\end{pmatrix}$; stretch by scale factor a^2 parallel to the x-axis and by scale factor b^2 parallel to the y-axis.

9 a $\begin{pmatrix}0 & -1\\1 & 0\end{pmatrix}$

b Rotation of 90° anticlockwise about (0, 0)

c Rotation of 45° anticlockwise about (0, 0)

d $\begin{pmatrix}1 & 0\\0 & 1\end{pmatrix}$ (Identity matrix)

10 a $k = -3$ **b** $\theta = 45°$

11 3

12 a $\begin{pmatrix}-\frac{1}{\sqrt{2}} & \frac{1}{\sqrt{2}}\\-\frac{1}{\sqrt{2}} & -\frac{1}{\sqrt{2}}\end{pmatrix}$ **b** $\begin{pmatrix}0 & 1\\1 & 0\end{pmatrix}$ **c** $\begin{pmatrix}\frac{1}{\sqrt{2}} & -\frac{1}{\sqrt{2}}\\-\frac{1}{\sqrt{2}} & -\frac{1}{\sqrt{2}}\end{pmatrix}$

13 a $\begin{pmatrix}k^2 + 3 & 0\\0 & 3 + k^2\end{pmatrix}$

b Enlargement by scale factor $k^2 + 3$ about (0, 0).

14 $\mathbf{P}^2 = \begin{pmatrix}a & b\\b & -a\end{pmatrix}\begin{pmatrix}a & b\\b & -a\end{pmatrix} = \begin{pmatrix}a^2 + b^2 & ab - ba\\ab - ba & b^2 + a^2\end{pmatrix}$

$= \begin{pmatrix}a^2 + b^2 & 0\\0 & a^2 + b^2\end{pmatrix}$

This represents an enlargement about the origin, scale factor $a^2 + b^2$.

Challenge

a $\mathbf{P}^2 = \begin{pmatrix}\cos\theta & -\sin\theta\\\sin\theta & \cos\theta\end{pmatrix}\begin{pmatrix}\cos\theta & -\sin\theta\\\sin\theta & \cos\theta\end{pmatrix}$

$= \begin{pmatrix}\cos^2\theta - \sin^2\theta & -2\sin\theta\cos\theta\\2\sin\theta\cos\theta & \cos^2\theta - \sin^2\theta\end{pmatrix} = \begin{pmatrix}\cos 2\theta & -\sin 2\theta\\\sin 2\theta & \cos 2\theta\end{pmatrix}$

b Two successive anticlockwise rotations about the origin by an angle θ are equivalent to a single anticlockwise rotation by an angle 2θ.

Exercise 6E

1 a Rotation of 90° anticlockwise about (0, 0)

b $\begin{pmatrix}0 & 1\\-1 & 0\end{pmatrix}$

c Rotation of 270° anticlockwise about (0, 0)

2 a i Rotation of 180° about (0, 0)

ii $\mathbf{S}^2 = \begin{pmatrix}-1 & 0\\0 & -1\end{pmatrix}\begin{pmatrix}-1 & 0\\0 & -1\end{pmatrix} = \begin{pmatrix}1+0 & 0+0\\0+0 & 0+1\end{pmatrix}$

$= \begin{pmatrix}1 & 0\\0 & 1\end{pmatrix} = \mathbf{I}$

iii Rotation of 180° about (0, 0)

b i Reflection in $y = -x$

ii $\mathbf{T}^2 = \begin{pmatrix}0 & -1\\-1 & 0\end{pmatrix}\begin{pmatrix}0 & -1\\-1 & 0\end{pmatrix} = \begin{pmatrix}0+1 & 0+0\\0+0 & 1+0\end{pmatrix}$

$= \begin{pmatrix}1 & 0\\0 & 1\end{pmatrix} = \mathbf{I}$

iii Reflection in $y = -x$

c $\det\mathbf{S} = 1$; the area of a shape is unchanged by a rotation.
$\det\mathbf{T} = -1$; the area of a shape is unchanged by a reflection (the minus sign indicating that it has been reflected).

3 a $\begin{pmatrix}1 & 0\\0 & -1\end{pmatrix}$; reflection in $y = 0$
b $\begin{pmatrix}1 & 0\\0 & -1\end{pmatrix}$; reflection in $y = 0$
c $\begin{pmatrix}-1 & 0\\0 & 1\end{pmatrix}$; reflection in $x = 0$
d $\begin{pmatrix}-1 & 0\\0 & 1\end{pmatrix}$; reflection in $x = 0$

4 a $\left(-\frac{13}{2}, \frac{23}{4}\right)$ b $\begin{pmatrix}-\frac{3}{4} & -\frac{1}{4}\\ \frac{1}{2} & \frac{1}{2}\end{pmatrix}$

5 a Enlargement, scale factor 4, centre (0, 0)
b $\begin{pmatrix}\frac{1}{4} & 0\\0 & \frac{1}{4}\end{pmatrix}$ c $\left(1, \frac{3}{2}\right)$, $\left(\frac{9}{4}, \frac{7}{4}\right)$, $\left(\frac{3}{4}, \frac{1}{4}\right)$

6 a $\begin{pmatrix}\frac{1}{a} & 0\\0 & \frac{1}{b}\end{pmatrix}$ b $\left(\frac{-6}{a}, \frac{8}{b}\right)$

7 a Rotation of 330° anticlockwise about (0, 0)
b $p = -2, q = 1$

8 $\begin{pmatrix}\frac{1}{2} & -1\\-\frac{3}{2} & 2\end{pmatrix}$

9 $\left(-\frac{a}{2} - b, -2a - 3b\right)$

Chapter review 6

1 a $\begin{pmatrix}0 & -1\\1 & 0\end{pmatrix}$ b $\begin{pmatrix}-2 & 3\\-1 & 2\end{pmatrix}$
c $\begin{pmatrix}1 & 0\\0 & 1\end{pmatrix}$ (Identity matrix); four successive anticlockwise rotations of 90° about (0, 0).

2 a $\begin{pmatrix}2 & 0\\0 & -2\end{pmatrix}$; reflection in x-axis and enlargement s.f. 2, centre (0, 0)
b $\begin{pmatrix}\frac{1}{2} & 0\\0 & -\frac{1}{2}\end{pmatrix}$; reflection in x-axis and enlargement s.f. 2, centre (0, 0)

3 a $\begin{pmatrix}k & 0\\0 & k\end{pmatrix}$ b $k = -3$ c $m = 1$ or $m = -1$

4 a 4
b 30° anticlockwise about (0, 0)
c $\left(\frac{\sqrt{3}}{8}a + \frac{b}{8}, -\frac{a}{8} + \frac{\sqrt{3}}{8}b\right)$

5 a **A** represents a reflection in the line $y = x$;
B represents a rotation through 270° anticlockwise about (0, 0)
b $(-p, q)$

6 a $k = -2.8$ or $k = 14.8$
b $\begin{pmatrix}-4 & 3\\1 & -2\end{pmatrix}\begin{pmatrix}x\\-\frac{1}{3}x\end{pmatrix} = \begin{pmatrix}-5x\\ \frac{5}{3}x\end{pmatrix}$; $(-5x) + 3\left(\frac{5}{3}x\right) = 0$ so the point satisfies the equation of the original line.

7 a $\begin{pmatrix}-4 & 0\\0 & 3\end{pmatrix}$ b 5

8 a $\begin{pmatrix}\frac{1}{a} & 0\\0 & \frac{1}{a}\end{pmatrix}$ b $\left(\frac{4}{a}, \frac{7}{a}\right)$

9 $\begin{pmatrix}1 & 4\\ \frac{1}{2} & \frac{5}{2}\end{pmatrix}$

10 a $\begin{pmatrix}-4 & 1\\-3 & 2\end{pmatrix}$ b 7 c $\begin{pmatrix}-\frac{2}{5} & \frac{1}{5}\\-\frac{3}{5} & \frac{4}{5}\end{pmatrix}$

Challenge

a Let the point be $P(a, b)$: $\begin{pmatrix}0 & 1\\0 & 1\end{pmatrix}\begin{pmatrix}a\\b\end{pmatrix} = \begin{pmatrix}0 + b\\0 + b\end{pmatrix} = \begin{pmatrix}b\\b\end{pmatrix}$
So P' is (b, b); its x- and y-coordinates are equal, so it is on $y = x$
b $\begin{pmatrix}0 & 1\\0 & m\end{pmatrix}$
c If $c = 0$, then the line $ax + by = c$ does not go through the origin. Hence the origin cannot be mapped to itself, and the transformation is not linear.

CHAPTER 7

Prior knowledge check

1 a $(x + 3)(x + 2)$
b $(x - 1)(x + 4)$
c $(2x + 3)(x + 2)$

2 a $(k + 1)(1 + k + 2) = (k + 1)(k + 3)$
b $\frac{1}{2}(k + 1)^2(1 + 2k^2)$
c $(2k - 1)(k^2 + 5)$

Exercise 7A

1 a 16 b 820 c 210 d 4950
e 775 f 15 150 g 610 h 3240

2 $n = 32$

3 $k = 14$

4 a $n(2n - 1)$
b $\sum_{r=n+1}^{2n-1} r = \sum_{r=1}^{2n-1} r - \sum_{r=1}^{n} r = n(2n - 1) - \frac{1}{2}n(n + 1)$
$= \frac{1}{2}n(4n - 2 - n - 1) = \frac{1}{2}n(3n - 3) = \frac{3}{2}n(n - 1)$

5 $\sum_{r=n-1}^{2n} r = \sum_{r=1}^{2n} r - \sum_{r=1}^{n-2} r = \frac{1}{2}(2n)(2n + 1) - \frac{1}{2}(n - 2)(n - 1)$
$= \frac{1}{2}(2n(2n + 1) - (n - 1)(n - 2)) = \frac{1}{2}(3n^2 + 5n - 2)$
$= \frac{1}{2}(n + 2)(3n - 1)$

6 a $\sum_{r=1}^{n^2} r - \sum_{r=1}^{n} r = \frac{1}{2}n^2(n^2 + 1) - \frac{1}{2}n(n + 1)$
$= \frac{1}{2}n(n(n^2 + 1) - (n + 1)) = \frac{1}{2}n(n^3 + n - n - 1)$
$= \frac{1}{2}n(n^3 - 1)$
b 3276

7 a 4565 b −28 485 c 2576

8 a $\sum_{r=1}^{n}(3r + 2) = 3\sum_{r=1}^{n} r + 2\sum_{r=1}^{n} 1 = \frac{3}{2}n(n + 1) + 2n$
$= \frac{1}{2}n(3(n + 1) + 4) = \frac{1}{2}n(3n + 7)$
b $\sum_{r=1}^{2n}(5r - 4) = 5\sum_{r=1}^{2n} r - 4\sum_{r=1}^{2n} 1 = \frac{5}{2}(2n)(2n + 1) - 4(2n)$
$= 5n(2n + 1) - 8n = 10n^2 - 3n = n(10n - 3)$
c $\sum_{r=1}^{n+2}(2r + 3) = 2\sum_{r=1}^{n+2} r + 3\sum_{r=1}^{n+2} 1 = (n + 2)(n + 3) + 3(n + 2)$
$= (n + 2)((n + 3) + 3) = (n + 2)(n + 6)$
d $\sum_{r=3}^{n}(4r + 5) = 4\sum_{r=3}^{n} r + 5\sum_{r=3}^{n} 1$
$= 4\left(\sum_{r=1}^{n} r - \sum_{r=1}^{2} r\right) + 5\left(\sum_{r=1}^{n} 1 - \sum_{r=1}^{2} 1\right)$
$= 4\left(\frac{1}{2}n(n + 1) - 3\right) + 5(n - 2) = 2n^2 + 2n - 12 + 5n - 10$
$= 2n^2 + 7n - 22 = (2n + 11)(n - 2)$

9 a $\sum_{r=1}^{k}(4r-5) = 4\sum_{r=1}^{k}r - 5\sum_{r=1}^{k}1 = \frac{4}{2}k(k+1) - 5k$
$= 2k^2 + 2k - 5k = 2k^2 - 3k$
b 51

10 $a = 7, b = -3$

11 a $\sum_{r=1}^{4n-1}(3r+1) = 3\sum_{r=1}^{4n-1}r + \sum_{r=1}^{4n-1}1 = \frac{3}{2}(4n-1)(4n) + (4n-1)$
$= 6n(4n-1) + (4n-1) = (4n-1)(6n+1)$
$= 24n^2 - 2n - 1$
b 14949

12 a $\sum_{r=1}^{2k+1}(4-5r) = -5\sum_{r=1}^{2k+1}r + 4\sum_{r=1}^{2k+1}1$
$= -\frac{5}{2}(2k+1)(2k+2) + 4(2k+1)$
$= -5(2k+1)(k+1) + 4(2k+1)$
$= (2k+1)(-5(k+1)+4) = (2k+1)(-5k-1)$
$= -(2k+1)(5k+1)$
b -1525
c $\sum_{r=1}^{15}(5r-4) = -\sum_{r=1}^{15}(4-5r) = -(-540) = 540$

13 If $g(r) = 2r$, then $\sum_{r=1}^{n}g(r) = n^2 + n$. Hence $f(r) = 2r + 3$.

14 a $f(r) = 4r - 1$ **b** 210

Challenge
$n = 4$

Exercise 7B

1 a 30 **b** 22140 **c** 19270
d 24502500 **e** 379507500 **f** 173880

2 a $\sum_{r=1}^{2n}r^2 = \frac{1}{6}(2n)(2n+1)(4n+1) = \frac{1}{3}n(2n+1)(4n+1)$
b $\sum_{r=1}^{2n-1}r^2 = \frac{1}{6}(2n-1)(2n)(2(2n-1)+1)$
$= \frac{1}{6}2n(2n-1)(4n-1) = \frac{1}{3}n(2n-1)(4n-1)$
c $\sum_{r=n}^{2n}r^2 = \sum_{r=1}^{2n}r^2 - \sum_{r=1}^{n-1}r^2$
$= \frac{1}{3}n(2n+1)(4n+1) - \frac{1}{6}(n-1)(n)(2(n-1)+1)$
$= \frac{1}{6}n(2(2n+1)(4n+1) - (n-1)(2n-1))$
$= \frac{1}{6}n(16n^2 + 12n + 2 - 2n^2 + 3n - 1)$
$= \frac{1}{6}n(n+1)(14n+1)$

3 $\sum_{r=1}^{m}r^3 = \frac{1}{4}m^2(m+1)^2$ and so $\sum_{r=1}^{n+k}r^3 = \frac{1}{4}(n+k)^2(n+k+1)^2$

4 a $\sum_{r=n+1}^{3n}r^3 = \sum_{r=1}^{3n}r^3 - \sum_{r=1}^{n}r^3 = \frac{1}{4}(3n)^2(3n+1)^2 - \frac{1}{4}n^2(n+1)^2$
$= \frac{1}{4}n^2(9(3n+1)^2 - (n+1)^2)$
$= \frac{1}{4}n^2(81n^2 - n^2 + 54n - 2n + 9 - 1)$
$= n^2(20n^2 + 13n + 2) = n^2(4n+1)(5n+2)$
b 213200

5 a $\sum_{r=n}^{2n}r^3 = \sum_{r=1}^{2n}r^3 - \sum_{r=1}^{n-1}r^3 = \frac{1}{4}(2n)^2(2n+1)^2 - \frac{1}{4}(n-1)^2n^2$
$= \frac{1}{4}n^2(4(2n+1)^2 - (n-1)^2)$
$= \frac{1}{4}n^2(16n^2 + 16n + 4 - n^2 + 2n - 1)$
$= \frac{3}{4}n^2(5n^2 + 6n + 1) = \frac{3}{4}n^2(n+1)(5n+1)$
b 3159675

6 a 9425 **b** 25420
c 10507320 **d** 393825

7 a $\sum_{r=1}^{n}(r+2)(r+5) = \sum_{r=1}^{n}r^2 + 7\sum_{r=1}^{n}r + 10\sum_{r=1}^{n}1$
$= \frac{1}{6}n(n+1)(2n+1) + \frac{7}{2}n(n+1) + 10n$
$= \frac{1}{6}n(2n^2 + 3n + 1 + 21n + 21 + 60)$
$= \frac{1}{3}n(n^2 + 12n + 41)$
b 51660

8 a $\sum_{r=1}^{n}(r^2 + 3r + 1) = \sum_{r=1}^{n}r^2 + 3\sum_{r=1}^{n}r + \sum_{r=1}^{n}1$
$= \frac{1}{6}n(n+1)(2n+1) + \frac{3}{2}n(n+1) + n$
$= \frac{1}{6}n(2n^2 + 3n + 1 + 9n + 9 + 6)$
$= \frac{1}{3}n(n^2 + 6n + 8) = \frac{1}{3}n(n+2)(n+4)$
$a = 2, b = 4$
b 22000

9 a $\sum_{r=1}^{n}r^2(r-1) = \sum_{r=1}^{n}r^3 - \sum_{r=1}^{n}r^2$
$= \frac{1}{4}n^2(n+1)^2 - \frac{1}{6}n(n+1)(2n+1)$
$= \frac{1}{12}n(n+1)(3n(n+1) - 2(2n+1))$
$= \frac{1}{12}n(n+1)(3n^2 - n - 2)$
b $\sum_{r=1}^{2n-1}r^2(r-1)$
$= \frac{1}{12}(2n-1)((2n-1)+1)(3(2n-1)^2 - (2n-1) - 2)$
$= \frac{1}{12}2n(2n-1)(12n^2 - 14n + 2)$
$= \frac{1}{3}n(2n-1)(6n^2 - 7n + 1)$

10 a $\sum_{r=1}^{n}(r+1)(r+3) = \sum_{r=1}^{n}r^2 + 4\sum_{r=1}^{n}r + 3\sum_{r=1}^{n}1$
$= \frac{1}{6}n(n+1)(2n+1) + 2n(n+1) + 3n$
$= \frac{1}{6}n(2n^2 + 3n + 1 + 12n + 12 + 18)$
$= \frac{1}{6}n(2n^2 + 15n + 31)$
b $\frac{1}{6}n(14n^2 + 45n + 31)$

11 a $\sum_{r=1}^{n}(r+3)(r+4) = \sum_{r=1}^{n}r^2 + 7\sum_{r=1}^{n}r + 12\sum_{r=1}^{n}1$
$= \frac{1}{6}n(n+1)(2n+1) + \frac{7}{2}n(n+1) + 12n$
$= \frac{1}{6}n(2n^2 + 3n + 1 + 21n + 21 + 72)$
$= \frac{1}{3}n(n^2 + 12n + 47)$
b $\frac{2}{3}n(13n^2 + 48n + 47)$

12 a $\sum_{r=1}^{n}r(r+3)^2 = \sum_{r=1}^{n}r^3 + 6\sum_{r=1}^{n}r^2 + 9\sum_{r=1}^{n}r$
$= \frac{1}{4}n^2(n+1)^2 + n(n+1)(2n+1) + \frac{9}{2}n(n+1)$
$= \frac{1}{4}n(n+1)(n(n+1) + 4(2n+1) + 18)$
$= \frac{1}{4}n(n+1)(n^2 + 9n + 22)$
b 59070

13 a $\sum_{r=1}^{kn}(2r-1) = 2\sum_{r=1}^{kn}r - \sum_{r=1}^{kn}1 = kn(kn+1) - kn = k^2n^2$
b $n = 9$

14 a $\sum_{r=1}^{n}(r^3 - r^2) = \sum_{r=1}^{n}r^3 - \sum_{r=1}^{n}r^2$
$= \frac{1}{4}n^2(n+1)^2 - \frac{1}{6}n(n+1)(2n+1)$
$= \frac{1}{12}n(n+1)(3n(n+1) - 2(2n+1))$
$= \frac{1}{12}n(n+1)(3n^2 - n - 2) = \frac{1}{12}n(n+1)(n-1)(3n+2)$
b $n = 4$

Challenge

a $f_1(x) = 1$, $f_2(x) = 2x - 1$, $f_3(x) = 3x^2 - 3x + 1$, $f_4(x) = 4x^3 - 6x^2 + 4x - 1$

b Given $h(x) = ax^3 + bx^2 + cx + d$,
$nh(n) = an^4 + bn^3 + cn^2 + dn$
$= \sum_{r=1}^{n}(af_4(r) + bf_3(r) + cf_2(r) + df_1(r)) = \sum_{r=1}^{n} g(r)$
for $g(r) = af_4(r) + bf_3(r) + cf_2(r) + df_1(r)$

Chapter review 7

1 **a** 55 **b** 1230 **c** 385 **d** 3025 **e** 37400 **f** 24001875 **g** 75640

2 **a** $\frac{3}{2}n^2 - \frac{7}{2}n$ **b** $\frac{1}{3}n(n + 1)(n + 2)$
c $n(n + 1)(n + 4)$ **d** $n(n + 1)(n^2 + 3n + 1)$
e $\frac{1}{6}n(n + 1)(2n - 5)$ **f** $\frac{n}{6}(n + 1)(4n + 7)$
g $\frac{1}{6}n(2n^2 + 3n - 29)$ **h** $\frac{1}{2}n(n^3 + 4n^2 + 5n + 10)$

3 27 900

4 **a** $\sum_{r=1}^{n} r^2(r - 3) = \sum_{r=1}^{n} r^3 - 3\sum_{r=1}^{n} r^2$
$= \frac{1}{4}n^2(n + 1)^2 - \frac{1}{2}n(n + 1)(2n + 1)$
$= \frac{1}{4}n(n + 1)(n(n + 1) - 2(2n + 1))$
$= \frac{1}{4}n(n + 1)(n^2 - 3n - 2)$
so $a = -3$, $b = -2$.
b 35 490

5 **a** $\sum_{r=1}^{n}(2r - 1)^2 = 4\sum_{r=1}^{n} r^2 - 4\sum_{r=1}^{n} r + \sum_{r=1}^{n} 1$
$= \frac{2}{3}n(n + 1)(2n + 1) - 2n(n + 1) + n$
$= \frac{1}{3}n(n + 1)(2(2n + 1) - 6) + n = \frac{1}{3}n(n + 1)(4n - 4) + n$
$= \frac{1}{3}n(4n^2 - 4 + 3) = \frac{1}{3}n(2n - 1)(2n + 1)$
b $\frac{2}{3}n(4n - 1)(4n + 1)$

6 **a** $\sum_{r=1}^{n} r(r + 2) = \sum_{r=1}^{n} r^2 + 2\sum_{r=1}^{n} r$
$= \frac{1}{6}n(n + 1)(2n + 1) + n(n + 1) = \frac{1}{6}n(n + 1)(2n + 7)$
b 9160

7 **a** $\sum_{r=n+1}^{2n} r^2 = \sum_{r=1}^{2n} r^2 - \sum_{r=1}^{n} r^2$
$= \frac{1}{6}(2n)((2n) + 1)(2(2n) + 1) - \frac{1}{6}n(n + 1)(2n + 1)$
$= \frac{1}{6}n(2n + 1)(2(4n + 1) - n - 1) = \frac{1}{6}n(2n + 1)(7n + 1)$
b 8215

8 **a** $\sum_{r=1}^{n}(r^2 - r - 1) = \sum_{r=1}^{n} r^2 - \sum_{r=1}^{n} r - \sum_{r=1}^{n} 1$
$= \frac{1}{6}n(n + 1)(2n + 1) - \frac{1}{2}n(n + 1) - n = \frac{1}{3}n(n^2 - 4)$
b 21 049 **c** $n = 7$

9 **a** $\sum_{r=1}^{n} r(2r^2 + 1) = 2\sum_{r=1}^{n} r^3 + \sum_{r=1}^{n} r = \frac{1}{2}n^2(n + 1)^2 + \frac{1}{2}n(n + 1)$
$= \frac{1}{2}n(n + 1)(n(n + 1) + 1) = \frac{1}{2}n(n + 1)(n^2 + n + 1)$
b $100\sum_{r=1}^{n} r^2 - \sum_{r=1}^{n} r = \frac{1}{6}n(n + 1)(200n + 97)$
Now if $\sum_{r=1}^{n} r(2r^2 + 1) = \sum_{r=1}^{n}(100r^2 - r)$, then
$= \frac{1}{2}n(n + 1)(n^2 + n + 1) = \frac{1}{6}n(n + 1)(200n + 97)$
$= \frac{1}{6}n(n + 1)(3(n^2 + n + 1) - (200n + 97)) = 0$
$= \frac{1}{6}n(n + 1)(3n^2 - 197n - 94) = 0.$
But $n \neq 0$, $n \neq -1$ and $3n^2 - 197n - 94$ has discriminant 39 937 which is not square.

10 **a** $\sum_{r=1}^{n} r(r + 1)^2 = \sum_{r=1}^{n} r^3 + 2\sum_{r=1}^{n} r^2 + \sum_{r=1}^{n} r$
$= \frac{1}{4}n^2(n + 1)^2 + \frac{1}{3}n(n + 1)(2n + 1) + \frac{1}{2}n(n + 1)$
$= \frac{1}{12}n(n + 1)(n + 2)(3n + 5)$
b $n = 10$

11 $n = 15$

Challenge

a $\sum_{i=1}^{n} \frac{1}{6}i(i + 1)(2i + 1) = \sum_{i=1}^{n}(\frac{1}{3}i^3 + \frac{1}{2}i^2 + \frac{1}{6}i)$
Then use the formulae for sums of i, i^2, i^3 to obtain the result.

b $\sum_{j=1}^{n}\left(\sum_{i=1}^{j}\left(\sum_{r=1}^{i} r\right)\right) = \sum_{j=1}^{n}\left(\frac{1}{2}\sum_{i=1}^{j} i(i + 1)\right)$
$= \frac{1}{2}\sum_{j=1}^{n}\left(\sum_{i=1}^{j} i^2\right) + \frac{1}{2}\sum_{j=1}^{n}\left(\sum_{i=1}^{j} i\right)$
$= \frac{1}{24}n(n + 1)^2(n + 2) + \frac{1}{4}\sum_{j=1}^{n} j^2 + \frac{1}{4}\sum_{j=1}^{n} j$
$= \frac{1}{24}n(n + 1)^2(n + 2) + \frac{1}{24}n(n + 1)(2n + 1) + \frac{1}{8}n(n + 1)$
$= \frac{1}{24}n(n + 1)(n^2 + 5n + 6)$
$= \frac{1}{24}n(n + 1)(n + 2)(n + 3)$

CHAPTER 8

Prior knowledge check

1 **a** $\frac{1}{2}n(n + 1)$ **b** $\frac{1}{6}(n + 1)(n + 2)(2n + 3)$

2 $3^{n+2} - 3^n = 3^n(9 - 1) = 3^n \times 8$

3 1, 4, 7, 10

4 $\begin{pmatrix} 5k & 4 - 8k \\ 5k + 5 & -8k - 14 \end{pmatrix}$

Exercise 8A

1 Basis: When $n = 1$: LHS = 1; RHS = $\frac{1}{2}(1)(1 + 1) = 1$
Assumption: $\sum_{r=1}^{k} r = \frac{1}{2}k(k + 1)$
Induction: $\sum_{r=1}^{k+1} r = \sum_{r=1}^{k} r + (k + 1) = \frac{1}{2}k(k + 1) + (k + 1)$
$= \frac{1}{2}(k + 1)(k + 2)$
So if the statement holds for $n = k$, it holds for $n = k + 1$.
Conclusion: The statement holds for all $n \in \mathbb{Z}^+$.

2 Basis step: When $n = 1$: LHS=1; RHS=$\frac{1}{4}(1)^2(1 + 1)^2 = 1$
Assumption: $\sum_{r=1}^{k} r^3 = \frac{1}{4}k^2(k + 1)^2$
Induction: $\sum_{r=1}^{k+1} r^3 = \sum_{r=1}^{k} r^3 + (k + 1)^3 = \frac{1}{4}k^2(k + 1)^2 + (k + 1)^3$
$= \frac{1}{4}(k + 1)^2(k^2 + 4(k + 1)) = \frac{1}{4}(k + 1)^2(k + 2)^2$
So if the statement holds for $n = k$, it holds for $n = k + 1$.
Conclusion: The statement holds for all $n \in \mathbb{Z}^+$.

3 **a** Basis: $n = 1$: LHS = 0; RHS = $\frac{1}{3}(1)(1 + 1)(1 - 1) = 0$
Assumption: $\sum_{r=1}^{k} r(r - 1) = \frac{1}{3}k(k + 1)(k - 1)$

Induction: $\sum_{r=1}^{k+1} r(r-1) = \sum_{r=1}^{k} r(r-1) + (k+1)k$
$= \frac{1}{3}k(k+1)(k-1) + k(k+1)$
$= \frac{1}{3}k(k+1)(k-1+3) = \frac{1}{3}k(k+1)(k+2)$
So if the statement holds for $n = k$, it holds for $n = k + 1$.
Conclusion: The statement holds for all $n \in \mathbb{Z}^+$.

b $\sum_{r=1}^{2n+1} r(r-1) = \frac{1}{3}(2n+1)((2n+1)+1)((2n+1)-1)$
$= \frac{4}{3}n(2n+1)(n+1)$

4 a Basis: $n = 1$: LHS = 2; RHS = $(1)^2(1+1) = 2$
Assumption: $\sum_{r=1}^{k} r(3r-1) = k^2(k+1)$
Induction: $\sum_{r=1}^{k+1} r(3r-1) = \sum_{r=1}^{k} r(3r-1) + (k+1)(3k+2)$
$= k^2(k+1) + (k+1)(3k+2)$
$= (k+1)^2(k+2)$
So if the statement holds for $n = k$, it holds for $n = k + 1$.
Conclusion: The statement holds for all $n \in \mathbb{Z}^+$.

b $n = 15$

5 a Basis: $n = 1$: LHS = $\frac{1}{2}$; RHS=$1 - \frac{1}{2} = \frac{1}{2}$
Assumption: $\sum_{r=1}^{k} \left(\frac{1}{2}\right)^r = 1 - \frac{1}{2^k}$
Induction: $\sum_{r=1}^{k+1} \left(\frac{1}{2}\right)^r = \sum_{r=1}^{k} \left(\frac{1}{2}\right) + \frac{1}{2^{k+1}} = 1 - \frac{1}{2^k} + \frac{1}{2^{k+1}}$
$= 1 - \frac{2}{2^{k+1}} + \frac{1}{2^{k+1}} = 1 - \frac{1}{2^{k+1}}$
So if the statement holds for $n = k$, it holds for $n = k + 1$.
Conclusion: The statement holds for all $n \in \mathbb{Z}^+$.

b Basis: $n = 1$: LHS = $1 \times 1! = 1$; RHS = $(1+1)! - 1 = 1$
Assumption: $\sum_{r=1}^{k} r(r!) = (n+1)! - 1$
Induction: $\sum_{r=1}^{k+1} r(r!) = \sum_{r=1}^{k} r(r!) + (k+1)(k+1)!$
$= (k+1)!-1 + (k+1)(k+1)!$
$= (k+1)!(k+2) - 1 = ((k+1)+1)! - 1$
So if the statement holds for $n = k$, it holds for $n = k + 1$.
Conclusion: The statement holds for all $n \in \mathbb{Z}^+$.

c Basis: $n = 1$: LHS = $\frac{4}{1 \times 3} = \frac{4}{3}$; RHS = $\frac{1 \times 8}{2 \times 3} = \frac{4}{3}$
Assumption: $\sum_{r=1}^{k} \frac{4}{r(r+2)} = \frac{k(3k+5)}{(k+1)(k+2)}$
Induction: $\sum_{r=1}^{k+1} \frac{4}{r(r+2)} = \sum_{r=1}^{k} \frac{4}{r(r+2)} + \frac{4}{(k+1)(k+3)}$
$= \frac{k(3k+5)}{(k+1)(k+2)} + \frac{4}{(k+1)(k+3)}$
$= \frac{k(3k+5)(k+3)}{(k+1)(k+2)(k+3)} + \frac{4(k+2)}{(k+1)(k+2)(k+3)}$
$= \frac{k(3k+5)(k+3) + 4(k+2)}{(k+1)(k+2)(k+3)} = \frac{(k+1)(3k+8)}{(k+2)(k+3)}$
$= \frac{(k+1)(3(k+1)+5)}{((k+1)+1)((k+1)+2)}$
So if the statement holds for $n = k$, it holds for $n = k + 1$.
Conclusion: The statement holds for all $n \in \mathbb{Z}^+$.

6 a The student has just stated and not shown that the statement is true for $n = k + 1$.
b e.g. $n = 2$: LHS = $(1+2)^2 = 9$; RHS = $1^2 + 2^2 \neq 9$, so that LHS $\neq$ RHS.

7 a The student has not completed the basis step.
b e.g. $n = 1$: LHS = 1; RHS=$\frac{1}{2}(1^2 + 1 + 1) = \frac{3}{2} \neq 1$

Challenge
Basis: $n = 1$: LHS = $(-1)^1 \times 1^2 = -1$;
RHS = $\frac{1}{2}(-1)^1(1)(1+1) = -1$
Assumption: $\sum_{r=1}^{k} (-1)^r r^2 = \frac{1}{2}(-1)^k k(k+1)$
Induction: $\sum_{r=1}^{k+1} (-1)^r r^2 = \sum_{r=1}^{k} (-1)^r r^2 + (-1)^{k+1}(k+1)^2$
$= \frac{1}{2}(-1)^k k(k+1) + (-1)^{k+1}(k+1)^2$
$= \frac{1}{2}(-1)^{k+1}(k+1)(-k + 2(k+1)) = \frac{1}{2}(-1)^{k+1}(k+1)(k+2)$
So if the statement holds for $n = k$, it holds for $n = k + 1$.
Conclusion: The statement holds for all $n \in \mathbb{Z}^+$.

Exercise 8B

1 a Let $f(n) = 8^n - 1$ where $n \in \mathbb{Z}^+$.
Basis: $n = 1$: $f(1) = 8 - 1 = 7$ is divisible by 7.
Assumption: $f(k)$ is divisible by 7.
Induction: $f(k+1) = 8^{k+1} - 1 = 8 \times 8^k - 1 = 8f(k) + 7$
So if the statement holds for $n = k$, it holds for $n = k + 1$.
Conclusion: The statement holds for al $n \in \mathbb{Z}^+$.

b Let $f(n) = 3^{2n} - 1$ where $n \in \mathbb{Z}^+$.
Basis: $n = 1$: $f(1) = 3^2 - 1 = 8$ is divisible by 8.
Assumption: f(k) is divisible by 8.
Induction: $f(k+1) = 3^{2(k+1)} - 1 = 3^{2k} \times 3^2 - 1$
$f(k+1) - f(k) = (3^{2k} \times 3^2 - 1) - (3^{2k} - 1) = 8 \times 3^{2k}$
So if the statement holds for $n = k$, it holds for $n = k + 1$.
Conclusion: The statement holds for all $n \in \mathbb{Z}^+$.

c Let $f(n) = 5^n + 9^n + 2$ where $n \in \mathbb{Z}^+$.
Basis: $n = 1$: $f(1) = 5 + 9 + 2 = 16$ is divisible by 4.
Assumption: $f(k)$ is divisible by 4.
Induction: $f(k+1) = 5^{k+1} + 9^{k+1} + 2$
$= 5 \times 5^k + 9 \times 9^k + 2$
$f(k+1) - f(k) = (5 \times 5^k + 9 \times 9^k + 2) - (5^k + 9^k + 2)$
$= 4 \times 5^k + 8 \times 9^k$
So if the statement holds for $n = k$, it holds for $n = k + 1$.
Conclusion: The statement holds for all $n \in \mathbb{Z}^+$.

d Let $f(n) = 2^{4n} - 1$ where $n \in \mathbb{Z}^+$.
Basis: $n = 1$: $f(1) = 2^4 - 1 = 15$ is divisible by 15.
Assumption: $f(k)$ is divisible by 15.
Induction: $f(k+1) = 2^{4(k+1)} - 1 = 16 \times 2^{4k} - 1$
$f(k+1) - f(k) = (16 \times 2^{4k} - 1) - (2^{4k} - 1)$
$= 15 \times 2^{4k}$
So if the statement holds for $n = k$, it holds for $n = k + 1$.
Conclusion: The statement holds for all $n \in \mathbb{Z}^+$.

e Let $f(n) = 3^{2n-1} + 1$ where $n \in \mathbb{Z}^+$
Basis: $n = 1$: $f(1) = 3^{2-1} + 1 = 4$ is divisible by 4.
Assumption: $f(k)$ is divisible by 4.
Induction: $f(k+1) = 3^{2(k+1)-1} + 1 = 3^2 \times 3^{2k-1} + 1$
$f(k+1) - f(k) = (3^2 \times 3^{2k-1} + 1) - (3^{2k-1} + 1)$
$= 8 \times 3^{2k-1}$
So if the statement holds for $n = k$, it holds for $n = k + 1$.
Conclusion: The statement holds for all $n \in \mathbb{Z}^+$.

f Let $f(n) = n^3 + 6n^2 + 8n$ where $n \in \mathbb{Z}^+$
Basis: $n = 1$: $f(1) = 1^3 + 6 \times 1^2 + 8 = 15$ is divisible by 3.
Assumption: $f(k)$ is divisible by 3.
Induction: $f(k + 1) = (k + 1)^3 + 6(k + 1)^2 + 8(k + 1)$
$= k^3 + 9k^2 + 23k + 15$
$f(k + 1) - f(k) = 3(k^2 + 5k + 5)$
So if the statement holds for $n = k$, it holds for $n = k + 1$.
Conclusion: The statement holds for all $n \in \mathbb{Z}^+$.

g Let $f(n) = n^3 + 5n$ where $n \in \mathbb{Z}^+$.
Basis: $n = 1$: $f(1) = 1^3 + 5 = 6$ is divisible by 6.
Assumption: $f(k)$ is divisible by 6.
Induction: $f(k + 1) = (k + 1)^3 + 5(k + 1)$
$= k^3 + 3k^2 + 8k + 6$
$f(k + 1) - f(k) = 3k^2 + 3k + 6$
$= 3k(k + 1) + 6$
where $3k(k + 1)$ is divisible by 3, and one of k and $k + 1$ must be even, so $3k(k + 1)$ is divisible by 6. Therefore, if the statement holds for $n = k$, it holds for $n = k + 1$.
Conclusion: The statement holds for all $n \in \mathbb{Z}^+$.

h Let $f(n) = 2^n \times 3^{2n} - 1$ where $n \in \mathbb{Z}^+$.
Basis: $f(1) = 2 \times 3^2 - 1 = 17$ is divisible by 17.
Assumption: $f(k)$ is divisible by 17.
Induction: $f(k + 1) = 2^{k+1} \times 3^{2(k+1)} - 1$
$= 18 \times 2^k \times 3^{2k} - 1$
$f(k + 1) - f(k) = (18 \times 2^k \times 3^{2k} - 1) - (2^k \times 3^{2k} - 1)$
$= 17 \times 2^k \times 3^{2k}$
So if the statement holds for $n = k$, it holds for $n = k + 1$.
Conclusion: The statement holds for all $n \in \mathbb{Z}^+$.

2 a $f(k + 1) = 13^{k+1} - 6^{k+1} = 13 \times 13^k - 6 \times 6^k$
$= 6(13^k - 6^k) + 7 \times 13^k = 6f(k) + 7(13^k)$

b Basis: $n = 1$: $f(1) = 13 - 6 = 7$ is divisible by 7.
Assumption: $f(k)$ is divisible by 7.
Induction: $f(k + 1) = 6f(k) + 7(13^k)$ by part **a**.
So if the statement holds for $n = k$, it holds for $n = k + 1$.
Conclusion: The statement holds for all $n \in \mathbb{Z}^+$.

3 a $g(k + 1) = 5^{2(k+1)} - 6(k + 1) + 8 = 25 \times 5^{2k} - 6k + 2$
$= 25(5^{2k} - 6k + 8) + 144k - 198$
$= 25g(k) + 9(16k - 22)$

b Basis: $n = 1$: $g(1) = 5^2 - 6 + 8 = 27$ is divisible by 9.
Assumption: $g(k)$ is divisible by 9.
Induction: $g(k + 1) = 25\,g(k) + 9(16k - 22)$ by part **a**.
So if the statement holds for $n = k$, it holds for $n = k + 1$.
Conclusion: The statement holds for all $n \in \mathbb{Z}^+$.

4 Let $f(n) = 8^n - 3^n$ where $n \in \mathbb{Z}^+$.
Basis: $n = 1$: $f(1) = 8 - 3 = 5$ is divisible by 5.
Assumption: $f(k)$ is divisible by 5.
Induction: $f(k + 1) = 8^{k+1} - 3^{k+1} = 8 \times 8^k - 3 \times 3^k$
$f(k + 1) - 3f(k) = (8 \times 8^k - 3 \times 3^k) - 3(8^k - 3^k)$
$= 5 \times 8^k$
So if the statement holds for $n = k$, it holds for $n = k + 1$.
Conclusion: The statement holds for all $n \in \mathbb{Z}^+$.

5 Let $f(n) = 3^{2n+2} + 8n - 9$ where $n \in \mathbb{Z}^+$.
Basis: $n = 1$: $f(1) = 3^4 + 8 - 9 = 80$ is divisible by 8.
Assumption: $f(k)$ is divisible by 8.
Induction: $f(k + 1) = 3^{2(k+1)+2} + 8(k + 1) - 9 = 9 \times 3^{2k+2} + 8k-1$
$f(k + 1) - f(k) = (9 \times 3^{2k+2} + 8k - 1) - (3^{2k+2} + 8k - 9)$
$= 8 \times 3^{2k+2} + 8 = 8(3^{2k+2} + 1)$
So if the statement holds for $n = k$, it holds for $n = k + 1$.
Conclusion: By induction, the statement holds for all $n \in \mathbb{Z}^+$.

6 Let $f(n) = 2^{6n} + 3^{2n-2}$ where $n \in \mathbb{Z}^+$
Basis: $n = 1$: $f(1) = 2^6 + 3^0 = 65$ is divisible by 5.
Assumption: $f(k)$ is divisible by 5.
Induction: $f(k + 1) = 2^{6(k+1)} + 3^{2(k+1)-2} = 64 \times 2^{6k} + 9 \times 3^{2k-2}$
$f(k + 1) + f(k) = (64 \times 2^{6k} + 9 \times 3^{2k-2}) + (2^{6k} + 3^{2k-2})$
$= 65 \times 2^{6k} + 10 \times 3^{2k-2}$
So if the statement holds for $n = k$, it holds for $n = k + 1$.
Conclusion: The statement holds for all $n \in \mathbb{Z}^+$.

Exercise 8C

1 Basis: $u_n = 5^n - 1 \Rightarrow n = 1, u_1 = 5^1 - 1 = 4$ as given
$n = 2, u_2 = 5^2 - 1 = 24$ from the general statement
and $u_2 = 5 \times 4 + 4 = 24$ from the recurrence relationship.
So u_n is true when $n = 1$ and is also true when $n = 2$
Assumption: $n = k \Rightarrow u_k = 5^k - 1$,
Then $n = k + 1 \Rightarrow u_{k+1} = 5^{k+1} - 1$ for all $k \in \mathbb{Z}^+$
Induction: $u_{k+1} = 5\left(5^{k+1} - 1\right) + 4 = 5^{k+1} - 1$
Conclusion: If u_n is true when $n = k$, then it has been shown that u_n is also true when $n = k + 1$.
As u_n is true for $n = 1$ and $n = 2$ then u_n is true for all $n \geq 1$ and $n \in \mathbb{Z}^+$ by mathematical induction.

2 Basis: $u_n = 2^{n+2} - 5 \Rightarrow n = 1$, $u_1 = 2^3 - 5 = 3$, as given
$n = 2$, $u_2 = 2^4 - 5 = 11$ from the general statement
and $u_2 = 2 \times 3 + 5 = 11$ from the recurrence relationship.
So u_n is true when $n = 1$ and is also true when $n = 2$
Assumption: $n = k \Rightarrow u_k = 2^{k+2} - 5$
Then $n = k + 1 \Rightarrow u_{k+1} = 2^{k+3} - 5$ for all $k \in \mathbb{Z}^+$
Induction: $u_{k+1} = 2(2^{k+2} - 5) + 5 = 2^{k+3} - 10 + 5 = 2^{k+3} - 5$
Conclusion: If u_n is true when $n = k$, then it has been shown that u_n is also true when $n = k+1$. As u_n is true for $n = 1$ and $n = 2$ then u_n is true for all $n \geq 1$ and $n \in \mathbb{Z}^+$ by mathematical induction.

3 Basis: $u_n = 5^{n-1} + 2 \Rightarrow n = 1, u_1 = 5^0 + 2 = 3$, as given
$n = 2, u_2 = 5^1 + 2 = 7$ from the general statement
and $u_2 = 5 \times 3 - 8 = 17$ from the recurrence relationship.
So u_n is true when $n = 1$ and is also true when $n = 2$
Assumption: $n = k \Rightarrow u_k = 5^{k-1} + 2$,
Then $n = k + 1 \Rightarrow u_{k+1} = 5^k + 2$ for all $k \in \mathbb{Z}^+$
Induction: $u_{k+1} = 5(5^{k-1} + 2) - 8 = 5^k + 10 - 8 = 5^k + 2$
Conclusion: If u_n is true when $n = k$, then it has been shown that u_n is also true when $n = k + 1$. As u_n is true for $n = 1$ and $n = 2$ then u_n is true for all $n \geq 1$ and $n \in \mathbb{Z}^+$ by mathematical induction.

4 Basis: $u_n = 3^n - 2^n \Rightarrow n = 1, u_1 = 3^1 - 2^1 = 1$ as given,

$u_2 = 3^2 - 2^2 = 5$

$n = 2, u_2 = 3^2 - 2^2 = 5$ from the general statement

$n = 3, u_3 = 3^3 - 2^3 = 19$ from the general statement

$n = 3, u_3 = 5 \times 5 - 6 \times 1 = 19$ from the recurrence relationship So u_n is true when $n = 1$ and is also true when $n = 2$ and $n = 3$

Assumption: $n = k \Rightarrow u_k = 3^k - 2^k$,

$n = k + 1 \Rightarrow u_{k+1} = 3^{k+1} - 2^{k+1}$,

$u_{k+2} = 3^{k+2} - 2^{k+2}$ for all $k \in \mathbb{Z}^+$

Induction: $u_{k+2} = 5(3^{k+1} - 2^{k+1}) - 6(3^k - 2^k)$

$= 5 \times 3^{k+1} - 2 \times 2^{k+1} - 2 \times 3^{k+1} + 2 \times 2^{k+1}$

$u_{k+2} = 3^{k+2} - 2^{k+2}$

Conclusion: If u_n is true when $n = k$, then it has been shown that u_n is also true when $n = k + 2$.

As u_n is true for $n = 1, n = 2$ and $n = 3$ then u_n is true for all $n \geq 1$ and $n \in \mathbb{Z}^+$ by mathematical induction.

5 Basis: $u_n = (n - 2)3^{n-1} \Rightarrow n = 1, u_1 = (1 - 2)3^{1-1} = -1$, as given

$n = 2, u_2 = (2 - 2)3^{2-1} = 0$ from the general statement

$u_3 = (3 - 2)3^{3-1} = 9$ from the general statement

and from the recurrence relation

$u_3 = 6 \times 0 - 9 \times -1 = 9$

Assumption: $n = k \Rightarrow u_k = (k - 2)3^{k-1}$,

Then $n = k + 1 \Rightarrow u_{k+1} = (k - 1)\,3^k$ for all $k \in \mathbb{Z}^+$

$[u_{k+2} = (k)3^{k+1}]$

Induction: $u_{k+1} = 6(k - 1)3^k - 9(k - 2)3^{k-1}$

$= 2k \times 3^{k+1} - 2 \times 3^{k+1} - k \times 3^{k+1} + 2 \times 3^{k+1}$

$u_{k+2} = (k)3^{k+1} \Rightarrow u_k = (k - 2)3^{k-1}$

Conclusion: If u_n is true when $n = k$, then it has been shown that u_n is also true when $n = k + 2$.

As u_n is true for $n = 1$, $n = 2$ and $n = 3$ then u_n is true for all $n \geq 1$ and $n \in \mathbb{Z}^+$ by mathematical induction.

6 Basis: $u_n = 2(5^{n-1}) - 2^{n-1} \Rightarrow n = 1, u_1 = 2(5^{1-1}) - 2^{1-1} = 1$, as given

$n = 2, u_2 = 2(5^{2-1}) - 2^{2-1} = 8$ from the general statement

$u_3 = 2(5^{3-1}) - 2^{3-1} = 46$ from the general statement

and from the recurrence relation $u_3 = 7 \times 8 - 10 \times 1 = 46$

Assumption: $n = k \Rightarrow u_k = 2(5^{k-1}) - 2^{k-1}$,

Then $n = k + 1 \Rightarrow u_{k+1} = 2(5^k) - 2^k$, for all $k \in \mathbb{Z}^+$

$[u_{k+2} = 2(5^{k+1}) - 2^{k+1}]$

Induction: $u_{k+2} = 7(2(5^k) - 2^k) - 10(2(5^{k-1}) - 2^{k-1})$

$= 14 \times 5^k - 7 \times 2^k - 20 \times 5^{k-1} + 10 \times 2^{k-1}$

$u_{k+2} = 14 \times 5^k - 7 \times 2^k - 4 \times 5^k + 5 \times 2^k$

$= 10 \times 5^k - 2 \times 2^k$

$u_{k+2} = 2(5^{k+1}) - 2^{k+1}$

Conclusion

If u_n is true when $n = k$, then it has been shown that u_n is also true when $n = k + 2$.

As u_n is true for $n = 1$, $n = 2$ and $n = 3$ then u_n is true for all $n \geq 1$ and $n \in \mathbb{Z}^+$ by mathematical induction.

7 Basis: $u_n = (3n - 2)3^n \Rightarrow n = 1,\ u_1 = (3 - 2)3^1 = 3$ as given

$n = 2,\ u_2 = (3 \times 2 - 2)3^2 = 36$ from the general statement

$u_3 = (3 \times 3 - 2)3^3 = 189$ from the general statement

and from the recurrence relation $u_3 = 6 \times 36 - 9 \times 3 = 189$

Assumption: $n = k \Rightarrow u_k = (3k - 2)3^k$,

Then $n = k + 1 \Rightarrow u_{k+1} = (3k - 1)3^{k+1}$ for all $k \in \mathbb{Z}^+$

$[u_{k+2} = (3k)3^{k+2}]$

Induction: $u_{k+2} = 6((3k - 1)3^{k+1}) - 9((3k - 2)3^k)$

$= 18k \times 3^{k+1} - 6 \times 3^{k+1} - 27k \times 3^k + 18 \times 3^k$

$u_{k+2} = 6k \times 3^{k+2} - 2 \times 3^{k+2} - 3k \times 3^{k+2} + 2 \times 3^{k+2}$

$u_{k+2} = 3k(3^{k+2})$

Conclusion: If u_n is true when $n = k$, then it has been shown that u_n is also true when $n = k + 2$.

As u_n is true for $n = 1, n = 2$ and $n = 3$ then u_n is true for all $n \geq 1$ and $n \in \mathbb{Z}^+$ by mathematical induction.

Exercise 8D

1 Basis: $n = 1$: LHS = RHS = $\begin{pmatrix} 1 & 2 \\ 0 & 1 \end{pmatrix}$

Assumption: $\begin{pmatrix} 1 & 2 \\ 0 & 1 \end{pmatrix}^k = \begin{pmatrix} 1 & 2k \\ 0 & 1 \end{pmatrix}$

Induction: $\begin{pmatrix} 1 & 2 \\ 0 & 1 \end{pmatrix}^{k+1} = \begin{pmatrix} 1 & 2 \\ 0 & 1 \end{pmatrix}^k \begin{pmatrix} 1 & 2 \\ 0 & 1 \end{pmatrix} = \begin{pmatrix} 1 & 2k \\ 0 & 1 \end{pmatrix}\begin{pmatrix} 1 & 2 \\ 0 & 1 \end{pmatrix}$

$= \begin{pmatrix} 1+0 & 2+2k \\ 0+0 & 0+1 \end{pmatrix} = \begin{pmatrix} 1 & 2(k+1) \\ 0 & 1 \end{pmatrix}$

So if the statement holds for $n = k$, it holds for $n = k + 1$.

Conclusion: The statement holds for all $n \in \mathbb{Z}^+$.

2 Basis: $n = 1$: LHS = RHS = $\begin{pmatrix} 3 & -4 \\ 1 & -1 \end{pmatrix}$

Assumption: $\begin{pmatrix} 3 & -4 \\ 1 & -1 \end{pmatrix}^k = \begin{pmatrix} 2k+1 & -4k \\ k & -2k+1 \end{pmatrix}$

Induction: $\begin{pmatrix} 3 & -4 \\ 1 & -1 \end{pmatrix}^{k+1} = \begin{pmatrix} 3 & -4 \\ 1 & -1 \end{pmatrix}^k \begin{pmatrix} 3 & -4 \\ 1 & -1 \end{pmatrix}$

$= \begin{pmatrix} 2k+1 & -4k \\ k & -2k+1 \end{pmatrix}\begin{pmatrix} 3 & -4 \\ 1 & -1 \end{pmatrix}$

$= \begin{pmatrix} 6k+3-4k & -8k-4+4k \\ 3k-2k+1 & -4k+2k-1 \end{pmatrix}$

$= \begin{pmatrix} 2(k+1)+1 & -4(k+1) \\ k+1 & -2(k+1)+1 \end{pmatrix}$

So if the statement holds for $n = k$, it holds for $n = k + 1$.

Conclusion: The statement holds for all $n \in \mathbb{Z}^+$.

3 Basis: $n = 1$: LHS = RHS = $\begin{pmatrix} 2 & 0 \\ 1 & 1 \end{pmatrix}$

Assumption: $\begin{pmatrix} 2 & 0 \\ 1 & 1 \end{pmatrix}^k = \begin{pmatrix} 2^k & 0 \\ 2^k - 1 & 1 \end{pmatrix}$

Induction: $\begin{pmatrix} 2 & 0 \\ 1 & 1 \end{pmatrix}^{k+1} = \begin{pmatrix} 2 & 0 \\ 1 & 1 \end{pmatrix}^k \begin{pmatrix} 2 & 0 \\ 1 & 1 \end{pmatrix} = \begin{pmatrix} 2^k & 0 \\ 2^k - 1 & 1 \end{pmatrix}\begin{pmatrix} 2 & 0 \\ 1 & 1 \end{pmatrix}$

$= \begin{pmatrix} 2^{k+1}+0 & 0+0 \\ 2^{k+1}-2+1 & 0+1 \end{pmatrix}$

$= \begin{pmatrix} 2^{k+1} & 0 \\ 2^{k+1}-1 & 1 \end{pmatrix}$

So if the statement holds for $n = k$, it holds for $n = k + 1$.

Conclusion: The statement holds for all $n \in \mathbb{Z}^+$.

4 a Basis: $n = 1$: LHS = RHS = $\begin{pmatrix} 5 & -8 \\ 2 & -3 \end{pmatrix}$

Assumption: $\begin{pmatrix} 5 & -8 \\ 2 & -3 \end{pmatrix}^k = \begin{pmatrix} 4k+1 & -8k \\ 2k & 1-4k \end{pmatrix}$

Induction: $\begin{pmatrix} 5 & -8 \\ 2 & -3 \end{pmatrix}^{k+1} = \begin{pmatrix} 5 & -8 \\ 2 & -3 \end{pmatrix}^k \begin{pmatrix} 5 & -8 \\ 2 & -3 \end{pmatrix}$

$= \begin{pmatrix} 4k+1 & -8k \\ 2k & 1-4k \end{pmatrix}\begin{pmatrix} 5 & -8 \\ 2 & -3 \end{pmatrix}$

$= \begin{pmatrix} 20k+5-16k & -32k-8+24k \\ 10k+2-8k & -16k-3+12k \end{pmatrix}$

$= \begin{pmatrix} 4(k+1)+1 & -8(k+1) \\ 2(k+1) & 1-4(k+1) \end{pmatrix}$

So if the statement holds for $n = k$, it holds for $n = k + 1$.

Conclusion: The statement holds for all $n \in \mathbb{Z}^+$.

b $n = 6$

5 a Basis: $n = 1$: LHS = RHS = $\begin{pmatrix} 2 & 5 \\ 0 & 1 \end{pmatrix}$

Assumption: $\mathbf{M}^k = \begin{pmatrix} 2^k & 5(2^k-1) \\ 0 & 1 \end{pmatrix}$

Induction:

$\mathbf{M}^{k+1} = \mathbf{M}^k\begin{pmatrix} 2 & 5 \\ 0 & 1 \end{pmatrix} = \begin{pmatrix} 2^k & 5(2^k-1) \\ 0 & 1 \end{pmatrix}\begin{pmatrix} 2 & 5 \\ 0 & 1 \end{pmatrix}$

$= \begin{pmatrix} 2^{k+1} & 5\times 2^k + 5(2^k-1) \\ 0 & 1 \end{pmatrix} = \begin{pmatrix} 2^{k+1} & 5(2^{k+1}-1) \\ 0 & 1 \end{pmatrix}$

So if the statement holds for $n = k$, it holds for $n = k + 1$.

Conclusion: The statement holds for all $n \in \mathbb{Z}^+$.

b $\begin{pmatrix} 2^{-n} & 5(2^{-n}-1) \\ 0 & 1 \end{pmatrix}$

Chapter review 8

1 Let $f(n) = 9^n - 1$ where $n \in \mathbb{Z}^+$.

Basis: $f(1) = 9^1 - 1 = 8$ is divisible by 8.

Assumption: $f(k)$ is divisible by 8.

Induction: $f(k + 1) = 9^{k+1} - 1 = 9 \times 9^k - 1$

$f(k+1) - f(k) = (9 \times 9^k - 1) - (9^k - 1) = 8 \times 9^k$

So if the statement holds for $n = k$, it holds for $n = k + 1$.

Conclusion: The statement holds for all $n \in \mathbb{Z}^+$.

2 a $\mathbf{B}^2 = \begin{pmatrix} 1 & 0 \\ 0 & 9 \end{pmatrix}$, $\mathbf{B}^3 = \begin{pmatrix} 1 & 0 \\ 0 & 27 \end{pmatrix}$

b $\mathbf{B}^n = \begin{pmatrix} 1 & 0 \\ 0 & 3^n \end{pmatrix}$

c Basis: $n = 1$: LHS= RHS = $\begin{pmatrix} 1 & 0 \\ 0 & 3 \end{pmatrix}$

Assumption: $\mathbf{B}^k = \begin{pmatrix} 1 & 0 \\ 0 & 3^k \end{pmatrix}$

Induction: $\mathbf{B}^{k+1} = \mathbf{B}^k\begin{pmatrix} 1 & 0 \\ 0 & 3 \end{pmatrix} = \begin{pmatrix} 1 & 0 \\ 0 & 3^k \end{pmatrix}\begin{pmatrix} 1 & 0 \\ 0 & 3 \end{pmatrix}$

$= \begin{pmatrix} 1+0 & 0+0 \\ 0+0 & 0+3^{k+1} \end{pmatrix} = \begin{pmatrix} 1 & 0 \\ 0 & 3^{k+1} \end{pmatrix}$

So if the statement holds for $n = k$, it holds for $n = k + 1$.

Conclusion: The statement holds for all $n \in \mathbb{Z}^+$.

3 Basis: $n = 1$: LHS = $3 \times 1 + 4 = 7$;

RHS = $\frac{1}{2} \times 1(3 \times 1 + 11) = 7$

Assumption: $\sum_{r=1}^{k}(3r + 4) = \frac{1}{2}k(3k + 11)$

Induction: $\sum_{r=1}^{k+1}(3r + 4) = \sum_{r=1}^{k}(3r + 4) + 3(k + 1) + 4$

$= \frac{1}{2}k(3k + 11) + 3(k + 1) + 4 = \frac{1}{2}(3k^2 + 17k + 14)$

$= \frac{1}{2}(k + 1)(3(k + 1) + 11)$

So if the statement holds for $n = k$, it holds for $n = k + 1$.

Conclusion: The statement holds for all $n \in \mathbb{Z}^+$.

4 a Basis: $n = 1$: LHS = RHS = $\begin{pmatrix} 9 & 16 \\ -4 & -7 \end{pmatrix}$

Assumption: $\mathbf{A}^k = \begin{pmatrix} 8k+1 & 16k \\ -4k & 1-8k \end{pmatrix}$

Induction:

$\mathbf{A}^{k+1} = \mathbf{A}^k\begin{pmatrix} 9 & 16 \\ -4 & -7 \end{pmatrix} = \begin{pmatrix} 8k+1 & 16k \\ -4k & 1-8k \end{pmatrix}\begin{pmatrix} 9 & 16 \\ -4 & -7 \end{pmatrix}$

$= \begin{pmatrix} 72k+9-64k & 128k+16-112k \\ -36k-4+32k & -64k-7+56k \end{pmatrix}$

$= \begin{pmatrix} 8(k+1)+1 & 16(k+1) \\ -4(k+1) & 1-8(k+1) \end{pmatrix}$

So if the statement holds for $n = k$, it holds for $n = k + 1$.

Conclusion: The statement holds for all $n \in \mathbb{Z}^+$.

b $\begin{pmatrix} 1-8n & -16n \\ 4n & 8n+1 \end{pmatrix}$

5 a $f(n + 1) = 5^{2(n+1)-1} + 1 = 25 \times 5^{2n-1} + 1$

$f(n + 1) - f(n) = (25 \times 5^{2n-1} + 1) - (5^{2n-1} + 1)$

$= 24 \times 5^{2n-1}$; $\mu = 24$

b Let $f(n) = 5^{2n-1} + 1$ where $n \in \mathbb{Z}^+$.

Basis: $n = 1$: $f(1) = 5^{2-1} + 1 = 6$ is divisible by 6.

Assumption: $f(k)$ is divisible by 6.

Induction: $f(k + 1) - f(k) = 24 \times 5^{2k-1} = 6 \times 4 \times 5^{2k-1}$

So if the statement holds for $n = k$, it holds for $n = k + 1$.

Conclusion: The statement holds for all $n \in \mathbb{Z}^+$.

6 Let $f(n) = 7^n + 4^n + 1$ where $n \in \mathbb{Z}^+$.

Basis: $n = 1$: $f(1) = 7^1 + 4^1 + 1 = 12$ is divisible by 6.

Assumption: $f(k)$ is divisible by 6.

Induction: $f(k + 1) = 7^{k+1} + 4^{k+1} + 1 = 7 \times 7^k + 4 \times 4^k + 1$

$f(k + 1) - f(k) = (7 \times 7^k + 4 \times 4^{k+1} + 1) - (7^k + 4^k + 1)$

$= 6 \times 7^k + 3 \times 4^k$

where both 6×7^k and 3×4^k are divisible by 6, since 4 is even.

So if the statement holds for $n = k$, it holds for $n = k + 1$.

Conclusion: The statement holds for all $n \in \mathbb{Z}^+$.

7 Basis:

$n = 1$: LHS = $1 \times 5 = 5$; RHS = $\frac{1}{6} \times 1 \times 2 \times (2 + 13) = 5$

Assumption: $\sum_{r=1}^{k} r(r + 4) = \frac{1}{6}k(k + 1)(2k + 13)$

Induction: $\sum_{r=1}^{k+1} r(r + 4) = \sum_{r=1}^{k} r(r + 4) + (k + 1)(k + 5)$

$= \frac{1}{6}k(k + 1)(2k + 13) + (k + 1)(k + 5)$

$= \frac{1}{6}(2k^3 + 21k^2 + 49k + 30) = \frac{1}{6}(k + 1)(k + 2)(2(k + 1) + 13)$

So if the statement holds for $n = k$, it holds for $n = k + 1$.

Conclusion: The statement holds for all $n \in \mathbb{Z}^+$.

8 a Basis: $n = 1$: LHS = $1 + 4 = 5$; RHS = $\frac{1}{3} \times 1 \times 3 \times 5 = 5$

Assumption: $\sum_{r=1}^{2k} r^2 = \frac{1}{3}k(2k + 1)(4k + 1)$

Induction: $\sum_{r=1}^{2(k+1)} r^2 = \sum_{r=1}^{2k} r^2 + (2k + 1)^2 + (2k + 2)^2$

$= \frac{1}{3}k(2k + 1)(4k + 1) + (2k + 1)^2 + (2k + 2)^2$

$= \frac{1}{3}(8k^3 + 30k^2 + 37k + 15)$

$= \frac{1}{3}(k + 1)(2(k + 1) + 1)(4(k + 1) + 1)$

So if the statement holds for $n = k$, it holds for $n = k + 1$.

Conclusion: The statement holds for all $n \in \mathbb{Z}^+$.

b Using **a** and the formula for $\sum_{r=1}^{n} r^2$,

$\frac{1}{6} \times 2n(2n + 1)(4n + 1) = \frac{1}{6}kn(n + 1)(2n + 1)$

$2n(2n+1)(4n+1) = kn(n+1)(2n+1)$
$16n^3 + 12n^2 + 2n = k(2n^3 + 3n^2 + n)$
$\Rightarrow k = \dfrac{2n(8n^2+6n+1)}{n(2n^2+3n+1)} = \dfrac{2(2n+1)(4n+1)}{(2n+1)(n+1)} = \dfrac{8n+2}{n+1}$
$\Rightarrow kn + k = 8n + 2 \Rightarrow n(k-8) = 2 - k \Rightarrow n = \dfrac{2-k}{k-8}$

9 Basis: $u_n = \dfrac{3^n - 1}{2} \Rightarrow n = 1,\ u_1 = \dfrac{3^1-1}{2} = 1$ as given
$n = 2,\ u_2 = \dfrac{3^2-1}{2} = 4$ from the general statement
and $u_2 = 3 \times 1 + 1 = 4$ from the recurrence relationship.
So u_n is true when $n = 1$ and is also true when $n = 2$
Assumption: $n = k \Rightarrow u_k = \dfrac{3^k - 1}{2}$,
Then $n = k+1 \Rightarrow u_{k+1} = \dfrac{3^{k+1}-1}{2}$ for all $k \in \mathbb{Z}^+$
Induction: $u_{k+1} = \left(3\ \dfrac{3^k-1}{2}\right) + 1 = \dfrac{3^{k+1}}{2} - \dfrac{3}{2} + 1$
$= \dfrac{3^{k+1}-1}{2}$
Conclusion: If u_n is true when $n = k$, then it has been shown that u_n is also true when $n = k+1$. As u_n is true for $n = 1$ and $n = 2$ then u_n is true for all $n \geq 1$ and $n \in \mathbb{Z}^+$ by mathematical induction.

10 a $u_1 = 2,\ u_2 = \dfrac{5}{4},\ u_3 = \dfrac{11}{16},\ u_4 = \dfrac{17}{64},\ u_1 = -\dfrac{13}{256}$

b Basis: $u_n = 4\left(\dfrac{3}{4}\right)^n - 1 \Rightarrow n = 1,\ u_1 = 4\left(\dfrac{3}{4}\right)^1 - 1 = 2$,
as given
$n = 2,\ u_2 = 4\left(\dfrac{3}{4}\right)^2 - 1 = \dfrac{5}{4}$ from the general statement
and $u_2 = \dfrac{3 \times 2 - 1}{4} = \dfrac{5}{4}$ from the recurrence relationship
Assumption: $n = k \Rightarrow u_k = 4\left(\dfrac{3}{4}\right)^k - 1$,
Then $n = k+1 \Rightarrow u_{k+1} = 4\left(\dfrac{3}{4}\right)^{k+1} - 1$ for all $k \in \mathbb{Z}^+$
Induction: $u_{k+1} = \dfrac{3\left[4\left(\frac{3}{4}\right)^n - 1\right] - 1}{4} = \dfrac{3 \times 4\left(\frac{3}{4}\right)^n}{4} - \dfrac{3}{4} - \dfrac{1}{4}$
$= 4\left(\dfrac{3}{4}\right)^{k+1} - 1$
Conclusion: If u_n is true when $n = k$, then it has been shown that u_n is also true when $n = k+1$. As u_n is true for $n = 1$, $n = 2$ then u_n is true for all $n \geq 1$ and $n \in \mathbb{Z}^+$ by mathematical induction.

11 a Basis: $n = 1$: LHS = RHS = $\begin{pmatrix} 2c & 1 \\ 0 & c \end{pmatrix}$
Assumption: $\mathbf{M}^k = c^k\begin{pmatrix} 2^k & \frac{2^k-1}{c} \\ 0 & 1 \end{pmatrix}$
Induction: $\mathbf{M}^{k+1} = \mathbf{M}^k\begin{pmatrix} 2c & 1 \\ 0 & c \end{pmatrix}$
$= c^k\begin{pmatrix} 2^k & \frac{2^k-1}{c} \\ 0 & 1 \end{pmatrix}\begin{pmatrix} 2c & 1 \\ 0 & c \end{pmatrix} = c^{k+1}\begin{pmatrix} 2^k & \frac{2^k-1}{c} \\ 0 & 1 \end{pmatrix}\begin{pmatrix} 2 & \frac{1}{c} \\ 0 & 1 \end{pmatrix}$
$= c^{k+1}\begin{pmatrix} 2^{k+1} & \frac{2^k+2^k-1}{c} \\ 0 & 1 \end{pmatrix} = c^{k+1}\begin{pmatrix} 2^{k+1} & \frac{2^{k+1}-1}{c} \\ 0 & 1 \end{pmatrix}$
So if the statement holds for $n = k$, it holds for $n = k+1$.
Conclusion: The statement holds for all $n \in \mathbb{Z}^+$.

b Consider $n = 1$: $\det\mathbf{M} = 50 \Rightarrow 2c^2 = 50$
So $c = 5$, since c is +ve.

Challenge

a Basis: $n = 1$: LHS = RHS = $\begin{pmatrix} \cos\theta & -\sin\theta \\ \sin\theta & \cos\theta \end{pmatrix}$
Assumption: $\mathbf{M}^k = \begin{pmatrix} \cos k\theta & -\sin k\theta \\ \sin k\theta & \cos k\theta \end{pmatrix}$
Induction:
$\mathbf{M}^{k+1} = \mathbf{M}^k\begin{pmatrix} \cos\theta & -\sin\theta \\ \sin\theta & \cos\theta \end{pmatrix} = \begin{pmatrix} \cos k\theta & -\sin k\theta \\ \sin k\theta & \cos k\theta \end{pmatrix}\begin{pmatrix} \cos\theta & -\sin\theta \\ \sin\theta & \cos\theta \end{pmatrix}$
$= \begin{pmatrix} \cos k\theta\cos\theta - \sin k\theta\sin\theta & -\cos k\theta\sin\theta - \sin k\theta\cos\theta \\ \sin k\theta\cos\theta + \cos k\theta\sin\theta & -\sin k\theta\sin\theta + \cos k\theta\cos\theta \end{pmatrix}$
$= \begin{pmatrix} \cos((k+1)\theta) & -\sin((k+1)\theta) \\ \sin((k+1)\theta) & \cos((k+1)\theta) \end{pmatrix}$
So if the statement holds for $n = k$, it holds for $n = k+1$.
Conclusion: The statement holds for all $n \in \mathbb{Z}^+$.

b The matrix $\mathbf{M}$ represents a rotation through angle θ, and so $\mathbf{M}^n$ represents a rotation through angle $n\theta$.

Review exercise 2

1 a Does not exist: **B** doesn't have 3 rows.
b $\begin{pmatrix} 3q & 2q & pq \\ 9 & 4 & 3p+1 \end{pmatrix}$
c $\begin{pmatrix} 6q + pq \\ 3p + 25 \end{pmatrix}$
d Does not exist: **C** doesn't have 2 columns.

2 a $a = -2, b = 3$

3 $bc - ad$

4 a $-\frac{2}{3}$ **b** -2 **c** -4

5 a $\begin{pmatrix} 1 & \frac{1}{2} \\ 3 & 2 \end{pmatrix}$ **b** $\begin{pmatrix} 2 & 1 \\ 3p+3 & 2p+\frac{3}{2} \end{pmatrix}$ **c** $-\frac{1}{2}$

6 a $\dfrac{1}{pq}\begin{pmatrix} q & q \\ 3p & 4p \end{pmatrix}$ **b** $\dfrac{1}{pq}\begin{pmatrix} pq & 4q^2 \\ 2p^2 & 13pq \end{pmatrix}$

7 a $\begin{pmatrix} -\frac{1}{\sqrt{2}} & -\frac{1}{\sqrt{2}} \\ -\frac{1}{\sqrt{2}} & \frac{1}{\sqrt{2}} \end{pmatrix}$

b $\begin{pmatrix} -\frac{1}{\sqrt{2}} & -\frac{1}{\sqrt{2}} \\ -\frac{1}{\sqrt{2}} & \frac{1}{\sqrt{2}} \end{pmatrix}\begin{pmatrix} -\frac{1}{\sqrt{2}} & -\frac{1}{\sqrt{2}} \\ -\frac{1}{\sqrt{2}} & \frac{1}{\sqrt{2}} \end{pmatrix}$
$= \begin{pmatrix} \frac{1}{2}+\frac{1}{2} & \frac{1}{2}-\frac{1}{2} \\ \frac{1}{2}-\frac{1}{2} & \frac{1}{2}+\frac{1}{2} \end{pmatrix} = \begin{pmatrix} 1 & 0 \\ 0 & 1 \end{pmatrix}$

8 a $a = 3, b = -4, c = 2, d = -3$
b $\begin{pmatrix} 3 & -4 \\ 2 & -3 \end{pmatrix}\begin{pmatrix} 3 & -4 \\ 2 & -3 \end{pmatrix} = \begin{pmatrix} 9-8 & -12+12 \\ 6-6 & -8+9 \end{pmatrix} = \begin{pmatrix} 1 & 0 \\ 0 & 1 \end{pmatrix}$
c $p = 36, q = 25$

9 **a** $\begin{pmatrix} -1 & 2 \\ 0 & 3 \end{pmatrix}$

b $\begin{pmatrix} -1 & 2 \\ 0 & 3 \end{pmatrix}\begin{pmatrix} x \\ 2x \end{pmatrix} = \begin{pmatrix} 3x \\ 6x \end{pmatrix}$; $6x = 2(3x)$ so the point satisfies the equation of the original line.

c $A(2, 1), B(0, 5), C(-2, 4)$

d

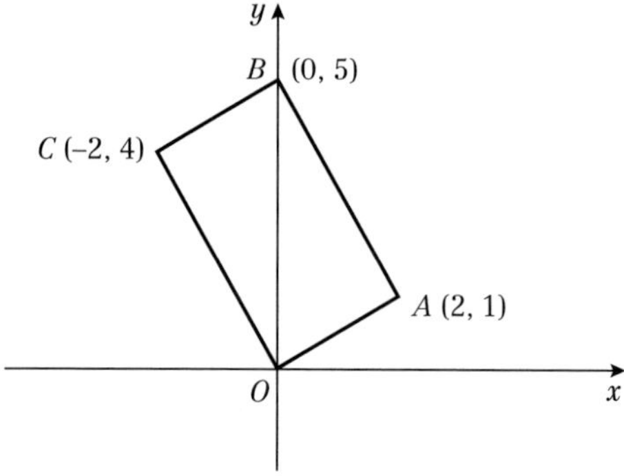

10 **a** $2k^2 + 3k - 3$

b $-\frac{7}{2}$ or 2

11 **a** Scale factor 3

b 45° anti-clockwise about (0, 0)

c $\left(\frac{p+q}{3\sqrt{2}}, \frac{-p+q}{3\sqrt{2}}\right)$

12 $\sum_{r=1}^{n}(2r-1)^2 = \sum_{r=1}^{n}(4r^2 - 4r + 1) = 4\sum_{r=1}^{n}r^2 - 4\sum_{r=1}^{n}r + n$

$= \frac{1}{3}n(n+1)(2n+1) - 2n(n+1) + n = \frac{1}{3}n(4n^2 - 1)$

13 $\sum_{r=1}^{n}r(r^2-3) = \sum_{r=1}^{n}r^3 - 3\sum_{r=1}^{n}r = \frac{1}{4}n^2(n+1)^2 - \frac{3}{2}n(n+1)$

$= \frac{1}{4}n(n+1)(n-2)(n+3)$

14 **a** $\sum_{r=1}^{n}r(2r-1) = 2\sum_{r=1}^{n}r^2 - \sum_{r=1}^{n}r$

$= \frac{1}{3}n(n+1)(2n+1) - \frac{1}{2}n(n+1) = \frac{1}{6}n(n+1)(4n-1)$

b 17 730

15 **a** $\sum_{r=1}^{n}(6r^2 + 4r - 5) = 6\sum_{r=1}^{n}r^2 + 4\sum_{r=1}^{n}r - 5n$

$= n(n+1)(2n+1) + 2n(n+1) - 5n = n(2n^2 + 5n - 2)$

b 32 480

16 **a** $\sum_{r=1}^{n}r(r+1) = \sum_{r=1}^{n}r^2 + \sum_{r=1}^{n}r$

$= \frac{1}{6}n(n+1)(2n+1) + \frac{1}{2}n(n+1) = \frac{1}{3}n(n+1)(n+2)$

b $\sum_{r=n}^{3n}r(r+1) = \sum_{r=1}^{3n}r(r+1) - \sum_{r=1}^{n-1}r(r+1)$

$= \frac{1}{3}(3n)(3n+1)(3n+2) - \frac{1}{3}(n-1)n(n+1)$

$= \frac{1}{3}n(26n^2 + 27n + 7) = \frac{1}{3}n(2n+1)(13n+7)$

$p = 13, q = 7$

17 **a** $p = 3, q = -1, r = -2$

b 23 703 950

18 Basis: When $n = 1$, LHS = RHS = 4

Assumption: $\sum_{r=1}^{k}r(r+3) = \frac{1}{3}k(k+1)(k+5)$

Induction: $\sum_{r=1}^{k+1}r(r+3) = \sum_{r=1}^{k}r(r+3) + (k+1)(k+4)$

$= \frac{1}{3}k(k+1)(k+5) + (k+1)(k+4)$

$= (k+1)\left(\frac{1}{3}k(k+5) + (k+4)\right)$

$= \frac{1}{3}(k+1)(k+2)(k+6)$

So if the statement holds for $n = k$, it holds for $n = k + 1$.

Conclusion: The statement holds for all $n \geqslant 1$.

19 Basis: When $n = 1$, LHS = RHS = 1

Assumption: $\sum_{r=1}^{k}(2r-1)^2 = \frac{1}{3}k(2k-1)(2k+1)$

Induction: $\sum_{r=1}^{k+1}(2r-1)^2 = \sum_{r=1}^{k}(2r-1)^2 + (2k+2-1)^2$

$= \frac{1}{3}k(2k-1)(2k+1) + (2k+1)^2$

$= (2k+1)(\frac{1}{3}k(2k-1) + 2k + 1)$

$= \frac{1}{3}(k+1)(2(k+1)-1)(2(k+1)+1)$

So if the statement holds for $n = k$, it holds for $n = k + 1$.

Conclusion: The statement holds for all $n \in \mathbb{Z}^+$.

20 Basis: $S_1 = a_1 = 6 = \frac{1}{2} \times 1 \times (1+1)^2 \times (1+2)$

Assumption: $S_k = \frac{1}{2}k(k+1)^2(k+2)$

Induction:

$S_{k+1} = S_k + a_{k+1} = \frac{1}{2}k(k+1)^2(k+2) + (k+1)(k+2)(2k+3)$

$= (k+1)(k+2)(\frac{1}{2}k(k+1) + (2k+3))$

$= \frac{1}{2}(k+1)(k+2)^2(k+3)$

So if the statement holds for $n = k$, it holds for $n = k + 1$.

Conclusion: The statement holds for all $n \geq 1$.

21 Basis: When $n = 1$, LHS = RHS = 0

Assumption: $\sum_{r=1}^{k}r^2(r-1) = \frac{1}{12}k(k-1)(k+1)(3k+2)$

Induction: $\sum_{r=1}^{k+1}r^2(r-1) = \sum_{r=1}^{k}r^2(r-1) + (k+1)^2k$

$= \frac{1}{12}k(k-1)(k+1)(3k+2) + (k+1)^2k$

$= \frac{1}{12}(k+1)k(k+2)(3(k+1)+2)$

So if the statement holds for $n = k$, it holds for $n = k + 1$.

Conclusion: The statement holds for all $n \geqslant 1$.

22 **a** $f(k+1) - f(k) = 3^{4k+4} + 2^{4k+6} - 3^{4k} - 2^{4k+2}$

$= 3^{4k}(3^4 - 1) + 2^{4k+2}(2^4 - 1) = 80 \times 3^{4k} + 15 \times 2^{4k+2}$

The first term is divisible by 15 since it is clearly divisible by 3, and 5 divides 80. Therefore $f(k+1) - f(k)$ is divisible by 15.

b Basis: When $n = 1$, $f(n) = 3^4 + 2^6 = 145 = 5 \times 29$

Assumption: $f(k)$ is divisible by 5.

Induction: From part **a**, $f(k+1) - f(k)$ is divisible by 15 and hence also by 5. Since $f(k)$ is divisible by 5, $f(k+1)$ is also divisible by 5.

So if the statement holds for $n = k$, it holds for $n = k + 1$.

Conclusion: The statement holds for all $n \in \mathbb{Z}^+$.

23 **a** $24 \times 2^{4(n+1)} + 3^{4(n+1)} - 24 \times 2^{4n} - 3^{4n}$

b Basis: When $n = 1$, $f(n) = 24 \times 2^4 + 3^4 = 465 = 5 \times 93$

Assumption: $f(k)$ is divisible by 5.

Induction: From part **a**,

$f(k+1) - f(k) = 24 \times 2^{4k+4} + 3^{4k+4} - 24 \times 2^{4k} - 3^{4k}$

$= 24 \times 2^{4k}(2^4 - 1) + 3^{4k}(3^4 - 1)$

$= 5(72 \times 2^{4k} + 16 \times 3^{4k})$

So if $f(k)$ is divisible by 5, $f(k+1)$ is divisible by 5.

Conclusion: $f(n)$ is divisible by 5 for all $n \in \mathbb{Z}^+$.

24 Let $f(n) = 7^n + 4^n + 1$.
Basis: $f(1) = 7 + 4 + 1 = 12$, which is divisible by 6.
Assumption: $f(k)$ is divisible by 6.
Induction: $f(k + 1) - f(k) = 7^{k+1} + 4^{k+1} + 1 - 7^k - 4^k - 1$
$= 7^k(7 - 1) + 4^k(4 - 1) = 6 \times 7^k + 3 \times 4^k$
The first term is divisible by 6, and since 4^k is even, the second term is divisible by 6. So if $f(k)$ is divisible by 6, then $f(k + 1)$ is also divisible by 6.
Conclusion: $f(n)$ is divisible by 6 for all $n \in \mathbb{Z}^+$.

25 Let $f(n) = 4^n + 6n - 1$.
Basis: When $n = 1$, $f(n) = 4^1 + 6(1) - 1 = 9$, which is divisible by 9.
Assumption: $f(k)$ is divisible by 9.
Induction: $f(k + 1) - f(k) = 4^{k+1} + 6(k + 1) - 1 - 4^k - 6k + 1$
$= 4^k(4 - 1) + 6 = 3(4^k - 1) + 9$
$4^k - 1$ is divisible by $4 - 1 = 3$.
First term has two factors of 3 so is divisible by 9 and the second term is divisible by 9. So if $f(k)$ is divisible by 9, then $f(k + 1)$ is also divisible by 9
Conclusion: $f(n)$ is divisible by 9 for all $n \in \mathbb{Z}^+$.

26 Let $f(n) = 3^{4n-1} + 2^{4n-1} + 5$
Basis: $f(1) = 3^3 + 2^3 + 5 = 27 + 8 + 5 = 40 = 10 \times 4$
Assumption: $f(k)$ is divisible by 10.
Induction: $f(k + 1) - f(k) = 3^{4k+3} + 2^{4k+3} - 3^{4k-1} + 2^{4k-1}$
$= 3^{4k-1}(3^4 - 1) + 2^{4k-1}(2^4 - 1) = 80 \times 3^{4k-1} + 15 \times 2^{4k-1}$
This is divisible by 10: 15 is divisible by 5 and 2^{4k-1} is even.
So if $f(k)$ is divisible by 10, then $f(k + 1)$ is divisible by 10.
Conclusion: $f(n)$ is divisible by 10 for all positive integers, n.

27 Basis: When $n = 1$, $\mathbf{A}^1 = \begin{pmatrix} 1 & c \\ 0 & 2 \end{pmatrix} = \begin{pmatrix} 1 & (2^1 - 1)c \\ 0 & 2^1 \end{pmatrix}$
Assumption: $\mathbf{A}^k = \begin{pmatrix} 1 & (2^k - 1)c \\ 0 & 2^k \end{pmatrix}$
Induction: $\mathbf{A}^{k+1} = \mathbf{A}^k\begin{pmatrix} 1 & c \\ 0 & 2 \end{pmatrix} = \begin{pmatrix} 1 & (2^k - 1)c \\ 0 & 2^k \end{pmatrix}\begin{pmatrix} 1 & c \\ 0 & 2 \end{pmatrix}$
$= \begin{pmatrix} 1 & c + 2c(2^k - 1) \\ 0 & 2^{k+1} \end{pmatrix} = \begin{pmatrix} 1 & (2^{k+1} - 1)c \\ 0 & 2^{k+1} \end{pmatrix}$
So if the statement holds for $n = k$, it holds for $n = k + 1$.
Conclusion: The statement holds for all positive integers, n.

28 Basis:
When $n = 1$, $\mathbf{A}^1 = \begin{pmatrix} 3 & 1 \\ -4 & -1 \end{pmatrix} = \begin{pmatrix} 2 \times 1 + 1 & 1 \\ -4 \times 1 & -2 \times 1 + 1 \end{pmatrix}$
Assumption: $\mathbf{A}^k = \begin{pmatrix} 2k + 1 & k \\ -4k & -2k + 1 \end{pmatrix}$
Induction:
$\mathbf{A}^{k+1} = \mathbf{A}^k\begin{pmatrix} 3 & 1 \\ -4 & -1 \end{pmatrix} = \begin{pmatrix} 2k + 1 & k \\ -4k & -2k + 1 \end{pmatrix}\begin{pmatrix} 3 & 1 \\ -4 & -1 \end{pmatrix}$
$= \begin{pmatrix} 6k + 3 - 4k & 2k + 1 - k \\ -12k + 8k - 4 & -4k + 2k - 1 \end{pmatrix}$
$= \begin{pmatrix} 2(k + 1) + 1 & k + 1 \\ -4(k + 1) & -2(k + 1) + 1 \end{pmatrix}$
So if the statement holds for $n = k$, it holds for $n = k + 1$.
Conclusion: The statement holds for all positive integers, n.

29 **a** She has not shown it true for $k = 1$
b Let $f(n) = 2^{2n} - 1$
Basis: When $n = 1$, $f(n) = 2^2 - 1 = 3$
Assumption: $f(k) = 2^{2k} - 1$ is divisible by 3.
Induction: $f(k + 1) = 2^{2(k+1)} - 1 = 4f(k) + 3$
So if $f(k)$ is divisible by 3, then $f(k + 1)$ is divisible by 3.
Conclusion: $f(n)$ is divisible by 3 for all positive integers, n.

30 **a** $u_2 = 3$, $u_3 = 7$
b Basis: $u_n = 2^n - 1 \Rightarrow n = 1, u_1 = 2^1 - 1 = 1$, as given
$n = 2$, $u_2 = 2^2 - 1 = 3$ from the general statement and
$u_2 = 2 \times 1 + 1 = 3$ from the recurrence relationship.
$n = 3$, $u_3 = 2^3 - 1 = 7$ from the general statement
and $u_2 = 2 \times 3 + 1 = 7$ from the recurrence relationship.
So u_n is true when $n = 1$ and is also true when $n = 2$ and when $n = 3$
Assumption: $n = k \Rightarrow u_k = 2^k - 1$,
Then $n = k + 1 \Rightarrow u_{k+1} = 2^{k+1} - 1$ for all $k \in \mathbb{Z}^+$
Induction: $u_{k+1} = 2(2^k - 1) + 1 = 2^{k+1} - 2 + 1 = 2^{k+1} - 1$
Conclusion: If u_n is true when $n = k$, then it has been shown that u_n is also true when $n = k + 1$.
As u_n is true for $n = 1$, $n = 2$ then u_n is true for all $n \geq 1$ and $n \in \mathbb{Z}^+$ by mathematical induction.

31 Basis: $u_n = 5(6^{n-1}) + 1 \Rightarrow n = 1, u_1 = 5(6^0) + 1 = 6$ as given
$n = 2$, $u_2 = 5(6^1) + 1 = 31$ from the general statement
and $u_2 = 6 \times 6 - 5 = 31$ from the recurrence relationship.
$n = 3$, $u_3 = 5(6^2) + 1 = 181$ from the general statement
and $u_2 = 6 \times 31 - 5 = 181$ from the recurrence relationship.
So u_n is true when $n = 1$ and is also true when $n = 2$ and when $n = 3$
Assumption: $n = k \Rightarrow u_k = 5(6^{k-1}) + 1$,
Then $n = k + 1 \Rightarrow u_{k+1} = 5(6^k) + 1$ for all $k \in \mathbb{Z}^+$
Induction: $u_{k+1} = 6(5(6^{k-1}) + 1) - 5 = 5(6^k) + 6 - 5 = 5(6^k) + 1$
Conclusion: If u_n is true when $n = k$, then it has been shown that u_n is also true when $n = k + 1$.
As u_n is true for $n = 1$, $n = 2$ then u_n is true for all $n \geq 1$ and $n \in \mathbb{Z}^+$ by mathematical induction.

Challenge

1 Basis: When $n = 1$, $r = 2 \Rightarrow 2(1) \leqslant 2 \leqslant \frac{1}{2}(1^2 + 1 + 2)$
Assumption: $2k \leqslant r \leqslant \frac{1}{2}(k^2 + k + 2)$
Induction: Lower bound – all lines pass through a single point, $r_k = 2k$
One more line added $\Rightarrow$ two more regions.
$r_{k+1} = 2k + 2 = 2(k + 1)$
Upper bound – lines do not pass through the intersection of any other pairs of lines, $r_k = \frac{1}{2}(k^2 + k + 2)$.
One more line added $\Rightarrow k + 1$ more regions.
$r_{k+1} = \frac{1}{2}(k^2 + k + 2) + k + 1 = \frac{1}{2}(k^2 + 3k + 4)$
$= \frac{1}{2}((k + 1)^2 + (k + 1) + 1)$
So if the statement holds for $n = k$, it holds for $n = k + 1$.
Conclusion: The statement holds for all positive integers, n.

2 $x = -1$, $y = 4$

Exam practice

1 **a** Det $\mathbf{P} = 2$ which is non-zero, hence non-singular
b $\mathbf{Q} = \begin{pmatrix} -3 \\ 5 \end{pmatrix}$

2 **a** $\arg(z_1) = -26.6°$
b $8 + \text{i}$
c $\frac{4 - 7\text{i}}{13}$

3 **a** $xy = 25$ $(\Rightarrow c = \pm 5)$
b midpoint is at $(15, 3)$

4 **a** $2y = -x + 9$ or equivalent
b $\left(-4, \frac{13}{2}\right)$ **c** $\left(\frac{169}{64}, \frac{13}{2}\right)$

5 **a** **i** Rotation, 30°, anticlockwise
ii Reflection in $y = -x$
b $\begin{pmatrix} -1.37 & -3.10 & -2.73 \\ -0.37 & 0.63 & -0.73 \end{pmatrix}$

6 **a** $f(-1) = 1$, $f(0) = -2$ sign change hence root.
b $\alpha = -0.5$ **c** $\beta = 2.21$
d $f(2.205) = -0.099...$, $f(2.215) = 0.0072888...$ sign change hence root

7 **a** $\frac{41}{4}$
b $\alpha + \beta = \frac{5}{2}$, $\alpha\beta = -2$

$$\alpha^3 + \beta^3 = (\alpha + \beta)^3 - 3\alpha\beta(\alpha + \beta)$$
$$= \left(\frac{5}{2}\right)^3 - 3 \times (-2) \times \left(\frac{5}{2}\right)$$
$$= \frac{245}{8}$$

Hence shown.
c $32x^2 - 162x + 189 = 0$

8 **a** $2 + \text{i}$, -2i
b $z^4 - 4z^3 + 9z^2 - 16z + 20 = 0$
$\Rightarrow a = -4, b = 9,$
$c = -16, d = 20$

9
$$\sum_{r=1}^{n}(r^3 + 6r - 3) = \frac{1}{4}n^2(n + 1)^2 + 6 \times \frac{1}{2}n(n + 1) - 3n$$
$$= \frac{1}{4}n^2(n + 1)^2 + 3n^2 + 3n - 3n$$
$$= \frac{1}{4}n^2[(n + 1)^2 + 12]$$
$$= \frac{1}{4}n^2(n^2 + 2n + 13)$$

10 **a** $f(n) = 2^n + 6^n$

$$6f(k) - 4(2^k) = 6(2^k + 6^k) - 4(2^k)$$
$$= 6(2^k) + 6(6^k) - 4(2^k)$$
$$= 2(2^k) + 6(6^k)$$
$$= (2^{k+1}) + (6^{k+1})$$
$$= f(k + 1)$$

Hence shown

b <u>Basis</u>: $n = 1, u_1 = 2^1 + 6^1 = 8$ (divisible by 8)
$n = 2, u_2 = 2^2 + 6^2 = 40$ (divisible by 8)
So when $n = 1$, u_1 is divisible by 8 and when $n = 2$, u_2 is divisible by 8
<u>Assumption</u>: $f(k)$ is divisible by 8 and
$f(k + 1) = 6f(k) - 4(2^k)$
<u>Induction</u>: $4(2^k) = 2^3(2^{k-1})$ and $2^3 = 8$
<u>Conclusion</u>: Result is implied for $n = k + 1$ and so is true for positive integers (for all $k \in \mathbb{Z}^+$) by mathematical induction.

Online Worked solutions are available in SolutionBank.

INDEX

M

N

P

Q

R

S

T

V

Z